Lecture
in Physi

Edited by J. Ehlers, München, K. Hepp, Zürich, and
H. A. Weidenmüller, Heidelberg
Managing Editor: W. Beiglböck, Heidelberg

29

Foundations of Quantum Mechanics and Ordered Linear Spaces

Advanced Study Institute
Marburg 1973

Edited by A. Hartkämper and H. Neumann
Fachbereich für Physik der Universität Marburg,
Marburg/BRD

Springer-Verlag
Berlin · Heidelberg · New York 1974

ISBN 3-540-06725-6 Springer-Verlag Berlin · Heidelberg · New York
ISBN 0-387-06725-6 Springer-Verlag New York · Heidelberg · Berlin

Offsetprinting and bookbinding: Julius Beltz, Hemsbach/Bergstr.

PREFACE

The Advanced Study Institute on "Foundations of Quantum Mechanics and Ordered Linear Spaces" has been held at Marburg, Federal Republic of Germany, from March 26th to April 6th 1973.

Mathematicians and physicists participated in the meeting. The lectures of the Institute were intended to prepare a common basis for discussions between mathematicians and physicists and for future research on foundations of quantum mechanics by ordered linear spaces. A series of lectures ("Course") provided a coherent introduction into the field of ordered normed vector spaces and their application to the foundation of quantum mechanics. Additional lectures treated special mathematical and physical topics, which were in more or less close connection to the lectures of the course.

The present volume contains the notes of the lectures revised by the authors.

The sponsorship of the Scientific Affairs Division of the North Atlantic Treaty Organization, of the Stiftung Volkswagenwerk and the University of Marburg is gratefully acknowledged.

A.Hartkämper H.Neumann

C O N T E N T S

Introduction	. . . .	1
COURSE		
Mathematics:		
H.H. Schaefer	Orderings of Vector Spaces	4
J. Mangold and R.J. Nagel	Duality of Cones in Locally Convex Spaces . .	11
R.J. Nagel	Order Unit and Base Norm Spaces	23
A.J. Ellis	Minimal Decompositions in Base Normed Spaces .	30
A. Goullet de Rugy	Simplex Spaces	33
A. Goullet de Rugy	Representation of Banach Lattices	41
A.J. Ellis	Order Ideals in Ordered Banach Spaces	47
W. Wils	Order Bounded Operators and Central Measures	54
G. Wittstock	Ordered Normed Tensor Products	67
E. Størmer	Positive Linear Maps of C^*-Algebras	85
Physics:		
A. Hartkämper	Axiomatics of Preparing and Measuring Procedures	107
H. Neumann	The Structure of Ordered Banach Spaces in Axiomatic Quantum Mechanics	116
G. Ludwig	Measuring and Preparing Processes	122
G.M. Prosperi	Models of the Measuring Process and of Macro Theories	163
C.M. Edwards	The Centre of a Physical System	199
K. Kraus	Operations and Effects in the Hilbert Space Formulation of Quantum Theory	206
D.J. Foulis and C.H. Randall	The Empirical Logic Approach to the Physical Sciences	230

SPECIAL TOPICS

M. Drieschner	The Structure of Quantum Mechanics: Suggestions for a Unified Physics	250
V. Gorini and E.C.G. Sudarshan	Irreversibility and Dynamical Maps of Statistical Operators	260
U. Krause	The Inner Orthogonality of Convex Sets in Axiomatic Quantum Mechanics	269
L. Lanz, L.A. Lugiato and G. Ramella	Reduced Dynamics in Quantum Mechanics. . . .	281
M. Mugur-Schachter	The Quantum Mechanical Hilbert Space Formalism and the Quantum Mechanical Probability Space of the Outcomes of Measurements . . .	288
R.J. Nagel	Mean Ergodic Semigroups and Invariant Ideals in Ordered Banach Spaces	309
H. Neumann	The Representation of Classical Systems in Quantum Mechanics	316
E. Prugovečki	Extended Hilbert Space Formulation of Dirac's Bra and Ket Formalism and its Applications to Abstract Stationary Scattering Theory	322
G.T. Rüttimann	Projections on Orthomodular Lattices	334
H.H. Schaefer	The Šilov Boundary of a Convex Cone	342
M. Wolff	A Radon-Nikodym-Theorem for Operators with an Application to Spectral Theory	345

INTRODUCTION

Since the physical contributions of the course are not so closely related to each other as the mathematical ones it seems useful to give some introductory remarks concerning the physical topics.

Since the early beginning of work with quantum mechanics physicists felt uncomfortable postulating the Hilbert space structure ad hoc. From this resulted attempts to deduce the Hilbert space structure by an axiomatic foundation of quantum mechanics. A further, far more ambitious aim of most of these attempts is to find structures of physical theories including more general theories than quantum mechanics. An axiomatic foundation of not only mathematical character usually starts with what could be called a pretheory, describing the physical notions and situations on which the final theory is based. This pretheory supplies the usual mathematical structures with an additional structure and simultaneously yields a particular interpretation of the final theory.

The attempts of an axiomatic foundation of quantum mechanics can be classified by the basic notions with which the pretheories are concerned and by the mathematical apparatus used in the sequel.

In contrast to the possibility of directly postulating properties of microsystems, the authors of this volume start from the macroscopic experimental situation.

The vector space structure enters the theory either in an early stage by embedding the basic statistical description into a dual pair of vector spaces, or in a later stage via a linear space of orthoadditive real valued functions on a "logic".

In accordance with the title of this volume very little will be found concerning the broad field of lattice theoretical approaches to an axiomatic foundation of quantum mechanics.

Several articles of this volume contribute to the discussion of the relation between quantum mechanics and the classical theories of macroscopic bodies. On one hand the latter seem intimately connected with quantum mechanics of many particles. On the other hand the classical theories to a certain extent enter quantum mechanics as a supposition, via axioms concerning the measurement with macroscopic apparata. In this sense the description of the measuring process appears as a problem of consistency of the theory.

Finally, it might be useful to give an elementary and short description of von Neumann's formulation, used throughout this volume, of the general frame of quantum mechanics in terms of ordered vector spaces. A quantum mechanical system is described in an infinite-dimensional, separable, complex Hilbert space H. If $L_h(H)$ denotes the set of bounded Hermitean operators of H the set

$K = \{ W \in L_h(H) \; / \; W \geqslant 0, \; \mathrm{tr}\, W = 1 \}$ is considered as the

set of ensembles of the system. K is the base of the positive cone of the base normed Banach space B of hermitean trace class operators:

$$B = \{ T \in L_h(H) \; / \; \| T \|_{tr} = \mathrm{tr}((T^* T)^{1/2}) < \infty \}$$

The set of self-adjoint operators in H is the set of observables of the system. By means of the spectral decomposition, the measurement of observables can be related to the measurement of compatible projection operators onto closed subspaces of H. The outcomes of the measurement of projection operators are 0 and 1 only, and, corresponding to various different interpretations, projection operators are called properties or events or decision effects. The projection operators form a complete orthocomplemented orthomodular lattice.

However, the analysis of the functioning and the statistics of realistic measurements with outcomes 0 and 1 and the analysis of the measuring process and of operations suggest that the whole set $L = \{ F \in L_h(H) \; / \; 0 \leq F \leq \mathbb{1} \}$ describes measurements with outcomes 0 and 1, $\mathrm{tr}(W \cdot F)$ being the probability for the outcome 1 in the ensemble W.

An element $F \in L$ is called effect, simple observable or test. $L_h(H)$ can be regarded as the dual Banach space B' of the space B of hermitean trace class operators, tr(TA) being the canonical bilinear form on $B \times B'$. $B' = L_h(H)$ is equipped with an order unit norm, where the unit operator $\mathbb{1}$ is the order unit of B'. L is the order interval $[0, \mathbb{1}]$ of B', and the set of projection operators is the set of extreme points of L.

ORDERINGS OF VECTOR SPACES

H.H. Schaefer
Mathematisches Institut der Universität Tübingen
Tübingen, Germany

What is meant by an ordered vector space, and what is the relevance of this concept in analysis ? In the following informal discussion, we try to give a first introduction and a few typical examples. For a closer study including detailed proofs, we must refer the reader to the references given at the end of the paper.

1. Vector Orderings. Let E denote a vector space over $\mathbb{R}$; a subset $C \subset$ satisfying $C + C \subset C$ and $\alpha C \subset C$ $(\alpha \geq 0)$ is called a cone; a cone C for which $C \cap -C = \{0\}$ is called a proper cone. An ordering (or partial ordering, semi-ordering), i.e., a reflexive, transitive, and anti-symmetric binary relation $\leq$ on E is called a vector ordering if it is invariant under translations and homothetic maps of ratio ≥ 0; that is, if $x \leq y$ implies $x + z \leq y + z$ and $\alpha x \leq \alpha y$ for all $x, y \in E$ and $0 \leq \alpha \in \mathbb{R}$. The pair $(E, \leq)$ is called an ordered vector space, and the set $E_+ := \{x \in E: x \geq 0\}$ is a proper cone called the positive cone of E. Conversely, each proper cone $C \subset E$ defines a (vector) ordering of E by virtue of "$x \leq y$ iff $y - x \in C$" (cf. [S_1], Chap. V, Exerc. 1-3).

Examples.

1. Let E be a Hausdorff locally convex space with topological dual E', let K be a $\sigma(E',E)$-compact subset of E' such that E' is the linear hull of K and K is contained in a $\sigma(E',E)$-closed hyperplane H, and suppose $0 \notin H$. Considering each $x \in E$ as an affine, continuou function on K, we can identify E with a space of affine continuous functions on K ([A], Chap.II, §2), ordered by the cone $E_+ := \{x \in E: \langle x, x'\rangle \geq 0 \text{ for all } x' \in K\}$. E is dense in the space A(K) of all affine continuous functions on K with respect to the uniform norm, and $E = A(K)$ if E is sequentially complete.

   E' is ordered by the cone $E'_+ := \bigcup_{\lambda \geq 0} \lambda K$, and these cones are dual to each other in the sense that $x' \in E'_+$ (respectively, $x \in E_+$) is equivalent to $\langle x, x'\rangle \geq 0$ for all $x \in E_+$ (respectively, $x' \in E'_+$).

2. Let X denote a (Hausdorff) compact space. The vector space of all continuous functions $X \to \mathbb{R}$ is a Banach space $C(X)$ under the supremum norm; the cone $C(X)_+ := \{ f: f(t) \geq 0 \text{ for all } t \in X \}$ defines the natural ordering of X. If X is finite (hence discrete) and contains precisely n elements, the ordered Banach space $C(X)$ can be identified with $\mathbb{R}^n$ under its usual (coordinatewise) ordering.

3. Let (X, Σ, μ) be a totally σ-finite measure space (see [DS]). The Banach spaces $L^p(\mu)$ $(1 \leq p \leq +\infty)$ of (equivalence classes of) measurable functions are ordered by defining an equivalence class to be ≥ 0 iff it contains an f such that $f(t) \geq 0$ everywhere on X. Recall that the Banach dual of $L^p(\mu)$ $(1 \leq p < +\infty)$ can be identified with $L^q(\mu)$, where $p^{-1} + q^{-1} = 1$.

Let E denote an ordered vector space, and let F denote a vector subspace of E. F is ordered by the cone $F_+ := F \cap E_+$ (induced ordering). If $q: E \to E/F$ denotes the quotient map, a natural ordering of E/F is defined iff the cone $q(E_+)$ is proper, which in turn is equivalent to F being an <u>order ideal</u> in E: $x,y \in F$, $z \in E$, and $x \leq z \leq y$ implies $z \in F$. (Notice that an order ideal J need not be positively generated even if E is; that is, $E = E_+ - E_+$ does not imply $J = J_+ - J_+$.)

If E,F are ordered vector spaces and L is a vector space of linear maps $E \to F$, consider the cone $L_+ := \{ T \in L: T(E_+) \subset F_+ \}$; whenever L_+ is a proper cone, it defines the <u>natural order</u> of L. (For L_+ to be a proper cone it is sufficient that $E = E_+ - E_+$.) Suppose $E = E_+ - E_+$ and let $F = \mathbb{R}$ in its standard ordering; the space E^* of all linear forms is ordered by the cone E^*_+ of positive linear functionals on E, and the vector subspace $E^*_+ - E^*_+$ of E^* is usually called the <u>order dual</u> of E. Examples of dual pairs of ordered vector spaces are contained in Examples 1, 3 above.

An element $x \in E_+ \setminus \{0\}$, where E is an ordered vector space, is called <u>extremal</u> [J] if $0 \leq y \leq x$, $y \in E$ implies that y is a scalar multiple of x. We point out this important fact:

> Let E be an ordered vector space such that $E = E_+ - E_+$. A positive linear form f on E is extremal in E^*_+ whenever its kernel $f^{-1}(0)$ is a maximal, positively generated ideal in E.

The proof depends on the fact that the annihilator $J^o \subset E^*$ of a positively generated ideal J is an ideal in E^*, and that $E/J \cong \mathbb{R}$ iff J is a maximal ideal.

An ordered vector space E is called <u>Archimedean</u> (<u>ordered</u>) if $x \leq 0$ whenever $nx \leq y$ for some $y \in E$ and all $n \in \mathbb{N}$. The order dual of an ordered vector space is Archimedean; moreover, if E is also a topological vecto space in which E_+ is sequentially closed then E is Archimedean (see Examples 1-3 above).

2. <u>Vector lattices</u>. An ordered vector space E is called a <u>vector lattic</u> if the greatest lower bound $x \wedge y := \inf\{x,y\}$ and the least upper boun $x \vee y := \sup\{x,y\}$ exist for each pair of elements $x,y \in E$. For the bas algebraic theory of vector lattices we refer to [J, Chap.II], [S_1, Chap.V, §1] or [S_2, Chap.II, §§1-4]. We recall in particular that

(1) $$x \vee y + x \wedge y = x + y$$

holds for all $x,y \in E$; defining $x^+ = x \vee 0$ and $x^- = (-x) \vee 0$ for $x \in E$, we obtain $x = x^+ - x^-$ by letting $y = 0$ in (1), and so $E = E_+ - E_+$. The element $|x| = x \vee (-x)$ is called the <u>modulus</u> of $x \in E$, and found to satisfy $|x| = x^+ + x^-$. Moreover, the modulus function $x \mapsto |x|$ satisfies the relations

(2) $$|x + y| \leq |x| + |y|, \quad |\alpha x| = |\alpha|\,|x|$$

for all $x,y \in E$, $\alpha \in \mathbb{R}$.

An important concept available in lattice ordered vector spaces is orthogonality: x,y are called (<u>lattice</u>) <u>orthogonal</u> if $|x| \wedge |y| = 0$; we write $A^\perp := \{y \in E: |y| \wedge |x| = 0 \text{ for all } x \in A\}$, where A is any subset of E. Orthogonality in vector lattices shows certain parallels to the like-named concept in Hilbert space theory; it is mainly useful for Archimedean vector lattices, however.

Examples of vector lattices are given above (Examples 2,3). The ordered space E' of Example 1 is a vector lattice only if K is a particular typ of set (precisely, iff K is a Choquet simplex; [A, Chap.II, §3]), while E is a vector lattice iff K is a Bauer simplex (l.c.) provided E is sequentially complete.

The property that the space E' of Example 1 be a vector lattice is equi valent to E having the <u>decomposition</u> (or <u>interpolation</u>) <u>property</u>, which is defined by the relation

(3) $$[0,x + y] = [0,x] + [0,y]$$

holding for all $x,y \in E_+$; as usual, $[0,z]$ denotes the order interval $\{w \in E: 0 \leq w \leq z\}$ of E. Every vector lattice satisfies (3), but not conversely; if E is a positively generated ordered vector space and satisfies (3), its order dual is a vector lattice.

If E is a vector lattice, J a vector subspace of E, then for E/J to be a vector lattice in its natural ordering (§1) with $q: E \to E/J$ preserving finite infima and suprema, it is necessary and sufficient that J be a positively generated order ideal. Such an ideal is called a lattice ideal (briefly, ideal). For each subset A of E, $A^{\perp}$ is a lattice ideal.

A vector lattice E such that each non-empty subset A which is order bounded (i.e., contained in some order interval) has a least upper bound sup A, is called order complete or Dedekind complete. The spaces of Example 3 are order complete; C(X) (Example 2) is order complete iff the compact space X is extremally disconnected.

An ideal $J \subset E$ (E any vector lattice) such that $A \subset J$ and $\sup A = x \in E$ implies $x \in J$, is called a band; for each $A \subset E$, $A^{\perp}$ is a band. If a band J has the property that $E = J + J^{\perp}$, then J is called a projection band, because $J \cap J^{\perp} = \{0\}$ forces this decomposition of E to be a direct sum, and the associated endomorphism $E \to J$ with kernel $J^{\perp}$ is a projection P_J (called the band projection associated with J). We can now state the famous Riesz decomposition theorem.

> Let E be an order complete vector lattice. Each band in E is a projection band, and the family $\mathcal{B}$ of all bands of E is a Boolean algebra with respect to the operations $B_1 \wedge B_2 := B_1 \cap B_2$ and $B_1 \vee B_2 := B_1 + B_2$ $(B_1,B_2 \in \mathcal{B})$. Accordingly, the family of all band projections $\mathcal{P}$ is a Boolean algebra with respect to the operations $P_1 \wedge P_2 := P_1 \cdot P_2$ and $P_1 \vee P_2 := P_1 + P_2 - P_1 \cdot P_2$ (in particular, commutative), and isomorphic to $\mathcal{B}$ under the mapping $B \mapsto P_B$.

For proofs of these results, see [LZ] or [S_2]. The Boolean algebras considered in this theorem can be represented in a particularly illuminating manner if we assume that E, in addition to being order complete, has a weak order unit (or Freudenthal unit) u; that is, an element $u \in E_+$ such that $u \wedge |x| = 0$ $(x \in E)$ implies $x = 0$. For this representation, we need the following result.

> Let E be any vector lattice, and let $e \in E_+$. The extreme points of the order interval $[0,e]$ form a Boolean algebra under the ordering induced by E.

In fact, it is not difficult to show that $x \in [0,e]$ is extreme iff $x \wedge (e-x) = 0$, and that these elements form a sublattice of $[0,e]$; since the lattice operations in E are distributive, the extreme boundary of $[0,e]$ is a Boolean algebra.

The announced result is as follows.

Let E be an order complete vector lattice possessing a weak order unit u. The mapping $P \mapsto u_P := Pu$ is an isomorphism of the Boolean algebra $\mathcal{P}$ onto the Boolean algebra of extreme points of $[0,u]$.

In fact, if P is a band projection on E then $0 = Pu \wedge (1_E - P)u = Pu \wedge (u-P$ and $0 \leq Pu \leq u$; therefore, Pu is an extreme point of $[0,u]$. Conversely, if $x \wedge (u-x) = 0$ and P is the band projection associated with the band generated by x, then $x = Pu$. Moreover, $P_1 u \geq P_2 u$ is equivalent with $P_1(E) \supset P_2(E)$ $(P_1, P_2 \in \mathcal{P})$; therefore, $P \mapsto u_P$ is an order isomorphism and hence, an isomorphism of Boolean algebras.

3. Order Unit and Base Norms. We consider very briefly two particular types of ordered vector spaces which, as the following lectures will show are dual to each other. We suppose throughout that E be an Archimedean ordered vector space.

Let e be an order unit of E, that is, let there exist an element $e \in E_+$ such that the ideal generated by e equals E (so that $E = \bigcup_{n \in \mathbb{N}} n[-e,e]$); the Minkowski functional (gauge) p_e of $[-e,e]$, defined by

$$p_e(x) = \inf \{ \lambda \in \mathbb{R} : -\lambda e \leq x \leq \lambda e \},$$

is a norm on E called an order unit norm; the interior points of E_+ in (E,p_e) are precisely the order units of E. (E,p_e) is a Banach space iff each increasing p_e-Cauchy sequence has a least upper bound. A typical (and the most general) example of an order-unit-normed Banach space is A(K) (§1, Example 1). If such a space is a vector lattice then it is isomorphic to C(X), where X is the extreme boundary of K (Kakutani-Krein).

Dually, if $E = E_+ - E_+$ and there exists a convex subset B of E_+ such that for each $x \in E_+$, there exists a unique number $f(x) \geq 0$ for which $x \in f(x)B$; then B is called a base of E_+. Let $q_B(z) = \inf f(x+y)$ where the infimum is taken over all pairs $(x,y) \in E_+ \times E_+$ such that $z = x-y$ $(z \in E)$; the set $J = q_B^{-1}(0)$ is an order ideal of E, and q_B defines a norm $\bar{q}_B$ on E/J which is called a <u>base norm</u>. The distinguishing property of q_B and $\bar{q}_B$ is additivity on E_+. A typical (but not the most general) example of an ordered vector space whose positive cone has a base, is the space E' of Example 1. (Here the base K is compact, and J = $\{0\}$.) A lattice ordered base normed Banach space is isomorphic to $L^1(\mu)$, where μ is a suitably chosen Radon measure on some locally compact space (Kakutani).

In conclusion, let us consider the following example which illuminates the entire preceding discussion. Let X be a set, Σ a σ-algebra of subsets, and let μ denote a finite positive measure on Σ. Then $L^1(\mu)$ is a base normed vector lattice with base $B = \{ f \geq 0 : \int f \, d\mu = 1 \}$, and its dual $L^\infty(\mu)$ is an order-unit-normed vector lattice with order unit e (the constant-one function on X). e is a weak order unit of $L^1(\mu)$, and $L^\infty(\mu)$ can be identified with the ideal of L^1 generated by e.

The extreme points of $[0,e]$ are the (equivalence classes of) characteristic functions of measurable sets $S \in \Sigma$; each projection band in L^1 (and L^∞) consists of the functions vanishing a.e. (μ) outside some fixed measurable set S, and the associated band projection is multiplication, $P_S : f \mapsto \chi_S f$, with the characteristic function of S. Thus the isomorphism of the last result of §2 is the mapping $P_S \mapsto \chi_S$. In particular, $\mathcal{P}$ is isomorphic with Σ/N, the Boolean algebra of measurable sets modulo μ-null sets.

On the other hand, the order unit normed space $L^\infty(\mu)$ can be identified with C(K), where K is the Stone representation space of the Boolean algebra Σ/N (see [V], [S_2]).

<u>References</u>

[A] Alfsen, E.M., Compact Convex Sets and Boundary Integrals. Springer-Verlag, Berlin-Heidelberg-New York 1971.

[DS] Dunford, N. and J.T. Schwartz, Linear Operators. Vol.I. Interscience Publ. 4th print, New York 1967.

[J] Jameson, G., Ordered Linear Spaces. Springer Lecture Notes No. 141, 1970.

[LZ] Luxemburg, W.A.J. and A.C. Zaanen, Riesz Spaces I, North-Holland Publ.Co., Amsterdam-London 1971.

[P] Peressini, A.L., Ordered Topological Vector Spaces.
Harper and Row, New York-Evanston-London 1967.

[S_1] Schaefer, H.H., Topological Vector Spaces. 3rd print.
Springer-Verlag, Berlin-Heidelberg-New York 1971

[S_2] Schaefer, H.H., Banach Lattices and Positive Operators.
Springer-Verlag (in preparation).

[V] Vulikh, B.Z., Introduction to the Theory of Partially Ordered Spaces. (Engl.Transl.)
Wolters-Noordhoff, Groningen 1967.

DUALITY OF CONES IN LOCALLY CONVEX SPACES

Jürgen Mangold and Rainer J. Nagel

Fachbereich Mathematik der Universität Tübingen

Tübingen, Germany

If a cone C in a locally convex space E is "big", its dual cone $C' := \{x' \in E' : \langle x, x' \rangle \geq 0 \text{ for all } x \in C\}$ is "small" and vice versa. This behavior will be expressed more precisely in section 2 by the duality theorem for normal and strict $\mathcal{S}$-cones due to S c h a e f e r ([4], V.3).

In normed vector spaces the normality resp. strictness of a cone can be measured by numerical constants. Again, a strong duality theorem is valid (section 3; see [2]). In section 4 we discuss a property of ordered Banach spaces which is motivated by the theory of Banach lattices (see [1]).

While all the results are more or less well known, we prove the main theorems by a new method (already used by K u n g - F u N g [3] in the normed case): By a consequent use of some basic properties of polars (see section 1), all proofs become simple and mechanical computations.

In general we follow the terminology of [4] and refer to [2] and [4] for additional information and historical comments.

1. Computation rules for polars

For a locally convex space E, we consider its dual space E' and the duality $\langle E,E'\rangle$. If M is a subset of E (resp. of E'), the polar of M is defined by

$$M^{\circ} := \{x' \in E' : \langle x,x'\rangle \leq 1 \quad \text{for all } x \in M\}$$

$$(\text{resp. } M^{\circ} := \{x \in E : \langle x,x'\rangle \leq 1 \quad \text{for all } x' \in M\}).$$

The basic properties of polars (especially the bipolar theorem) are proved in [4], IV.1. For our purposes we need some additional computation rules:

Let M and N be convex subsets of E (resp. E') containing the origin. The following inclusions are easy to prove:

(1) $(M + N)^{\circ} \supset 1/2\ (M^{\circ} \cap N^{\circ})$

(2) $(M \cap N)^{\circ} \supset 1/2\ (M^{\circ} + N^{\circ})$

(3) $(M + N)^{\circ} \subset M^{\circ} \cap N^{\circ}$

(4) if $\overline{M} \cap \overline{N} = \overline{M \cap N}$, then

$$(M \cap N)^{\circ} = \overline{co}\ (M^{\circ} \cup N^{\circ}) \subset \overline{M^{\circ} + N^{\circ}} \quad ^{1)}$$

<u>Remarks</u>: 1. For a convex subset of E, its closure is the same for all locally convex topologies consistent with $\langle E,E'\rangle$ and hence identical with the $\sigma(E,E')$-closure (see [4], IV.3.1).

2. Rule (4) follows from [4], IV.1.5, corollary 2, and the fact that $(M \cap N)^{\circ} = (\overline{M \cap N})^{\circ}$.

3. The assumption in (4) is satisfied if M and N are closed (trivially) or if the origin is an interior point of M and N (use [4], II.1.1 ; for a more detailed proof, see [3]).

If E is a normed vector space with closed unit ball U, we can deduce from (3) and (4) further computation rules for polars, which will be needed in section 3: Let C be a convex cone of vertex O such that $O \in C$. The dual cone is $C' = (-C)^{\circ}$. For a detailed proof of the following rules, see again [3] :

(5) $((U + C) \cap (U - C))^{\circ} = co((C' \cap U^{\circ}) \cup -(C' \cap U^{\circ}))$ 2)

---

1) On E' we consider (if not otherwise stated) the $\sigma(E',E)$-topology.

2) Since $C' \cap U^{\circ}$ is $\sigma(E',E)$-compact, $co((C' \cap U^{\circ}) \cup -(C' \cap U^{\circ}))$ is closed.

(6) if C is closed:

$$((U^{o} + C') \cap (U^{o} - C'))^{o} = \overline{co}((C \cap U) \cup -(C \cap U)) .$$

2. Duality of cones in locally convex spaces

In this section, we assume that E is a real locally convex space with 0-neighborhood base $\mathcal{U}$. Let C be a convex cone of vertex 0 in E such that $0 \in C$. The dual cone is

$$C' := \{x' \in E' : \langle x, x' \rangle \geq 0 \quad \text{for all } x \in C\} = (-C)^{o}$$

The following definition will help us to gauge the pointedness, resp. the bluntness of C and C'.

(2.1) DEFINITION: Let A be a subset of E. We call

(i) $[A] := (A + C) \cap (A - C)$ the C-saturated hull of A [3].

(ii) $]A[:= co((A \cap C) \cup -(A \cap C))$ the C-convex kernel of A.

For subsets A' of E' and the dual cone C', the sets $[A']$ and $]A'[$ are defined in an analogue way.

Example: Take $E = \mathbb{R}^2$, $C = \{(\lambda, \gamma) : \lambda \geq 0, \gamma \geq 0\}$. Let A be the euclidean unit ball :

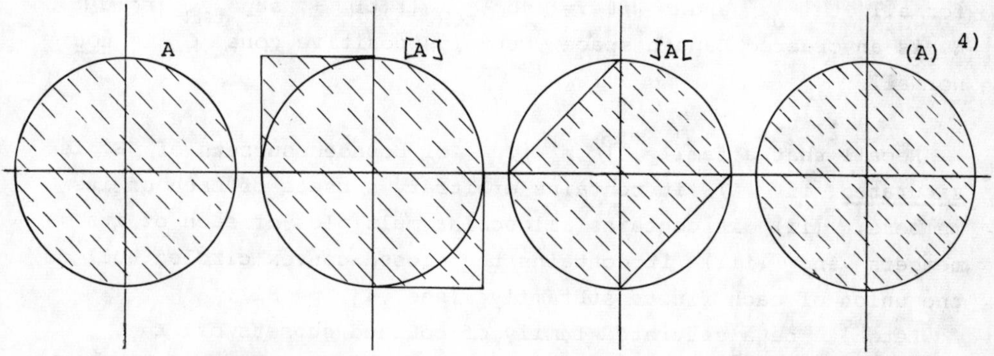

3) If C is a proper cone (and hence induces an ordering on E), we have $[A] = \bigcup_{x,y \in A} [x,y]$, where $[x,y] = \{z \in E: x \leq z \leq y\}$.

4) $(A) := \bigcup_{x \in A} [-x,x]$; see (4.2).

The family $[\mathfrak{U}] := \{[U] : U \in \mathfrak{U}\}$ defines a locally convex topology on E which is the coarser the 'blunter' the cone C is. Hence we can restrict the 'width' of C by requiring C to have the following property.

(2.2) <u>DEFINITION</u>: C <u>is a normal cone (with respect to</u> $\mathfrak{U}$) <u>if</u> $[\mathfrak{U}]$ <u>is equivalent to</u> $\mathfrak{U}$.

<u>Remarks</u>: 1. Every normal cone is proper (since E is Hausdorff) and the closure of a normal cone is normal. See [4], V.3.1 for further elementary properties.

2. J a m e s o n [2] , p. 88 , uses the term 'self-allied wedge'.

3. Let C be a cone in a normed vector space E with closed unit ball U . C <u>is normal iff there is</u> α, $1 \leq \alpha \in \mathbb{R}$ <u>such that</u> $[U] \subset \alpha U$ (or equivalently: there is α, $1 \leq \alpha \in \mathbb{R}$ such that $\|x\|$, $\|y\| \leq 1$ and $x \leq z \leq y$ implies $\|z\| \leq \alpha$).

<u>Examples</u>: 1. The positive cone in a locally convex vector lattice is normal ([4], V.7.1).

2. Let E be a vector space of bounded real-valued functions on a set X , endowed with the canonical ordering and the sup-norm. Then E is a normed vector space ordered by a normal cone.

3. Let E be the space of all real-valued functions on $\mathbb{R}$ having a bounded continous derivative. Define $C := \{f \in E : f(t) \geq 0$ for all $t \in \mathbb{R}\}$ and $\|f\| := \sup_{t\in\mathbb{R}} |f(t)| + \sup_{t\in\mathbb{R}} |f'(t)|$. E is an ordered Banach space , but its positive cone C is <u>not</u> normal.

Recall that a family $\mathfrak{S} \neq \{\emptyset\}$ of bounded subsets of E is <u>saturated</u> if (i) it contains arbitrary subsets of each of its members, (ii) it contains all scalar multiples of each of its members and (iii) it contains the closed convex circled hull of the union of each finite subfamily (see [4], p. 81).

Let $\mathfrak{S}$ be a saturated family of bounded subsets of E . Obviously, $]\mathfrak{S}[:= \{]S[: S \in \mathfrak{S} \}$ is a subfamily of $\mathfrak{S}$ whose elements are as 'small' as the cone C is 'small'. Hence the following property ensures that C is not too 'pointed'.

(2.3) <u>DEFINITION</u>: C <u>is a strict</u> $\mathfrak{S}$<u>-cone if</u> $]\mathfrak{S}[$ <u>is a fundamental subfamily of</u> $\mathfrak{S}$.[5)]

<u>Remarks</u>: 1. If $\mathfrak{S}'$ is a family of $\sigma(E',E)$-bounded subsets of E' , an analogous definition applies to the dual cone C'.

2. If $E = \bigcup\{S : S \in \mathfrak{S}\}$, every strict $\mathfrak{S}$-cone is generating, i.e., $E = C - C$.

3. J a m e s o n [2], p. 98, uses the term '(A)-decomposition'.

4. Let C be a cone in a normed vector space with closed unit ball U . Take for $\mathfrak{S}$ the family $\mathfrak{B}$ of <u>all</u> bounded subsets of E . Then C <u>is a strict</u> $\mathfrak{B}$ <u>-cone iff there is</u> $1 \leq \alpha \in \mathbb{R}$ <u>such that</u> $U \subset \alpha]U[$ (or equivalently: there is $1 \leq \alpha \in \mathbb{R}$ such that each $z \in E$ can be written as $z = x - y$, where $x,y \in C$ and $\|x\| + \|y\| \leq \alpha\|z\|$).

<u>Examples</u>: 1. The positive cone in a locally convex vector lattice E is a strict $\mathfrak{B}$-cone (for $\mathfrak{B}$ the family of all bounded subsets of E) (see [4], V.7.2).

2. Any closed and generating cone in an ordered Banach space is a strict $\mathfrak{B}$-cone (see (3.3) , or [4], V.3.5, corollary).

3. Further examples may be deduced from the subsequent duality theorems.

We are now able to formulate and prove S c h a e f e r ' s duality theorem for normal and strict $\mathfrak{S}$-cones in locally convex spaces (see [4], V.3.3). To this end we recall that for every saturated family $\mathfrak{S}$ of bounded[6)] subsets covering E (resp. of $\sigma(E',E)$-bounded subsets covering E') , the family of polars $\{S^o : S \in \mathfrak{S}\}$ is a 0-neighborhood base for the so-called $\mathfrak{S}$<u>-topology</u> on E' (resp. E). In particular: A locally convex topology on E (resp. on E') is consistent with the given duality $\langle E,E'\rangle$ if and only if it is the $\mathfrak{S}$-topology for a saturated family $\mathfrak{S}$, covering E' , of $\sigma(E',E)$-relatively compact subsets of E' (resp. covering E , of $\sigma(E,E')$-relatively compact subsets of E). For details and further results see [4], III.3 , IV.1.5 and IV.3 .

5) This means that each member of $\mathfrak{S}$ is contained in some member of $]\mathfrak{S}[$.

6) i.e. $\sigma(E,E')$-bounded (see [4], IV.3.2, corollary 2) .

(2.4) THEOREM: Let E be a locally convex space and let C be a cone in E with dual cone C' in E'.

(i) Let $\mathfrak{S}$ be a saturated family of $\sigma(E',E)$-relatively compact subsets covering E':
C' is a strict $\mathfrak{S}$-cone iff C is normal for the $\mathfrak{S}$-topology on E.

(ii) Let $\mathfrak{S}$ be a saturated family of $\sigma(E,E')$-relatively compact subsets covering E:
$\overline{C}$ is a strict $\mathfrak{S}$-cone iff C' is normal for the $\mathfrak{S}$-topology on E'.

Remark: For greater symmetry, we did not state the above theorem in its greatest possible generality. See [4], V.3 for additional results.

Proof: By using the rules (1) - (4) of section 1, all implications are proved easily. As an example we give an explicit proof of (ii):

'$\Longrightarrow$': Choose a 0-neighborhood S^o, $S \in \mathfrak{S}$, for the $\mathfrak{S}$-topology on E'. We have to show, that there is a $S_1 \in \mathfrak{S}$ such that $[S_1{}^o] \subset S^o$ (compare definition (2.2)).

By assumption, there is $S_1 \in \mathfrak{S}$, circled and convex, such that

$S \subset]1/4\, S_1[\; = \; co\,((1/4\, S_1 \cap \overline{C}) \cup -(1/4\, S_1 \cap \overline{C}))$.

A fortiori, we have

$S \subset 1/4\, (S_1 \cap \overline{C} \; - \; S_1 \cap \overline{C})$.

By taking polars we get

$S^o \supset 4(S_1 \cap \overline{C} \; - \; S_1 \cap \overline{C})^o$

by (1)

$\supset 2((S_1 \cap \overline{C})^o \cap (S_1 \cap (-\overline{C}))^o)$

by (2)

$\supset (S_1{}^o - C') \cap (S_1{}^o + C') \; = \; [S_1{}^o]$.

'$\Longleftarrow$': For $S \in \mathfrak{S}$ we have to show, that there exists $S_1 \in \mathfrak{S}$ such that $S \subset]S_1[$ (compare (2.3)). If C' is normal for the $\mathfrak{S}$-topology on E', there is a circled, convex, $\sigma(E,E')$-compact $S_1 \in \mathfrak{S}$ such that $[S_1{}^o] \subset S^o$.

Taking polars we get

$S \subset S^{oo} \subset [S_1{}^o]^o \; = \; ((S_1{}^o + C') \cap (S_1{}^o \; - \; C'))^o$.

Applying rule (4) and (3) yields

$S \subset \overline{co}\,((S_1^{\circ} + C')^{\circ} \cup (S_1^{\circ} - C')^{\circ})$
$\quad \subset \overline{co}\,((S_1^{\circ\circ} \cap (-C)^{\circ\circ}) \cup (S_1^{\circ\circ} \cap C^{\circ\circ}))$.

Since $S_1^{\circ\circ} = S_1$ and $C^{\circ\circ} = \overline{C}$ we get finally

$S \subset \overline{co}\,(-(S_1 \cap \overline{C}) \cup (S_1 \cap \overline{C})) = co\,((S_1 \cap \overline{C}) \cup -(S_1 \cap \overline{C}))$,

where the last equality holds since $S_1 \cap \overline{C}$ is $\sigma(E,E')$-compact.

3. Numerical duality theorems for cones in normed vector spaces

The previous duality theorem can be considerably strengthened for normal and strict $\mathcal{B}$ -cones in normed vector spaces ($\mathcal{B}$ the family of all bounded subsets). Not only will we obtain complete symmetry between normal and strict $\mathcal{B}$ -cones in the normed spaces E and E'_{β} (notice that the norm topology $\beta(E',E)$ on E' is not necessarily consistent with the duality $\langle E,E'\rangle$, hence theorem (2.4) does not apply), but we also can introduce a numerical constant which measures the normality resp. the strictness of the cone, and which is preserved under duality. This constant appeared already in (2.2), remark 3, and in (2.3), remark 4. We restate these properties, using the terminology of J a m e s o n [2], 3.6 .

(3.1) DEFINITION: Let E be a normed vector space with closed unit ball U and let C be a cone in E .

(i) C is α-normal if $[U] \subset \alpha U$ for $\alpha \in \mathbb{R}$.

(ii) C is α-generating if $U \subset \alpha\,]U[$ for $\alpha \in \mathbb{R}$.

Remarks: 1. C is normal iff C is α-normal for some $\alpha \in \mathbb{R}$.

2. C is a strict $\mathcal{B}$-cone iff C is α-generating for some $\alpha \in \mathbb{R}$.

Examples: 1. The positive cone in an order unit space is 1-normal and 2-generating.

2. The positive cone in a base norm space is 2-normal and 1-generating.

3. The positive cone in a Banach lattice is 2-normal and 2-generating.

(3.2) THEOREM: *Let E be a normed vector space and let C be a cone in E . Let E' have the norm topology.*

(i) *C is α-normal iff C' is α-generating.*

(ii) *If E is a Banach space and if C is closed:*
C' is α-normal iff C is $(\alpha+\varepsilon)$-generating for all $\varepsilon > 0$.

Proof: In fact, most of the proof is an easy computation using the rules (5) and (6) of section 1. Only for one implication in (ii) we have to do some extra work:

(i) With (5) it is clear that $(U+C) \cap (U-C) \subset \alpha U$ if and only if $\alpha^{-1}U^o \subset co((C' \cap U^o) \cup -(C' \cap U^o))$.

(ii) Similarly we have with the aid of (6):
$(U^o+C') \cap (U^o-C') \subset \alpha U^o$ if and only if
$\alpha^{-1}U \subset \overline{co}((C \cap U) \cup -(U \cap C))$.
The following lemma implies that this is equivalent to
$(\alpha+\varepsilon)^{-1} U \subset co((U \cap C) \cup -(U \cap C)) =]U[$.

LEMMA: *Let E be a Banach space, and let A, B be closed, convex, bounded subsets of E . The interior of $\overline{co}(A \cup B)$ is contained in co $(A \cup B)$.*

Proof: Let x be an interior point of $\overline{co}(A \cup B)$ .There exists $a_1 \in co(A \cup B)$ such that $b_1 := 2x - a_1 = x + (x - a_1)$ is contained in the interior of $\overline{co}(A \cup B)$. Therefore we have $x = 1/2\,(a_1 + b_1)$. Repeating this procedure for b_1 gives an $a_2 \in co(A \cup B)$ and b_2 in the interior of $\overline{co}(A \cup B)$ such that $x = 1/2\, a_1 + 1/4\, a_2 + 1/4\, b_2$. Finally we get $x = \sum_{n\in\mathbb{N}} 2^{-n} a_n$, where $a_n \in co(A \cup B)$. Since A and B are bounded and complete, we conclude $x \in co(A \cup B)$.

(3.3) COROLLARY: *Let E be a Banach space with closed cone C . The following are equivalent:*

(a) $E = C - C$ *and* $E' = C' - C'$.

(b) *C is a normal, strict $\mathcal{B}$-cone.*

(c) *C' is a normal, strict $\mathcal{B}$-cone in E'_β .*

(d) *C is α-normal and β-generating for some $\alpha, \beta \geq 1$.*

Proof: The equivalence of (b) and (d) follows from the remarks after (3.1), and the above theorem implies the equivalence of (d) and (c).

'(d) $\Rightarrow$ (a)': C is β-generating in E, and C' is α-generating in E'_β. A fortiori, C and C' are generating cones.

'(a) $\Rightarrow$ (b)': $M := \overline{]U[}$ is closed, convex, circled and such that $E = \bigcup_{n \in \mathbb{N}} nM$. E is a Baire space, hence M is a 0-neighborhood. By the above lemma, it follows again that $\varepsilon U \subset]U[$ for some $\varepsilon > 0$. As a consequence, we see that a closed generating cone in a Banach space is a strict $\mathcal{L}$-cone. Similarly, one can show that, under the assumptions of (a), C' is a strict $\mathcal{B}$-cone in E'_β, hence C is normal in E.

4. Regularly ordered normed vector spaces

In this section we assume that E is a normed vector space ordered by a closed positive cone C. If C is α-normal and β-generating, it is interesting to know, how small α and β can be chosen simultaneously. It turns out that α and β cannot simultaneously attain their minimum (which is 1).

(4.1) PROPOSITION: Let E be an ordered normed vector space with $\dim E \geqslant 2$. If its positive cone C is α-normal and β-generating, then $\alpha+\beta > 2$. If, in addition, E is a Banach lattice, $2\sqrt{2}$ is the best lower bound for $\alpha+\beta$.

Proof: Assume that C is 1-generating. Then exists $x \in E$ such that $x = x_1 - x_2$, $0 \neq x_1$, $x_2 \in C$, and $\|x\| = \|x_1\| + \|x_2\| = 1$. If C is also 1-normal, $-x_2 \leq x \leq x_1$ yields the contradiction $\|x\| \leq \max(\|x_1\|, \|x_2\|) < 1$. Therefore, $\alpha+\beta > 2$.

If E is a Banach lattice and C is α-normal, there exists $x \in E$ such that $-x^- \leq x \leq x^+$, $\|x^+\| = \|x^-\| = 1$ and $\|x\| \leq \alpha$. For $y := (\|x\|)^{-1} x$ we get the decomposition

$$y = (\|x\|)^{-1}(x^+ - x^-) \quad \text{and hence}$$

$$(\|x\|)^{-1}(\|x^+\| + \|x^-\|) = 2(\|x\|)^{-1} \geqslant 2/\alpha .$$

This shows that C is not β_0-generating for $\beta_0 < 2/\alpha$ and hence

$\alpha + \beta \geq \alpha + 2/\alpha$. The minimum of the right side is attained for $\alpha = \sqrt{2}$. The example in (2.1) shows that this bound can be attained.

The above proposition (and the example in (2.1)) motivates the search for a more symmetric property for cones in Banach spaces. The following results can be found in [1] or [2] , p.115 .

(4.2) DEFINITION: Let E be an ordered normed vector space with closed unit ball U .

(i) $(U) := \bigcup_{x \in U} [-x,x]$; $(U^o) := \bigcup_{x' \in U^o} [-x',x']$.

(ii) E is regularly ordered if

$(U) \subset U \subset (1+\varepsilon)(U)$ for all $\varepsilon > 0$.

Remarks: 1. $]U[\subset (U) \subset [U]$.

2. Every Banach lattice is regularly ordered.

3. E is regularly ordered iff $x,y \in E$ and $-x \leq y \leq x$ implies $\|y\| \leq \|x\|$, and for $x \in E$, $\|x\| < 1$, there exists $y \in E$ such that $-y \leq x \leq y$ and $\|y\| < 1$.

4. If E is an ordered Banach space with closed cone C , E is regularly ordered iff U is equal to the closure of (U) (use lemma (3.2)).

(4.3) LEMMA: $(U)^o = (U^o)$.

Proof: It is clear that

$\bigcup_{x' \in U^o} [-x',x'] \subset (\bigcup_{x \in U} [-x,x])^o$.

For $y' \in (U)^o$ define

$p(x) := \sup\{\langle y,y' \rangle : y \in [-x,x]\}$, $x \in E$.

Certainly, $p(x) \leq 1$ for all $x \in U \cap C$. Since p is superlinear on C [7] , $P := \{z \in C: p(z) > 1\}$ is convex.

If $P = \emptyset$, take $x' = 0$ and we get $y' \in [-x',x']$.

Assume now $P \neq \emptyset$: the first separation theorem ([4], II.9.1) implies the existence of $x' \in E'$ separating U and P , i.e.

$\langle y,x' \rangle \leq 1$ for all $y \in U$, and

$\langle y,x' \rangle \geq 1$ for all $y \in P$.

---

7) i.e. $p(x + y) \geq p(x) + p(y)$ and $p(\lambda x) = \lambda p(x)$ for all $x,y \in C$, $\lambda \in \mathbb{R}_+$.

From the first inequality follows $x' \in U^o$. Take $y \in C$. The second inequality implies

$p(y) \leq \langle y,x' \rangle$ if $p(y) \neq o$.

If $p(y) = o$, take $z \in P$. Then $(ny + z) \in P$ for all $n \in \mathbb{N}$, hence $n\langle y,x' \rangle + \langle z,x' \rangle \geq 1$ and $\langle y,x' \rangle \geq o$.

From both cases we conclude

$\pm\langle y,y' \rangle \leq p(y) \leq \langle y,x' \rangle$ for all $y \in C$,

hence $y' \in [-x',x'] \subset (U^o)$.

The following theorem expresses the symmetric behaviour of (U) and (U^o) , which is similar to that of $[U]$ and $]U^o[$ (see [2], 3.6.7 and 3.6.8) . The subsequent corollaries follow easily and express the main properties of regularly ordered Banach spaces.

(4.4) _THEOREM_: _Let_ E _be an ordered Banach space with closed positive cone and closed unit ball_ U . _Let_ $\alpha \in \mathbb{R}$.

(i) $(U) \subset \alpha U$ _iff_ $U^o \subset \alpha(U^o)$.

(ii) $(U^o) \subset \alpha U^o$ _iff_ $U \subset (\alpha+\varepsilon)(U)$ _for all_ $\varepsilon > o$.

Proof: (i) is trivial by the above lemma.

(ii) ' $\Leftarrow$ ' : From $U \subset (\alpha+\varepsilon)(U)$ we get by taking polars $(\alpha+\varepsilon)U^o \supset (U)^o = (U^o)$. Because this is valid for all $\varepsilon > o$, we conclude $\alpha U^o \supset (U^o)$.

' $\Rightarrow$ ' : Again taking polars we get

$\alpha\overline{(U)} = \alpha(U^o)^o \supset U^{oo} = U$.

Since $\sum_{n\in\mathbb{N}} 2^{-n} x_n \in (U)$ for any sequence $\{x_n\} \subset (U)$, one can show as in lemma (3.2) that the interior of $\overline{(U)}$ is contained in (U) . A fortiori,

$(\alpha+\varepsilon)(U) \supset U$ for all $\varepsilon > o$.

(4.5) _COROLLARY_: _Every ordered Banach space_ E _with closed generating positive cone and generating dual cone has an equivalent norm under which_ E _is regularly ordered_.

Proof: If U is the (original) unit ball, then (U) defines a new, but equivalent norm, for which E is regularly ordered.

(4.6) <u>COROLLARY</u>: <u>An ordered Banach space is regularly ordered iff its dual is regularly ordered</u>.

references:

[1] Davis, E.B.: The structure and ideal theory of the predual of a Banach lattice. Trans. Amer. Math. Soc. <u>131</u>, 544-555 (1968).

[2] Jameson, G.: Ordered Linear Spaces. Lecture Notes in Math. 141, Berlin-Heidelberg-New York: Springer 197o.

[3] Ng, Kung-Fu: On a computation rule for polars. Math. Scand. <u>26</u>, 14-16 (197o).

[4] Schaefer, H.H.: Topological Vector Spaces, 3rd print. Berlin-Heidelberg-New York: Springer 1971.

ORDER UNIT AND BASE NORM SPACES

Rainer J. Nagel

Fachbereich Mathematik der Universität Tübingen

Tübingen, Germany

The dual behavior of (AM)- and (AL)-spaces and the Kakutani represenatation theorems are well known (e.g. see: [6], V.8). Order unit and base norm spaces are the natural generalizations to the non-vector-lattice case, and we will see that similar results are valid.

In section 1 we represent every order unit Banach space as a space $A(K)$. The dual of an order unit space is a base norm space, whose basic properties are stated in section 2. Finally, the complete duality theory of order unit and base norm spaces will be developed in section 3.

We use the terminology of Schaefer [6]; the results can be found in Alfsen [1], II.1 or Jameson [2], 3.7 - 3.9.

1. Order unit spaces

Let E be an ordered vector space (over $\mathbb{R}$) with positive cone E_+ having an order unit u 1). If E is Archimedean,

$$p_u(x) := \inf\{\lambda \in \mathbb{R}_+ : x \in \lambda[-u,u]\} \quad , \quad x \in E ,$$

---

1) u is an order unit if $E = \bigcup_{n \in \mathbb{N}} n[-u,u]$.

is a norm on E, for which the unit ball $U = [-u,u]$ and the positive cone E_+ are closed (see [2], 3.7.2) .

(1.1) <u>DEFINITION</u>: <u>An Archimedean ordered vector space</u> E <u>with order unit</u> u <u>is called an order unit space</u> (<u>order unit Banach space</u>) <u>if it is endowed with the order unit norm</u> p_u (<u>and is complete</u>).

<u>Remarks</u>: 1. The positive cone E_+ in an order unit space is 1-normal, 2-generating and regular (see [3] for the definitions).

2. An ordered Banach space is isomorphic (as ordered topological vector space) to an order unit Banach space iff the positive cone is normal, closed and has interior points ([6], V.6.2, corollary 1).

<u>Examples</u>: 1. Let $\mathfrak{A}$ be a C^*-algebra (with algebraic unit). The subspace of all self-adjoint elements $\mathfrak{A}_s := \{x \in \mathfrak{A} : x = x^*\}$ ordered by the positive cone $\mathfrak{A}_+ := \{x \in \mathfrak{A} : x = yy^*\}$ is an order unit Banach space (see [5]).

2. Let K be a compact convex subset of a locally convex vector space. The space $A(K)$ of all continuous affine functions on K is an order unit Banach space for the natural positive cone and the sup-norm. We will see in (1.3), that every order unit Banach space can be represented as an $A(K)$.

3. Let E be an arbitrary Banach space. Take F a closed homogeneous hyperplane in E and write $E = F \oplus \mathbb{R}$. Define a cone $E_+ := \{(x,\lambda) : \|x\| \le \lambda\}$ and set $u = (o,1)$. The corresponding norm p_u is a new, but equivalent norm on E, which makes E an order unit Banach space.

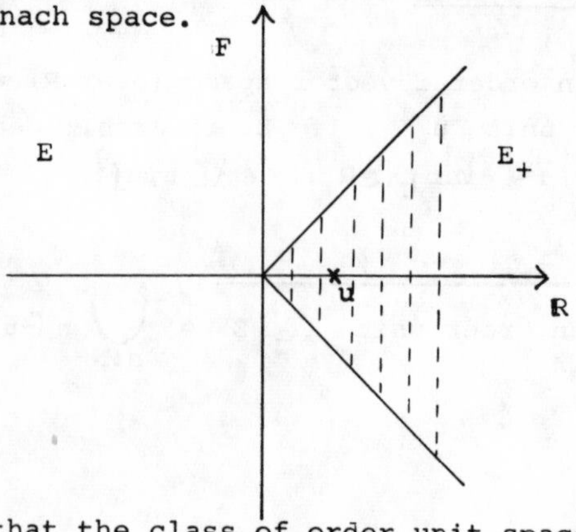

This example shows that the class of order unit spaces "contains" all Banach spaces.

(1.2) <u>DEFINITION</u>: <u>Let</u> (E,u,p_u) <u>be an order unit space</u>. <u>The set</u>

$$K := \{x' \in E' : o \leq x' \text{ and } \langle u,x'\rangle = 1\}$$

<u>is called the state space of</u> E .

<u>Remarks</u>: 1. K is a $\sigma(E',E)$-compact convex subset of the dual unit ball U^o .

2. K is a <u>base</u> of the dual cone E'_+ , i.e. for each $o \neq x' \in E'_+$ there exists a unique $\lambda \in \mathbb{R}_+$ such that $\lambda x' \in K$.

3. $x \in E$ is positive iff $\langle x,x'\rangle \geq o$ for all $x' \in K$: The second condition implies $\langle x,x'\rangle \geq o$ for all $x' \in E'_+$ (K is a base !) . Since E_+ is closed, this implies $x \in E_+ = (-E'_+)^o$.

4. $U^o = co(K \cup -K)$: This follows from the fact that K is a base of the 1-generating cone E'_+ , i.e. $U^o = co((U^o \cap E'_+) \cup \cup -(U^o \cap E'_+))$ (see [3], 3.2) .

5. $\|x\| = \sup\{|\langle x,x'\rangle| : x' \in K\}$ for all $x \in E$: This is immediate from 4 .

6. The dual cone E'_+ is regular , 2-normal and 1-generating (for the norm topology on E') . This again follows from [3] .

(1.3) <u>THEOREM</u>: <u>Every order unit Banach space</u> E <u>is norm and order isomorphic to the space</u> A(K) <u>of all continuous affine functions on the state space</u> K <u>of</u> E .

<u>Proof</u>: The evaluation $x \longrightarrow \langle x,x'\rangle$ for $x \in E$, $x' \in K$ defines a continuous affine function on K . By remark 3 and 5 this evaluation is an order and norm isomorphism and it remains to show , that every element $a \in A(K)$ can be linearly extended to a $\sigma(E',E)$-continuous linear form on E' , i.e. to an element of E .

First observe that the linear extension of $a \in A(K)$ to E' is a well defined linear form $\bar{a}$ on E' . For the $\sigma(E',E)$-continuity of $\bar{a}$ it suffices to show that $\{x' \in U^o : \langle \bar{a},x'\rangle = o\}$ is $\sigma(E',E)$-closed (use the Krein-Smulian theorem ([6], IV.6.4) and the completness of E):

Let $\{x'_\alpha\} \subset U^o = co(K \cup -K)$ be a net converging weakly to $x' \in E'$. Decompose $x'_\alpha = \lambda_\alpha x'_{\alpha 1} - (1 - \lambda_\alpha)x'_{\alpha 2}$ such that $\lambda_\alpha \in [o,1]$, $x'_{\alpha 1}$, $x'_{\alpha 2} \in K$. Since K is compact, we can choose a subnet $\{x'_\beta\}$ such that $\{\lambda_\beta\}$, $\{x'_{\beta 1}\}$ and $\{x'_{\beta 2}\}$ converge. But $\bar{a}$ is continuous on K and linear on U^o . Hence $\langle \bar{a},x'_\beta\rangle = o$ for all β implies $\langle \bar{a},x'\rangle = o$, qed.

Remark: Let X be the $\sigma(E',E)$-closure of the extreme points of K. The Krein-Milman theorem implies that E is also norm and order isomorphic to a closed subspace of $C(X)$, which is equal to $C(X)$ iff E is a vector lattice (K a k u t a n i's theorem; see [6], V.8.5 or the lecture of A. G o u l l e t de R u g y in this volume).

2. Base norm spaces

Some of the properties of the dual of an order unit space (as collected in the above remarks) are now used to define a new class of ordered normed vector spaces:

Let E be an ordered vector space with generating positive cone E_+. Assume that E_+ has a base K, i.e.

$$K = \{x \in E_+ : f(x) = 1\}$$

for some strictly positive linear form[2)] f on E. If

$$B := co(K \cup -K)$$

is linearly bounded[3)], then

$$p_B(x) := \inf \{\lambda \in \mathbb{R}_+ : x \in \lambda B\}$$

is a norm on E.

(2.1) DEFINITION: An ordered vector space E whose positive cone E_+ has a base K such that $B = co(K \cup -K)$ defines a norm p_B on E is called a base norm space.

Remarks: 1. If E is a base norm space, we have $K = \{x \in E_+ : \|x\| = 1\}$.

2. K is closed iff E_+ is closed ([2], 3.8.3).

3. There exists $f \in E'_+$ such that $\|f\| = 1$ and $f(x) = \|x\|$ for all $x \in E_+$, or: the norm of E is additive on E_+.

4. The positive cone in a base norm space is regular, 2-normal and α-generating for all $\alpha > 1$ (see [3]).

---

2) f is strictly positive if $f(x) > 0$ for all $0 \neq x \in E_+$.

3) B is linearly bounded (linearly compact) if $B \cap L$ is a bounded (bounded and closed) segment for every line L through the origin of E.

5. If B is linearly compact[3)], the closed unit ball U is equal to B. In this case E_+ is 1-generating.

6. If K is compact for some locally convex Hausdorff topology on E, then (E,p_B) is a Banach space ([1], II.1.12).

Examples: 1. Every (AL)-space, hence every $L^1(X,\mu)$ is a base norm space.

2. The dual of an order unit space is a base norm space by the remarks in 1.2.

3. The self-adjoint linear forms on a C^*-algebra form a base norm space. Moreover, the self-adjoint normal linear forms (i.e. the predual) of a W^*-algebra form a base norm space.

4. On every Banach space one can define an ordering and an equivalent norm under which it becomes a base norm space (compare 1.1, example 3).

(2.2) PROPOSITION: The dual of a base norm space is an order unit Banach space.

Proof: We have only to show that the linear form $f \in E'_+$ for which $K = \{x \in E_+ : f(x) = 1\}$, is an order unit in E' and that it determines the dual norm on E'. This is clear from the following computation:

$$\begin{aligned}
\|x'\| &= \sup\{|\langle x,x'\rangle| : x \in B\} \\
&= \sup\{|\langle x,x'\rangle| : x \in K\} \\
&= \inf\{\lambda \in \mathbb{R}_+ : -\lambda \le \langle x,x'\rangle \le \lambda \quad \text{for all} \quad x \in K\} \\
&= \inf\{\lambda \in \mathbb{R}_+ : -\lambda\langle x,f\rangle \le \langle x,x'\rangle \le \lambda\langle x,f\rangle \quad \text{for all} \quad x \in K\} \\
&= \inf\{\lambda \in \mathbb{R}_+ : x' \in \lambda[-f,f]\} \\
&= p_f(x') .
\end{aligned}$$

3. Duality of order unit and base norm spaces

(3.1) THEOREM: Let E be an ordered Banach space with closed and generating positive cone.

(i) E is an order unit space iff E' is a base norm space with $\sigma(E',E)$-compact base.

(ii) E is a base norm space iff E' is an order unit space.

Proof: Two implications were already proved in (1.2), remarks, and in (2.2). It remains to show the following:

(i) "$\Leftarrow$" : Let K be the $\sigma(E',E)$-compact base of E'_+ and let f be the linear functional on E' for which

$$K = \{x' \in E'_+ : f(x') = 1\} .$$

Since $\{x' \in E' : \|x'\| \le 1 \text{ and } f(x') = 0\}$ is equal to $1/2(K - K)$, it is $\sigma(E',E)$-closed. Hence f is $\sigma(E',E)$-continuous (again by [6], IV.6.4) and $f \in E_+$. As in (2.2), one shows that the unit ball U in E is equal to $[-f,f]$.

(ii) "$\Leftarrow$" : Let f be the order unit in E', so that the dual unit ball is $U^{\circ} = [-f,f]$. By [3], 3.2, the positive cone E_+ is α-generating for all $\alpha > 1$. Hence, the Minkowski functional of $B = co((E_+ \cap U) \cup (-E_+ \cap U))$, U the closed unit ball of E, defines the norm on E. Since

$$K = \{x \in E_+ : f(x) = 1\}$$

is a base of E_+, we get $B = co(K \cup -K)$ and consequently, E is a base norm space.

Remark: For E a vector lattice, we retrieve the classical duality theorems for (AM)- and (AL)-spaces ([6], V.8.4).

(3.2) The equivalence in (i) no longer holds if one drops the assumption that K be $\sigma(E',E)$-compact: The dual of c_o is ℓ^1 which is a base norm space, while c_o is not an order unit space. Without proof, we give a theorem of Kung-Fu Ng [4], which deals with this situation.

THEOREM: Let E be an ordered Banach space with closed and generating positive cone E_+:
E' is a base norm space iff E_+ is 1-normal and the open unit ball in E is directed upwards.

4. Problems

Let E be a base norm Banach space. Under which additional assumptions is it true, that

(i) E is an order ideal in E'' ?

(ii) each order intervall in E is weakly compact ?

(iii) E is the order ideal of all order continuous linear forms on some order complete order unit space ?

While none of these properties holds in the general case (use 2.1, example 4), they are true if E is a vector lattice ([6], V.8) or if E' is isomorphic (as an ordered Banach space) to the self-adjoint part of a C^*-algebra.

references

[1] Alfsen, E.M.: Compact Convex Sets and Boundary Integrals. Berlin-Heidelberg-New York: Springer 1971.

[2] Jameson, G.: Ordered Linear Spaces. Berlin-Heidelberg-New York: Springer 197o.

[3] Mangold, J. - Nagel, R.J.: Duality of cones in locally convex spaces. See the previous article.

[4] Ng, Kung-Fu: The duality of partially ordered Banach spaces. Proc. London Math. Soc. 19, 269-288 (1969).

[5] Sakai, S.: C^*-algebras and W^*-algebras. Berlin-Heidelberg-New York: Springer 1972.

[6] Schaefer, H.H.: Topological Vector Spaces, 3rd print. Berlin-Heidelberg-New York: Springer 1971.

MINIMAL DECOMPOSITIONS IN BASE NORMED SPACES

A.J. Ellis

Department of Pure Mathematics, University College of Swansea, Wales

Let E be a base normed Banach space with a closed positive cone E_+ and base K. Therefore, each $x \in E$ has, for each $\varepsilon > 0$, a positive decomposition $x = y - z$ such that $\|y\| + \|z\| \leq (1+\varepsilon)\|x\|$. If x has such a decomposition for $\varepsilon = 0$ then that decomposition is called minimal.

If E is a vector lattice, i.e. if K is a Choquet simplex, then if $x = y - z$, $y, z \geq 0$, we have $y \geq x^+$, $z \geq x^-$, so that $\|y\| + \|z\| \geq \|x^+\| + \|x^-\| = \|x\|$. In this case each x has a minimal decomposition.

Example (L. Asimow). There exists a base normed Banach space E and $x \in E$ such that x does not have a minimal decomposition: let E have the base norm induced by K where

$$E = \left\{\underline{x} \in c_0 : x_0 \in R\,,\ x_1 + x_2 = \frac{x_3}{2} + \frac{x_4}{2^2} + \ldots\right\},$$

$$K = \left\{\underline{x} \in E : x_i \geq 0\,\forall i\,,\ x_0 = 1 = \|\underline{x}\|_{c_0}\right\},\ \underline{x} = (0, \tfrac{1}{2}, -\tfrac{1}{2}, 0, 0, \ldots).$$

In this example the base norm is equivalent to the c_0-norm and $\underline{x} \in (1+\varepsilon)\mathrm{co}(K \cup -K)$ for all $\varepsilon > 0$, but $\underline{x}$ does not belong to $\mathrm{co}(K \cup -K)$.

For each strictly positive f on E, i.e. $f \in (E_+^*)^o$, we obtain a base $K_f = \{x \in E_+ : f(x) = 1\}$ for E_+, and an equivalent base norm, the Minkowski functional of $\mathrm{co}(K_f \cup -K_f)$ for E. Given an element $x \in E$ we consider the problem of finding a base norm for which x has a minimal decomposition.

Since if $x = y - z$ and $f \in E^*$ we have $f(y) + f(z) = 2f(y) - f(x)$ it is easy to show that $x = y - z$ is a minimal decomposition of x relative to $f \in (E_+^*)^o$ if and only if $y \in E_+ \cap (E_+ + x)$ and $f(y) = \inf\{f(u) : u \in E_+ \cap (E_+ + x)\}$. It is clear then that any

$f \in (E_+^*)^o$ which is a support functional for $E_+ \cap (E_+ + x)$ gives rise to a minimal decomposition of x. We have (cf. [4]):

Theorem 1. The strictly positive support functionals for $E_+ \cap (E_+ + x)$ are dense in $(E_+^*)^o$, and hence also in E_+^*.

Proof. If $f \in (E_+^*)^o$ then f is bounded below on $E_+ \cap (E_+ + x)$ and so, by a theorem of Bishop and Phelps [2], f can be approximated in the norm topology by support functionals for $E_+ \cap (E_+ + x)$. Since $f \in (E_+^*)^o$ it can be approximated by support functionals lying in $(E_+^*)^o$.

Although there are many support functionals for $E_+ \cap (E_+ + x)$ there may be only one support point, in fact if x^+ exists then x^+ will be the unique support point for all $f \in (E_+^*)^o$. Conversely, if only finitely many support points of $E_+ \cap (E_+ + x)$ exist (for $(E_+^*)^o$) then x^+ exists.

We now consider the existence of unique minimal decompositions. An element x will have a unique minimal decomposition $x = y - z$ relative to $f \in (E_+^*)^o$ if and only if f supports $E_+ \cap (E_+ + x)$ at the exposed point y. If $E_+ \cap (E_+ + x)$ is locally norm-compact, e.g. if E is finite-dimensional, then a result of Klee [7] shows that $E_+ \cap (E_+ + x)$ always has an exposed point y; it can be shown that y must be supported by some $f \in (E_+^*)^o$.

E will be said to have the unique minimal decomposition property (u.m.d.p.) if every $x \in E$ has a unique minimal decomposition relative to the given base K. This property of K can be interpreted [4] as an intersection property reminiscent of that defining Choquet simplexes:-

Theorem 2. E has the u.m.d.p. for K if and only if $co(K \cup -K)$ is closed and, for each $x \in E$, $K \cap (x + K)$ is either empty, a singleton, or contains a set of the form $y + \lambda K$, for some $y \in E$, $\lambda > 0$.

The AL-spaces have the u.m.d.p. for all bases, but then so does R^3 with a circular cone. An important example of a space E which possesses the u.m.d.p. for the given base is the space of Hermitian functionals on a B^*-algebra A with identity; here K is just the usual state-space for A, and $E = \text{lin} K$. This property of $\text{lin} K$ does not distinguish B^*-algebras amongst unital complex Banach algebras. In fact, if A is a Dirichlet algebra on a compact Hausdorff space Ω then the state-space K is precisely the probability Radon measures on Ω, so that $\text{lin} K$ is an AL-space.

Let A be a complex unital Banach algebra with identity e and let $K = \{\phi \in A^* : \phi(e) = 1 = \|\phi\|\}$ be the state-space of A. Then, using a result of Bohnenblust and Karlin (cf.[3]) and a construction of Asimow[1], the map $\theta : A \to A(Z) = A(co(K \cup -iK))$, $\theta a(z) = \mathrm{re}\, z(a)$ $\forall a \in A, z \in Z$, is a topological real-linear isomorphism of A onto A(Z), the Banach space of all continuous real-valued affine functions on Z with the supremum norm.

The dual space of A(Z) is the base normed space lin Z, with base Z. Using the Vidav-Palmer theorem (cf. [3]) the following characterization of B^*-algebras can be obtained [5].

<u>Theorem 3</u>. If A is a complex unital Banach algebra then A is a B^*-algebra if and only if $\mathrm{lin}\, Z (= A(Z)^*)$ has the u.m.d.p.

If lin Z is an AL-space then K is a simplex and it is well known that this implies that the B^*-algebra A is commutative. A related result is the following [5].

<u>Theorem 4</u>. If A is a complex unital Banach algebra such that A^* is isometrically isomorphic to a complex L^1-space then A is a commutative B^*-algebra.

Theorem 4 was first proved, for the case of a function algebra, by Hirsberg and Lazar [6].

<u>References</u>

1. L. ASIMOW, 'Decomposable compact convex sets and peak sets for function spaces', Proc. Amer. Math. Soc. 25(1)(1970) 75-9.
2. E. BISHOP and R.R. PHELPS, 'The support functionals of a convex set', Proc. Symp. Pure Mathematics VII(Convexity), Amer. Math. Soc. (1963), 27-35.
3. F.F. BONSALL and J. DUNCAN, '<u>Numerical ranges of operators on normed spaces and of elements of normed algebras</u>', Cambridge 1971.
4. A.J. ELLIS, 'Minimal decompositions in partially ordered normed vector spaces', Proc.Camb.Phil.Soc. 64(1968),989-100
5. A.J. ELLIS, 'Some applications of convexity theory to Banach algebras', (submitted for publication).
6. B. HIRSBERG and A.J. LAZAR, 'Complex Lindenstrauss spaces with extreme points', (to appear).
7. V.L. KLEE, 'Extremal structure of convex sets II', Math. Z. 69(1958), 90-104.

SIMPLEX SPACES

Alain GOULLET de RUGY

Equipe d'Analyse, Université de Paris VI

0. INTRODUCTION.

The aim of this lecture is to make a survey of the theory of integral representation on compact convex sets and its connexions with the theory of ordered Banach spaces.

As I am short of time and as there now exists the book of ALFSEN covering these matters, I shall not give full proofs. I shall simply try to give the major underlying ideas of the theory. Discussion of references is relegated to the last section.

1. KREIN-MILMAN THEOREM AND BAUER'S MAXIMUM PRINCIPLE.

1.0. NOTATIONS.

By a compact convex set, I shall always mean a compact convex set of a Hausdorff locally convex real topological vector space (HLCRTVS).

Let X be a compact convex set in some HLCRTVS E. Denote $A_c(X)$ the space of affine continuous functions on X and $Q_c(X)$ (resp. $Q_s(X)$) the space of convex continuous (resp. u.s.c.) functions on X.

An extreme point $x \in X$ is, by definition, a point such that $(X - \{x\})$ is still convex. Denote E(X) the set of extreme points of X.

1.1. THEOREM.-(BAUER's maximum principle).

Let f in $Q_s(X)$. Then f attains its maximum on E(X).

PROOF (sketch of).

Define a stable set of X to be a non empty closed subset S of X such that :

$$x, y \in X, \forall t \in]0,1[, \ tx + (1-t)y \in S \Longrightarrow x, y \in S.$$

Prove that the family of stable sets of X is downwards inductive and that the minimal stable sets are just the one point sets $\{x\}$ with x in E(X) using the following fact which relate stable sets and convex functions :

- If S is a stable set, $f \in Q_s(X)$ and $r = \max\{f(s) : s \in S\}$, then $S \cap F^{-1}(r)$ is a stable set.

1.2. COROLLARY (KREIN-MILMAN Theorem).

Let A be a closed subset of X. Then, the following are equivalent :

(a) X is equal to the closed convex hull $\overline{\text{conv}}(A)$ of A ;

(b) $A \supset E(X)$;

(c) A has the property : $\forall f \in A_c(X)$: $f \geq 0$ on $A \Longrightarrow f \geq 0$ on X ;

(d) A has the property : $\forall f \in Q_s(X)$: f attains its maximum on A.

PROOF.

By 1.1. (b) $\Longrightarrow$ (d) and it is clear that (d) $\Longrightarrow$ (c). The assertion (c) $\Longrightarrow$ (a) is a Hahn-Banach argument and (a) $\Longrightarrow$ (b) follows from :

1.3. LEMMA.

In a compact convex set X each extreme point x is strongly extreme in the sense that : For any neighbourhood V of x in X there exists $f \in E'$ and $r \in \underline{R}$ such that the slice $\{f < r\} \cap X$ contains x and is contained in V.

PROOF.

See H.H. SCHAEFER'lecture : The Silov boundary of a cone.

2. LINKS BETWEEN COMPACT CONVEX SETS AND ORDERED SPACES.

2.0. DEFINITIONS.

Let V be an order unit normed space. A state of V is a linear functional x on V such that : $x(V_+) \subset \underline{R}_+$ and $x(e) = 1$ where e denotes the order unit which define the norm in V. Clearly one can replace the assumption $x(e) = 1$ by $\|x\| = 1$. In particular, every state of V is continuous. The set X of all states of V is called the state space of V. Then, X is a non-empty convex and $\sigma(V',V)$-compact convex subset of the dual V' of V. For $x \in X$ and $a \in V$ let $\tilde{a}(x) = x(a)$. The map $a \mapsto \tilde{a}$, is called the KADISON map. It carries V into $A_c(X)$. The properties of this map are summarized in the following theorem, the proof of which reduces to the bipolar theorem. (See NAGEL'lecture : Order unit and base norm spaces).

2.1. THEOREM.

Let V be an order unit normed space and let X be the state space of V. Then the map $a \mapsto \tilde{a}$ from V into $A_c(X)$ is a bipositive linear isometry of V onto a dense subspace of the order unit Banach space $A_c(X)$ and $\tilde{e}(x) = 1$ for all $x \in X$, where e is the order unit of V. The map $a \mapsto \tilde{a}$ is onto $A_c(X)$ if an only if the normed space V is complete.

If we apply Kadison map to a space $A_c(X)$ for some compact convex set X, do we obtain the same space at the end? The answer is given by the following.

2.2. PROPOSITION.

Let X be a non-empty compact convex set and for every $x \in X$ let us denote by e(x) the evaluation map defined by $e(x)f = f(x)$ for any $f \in A_c(X)$. Then the map $x \mapsto e(x)$

is an affine homeomorphism of X onto the state space of $A_c(X)$.

2.3. COROLLARY.

Let X be a non-empty compact convex set and $\theta \in \mathcal{M}_+^1(X)$. Then, there exists a point x in X, denoted $r(\theta)$ and called the resultant or barycenter of θ such that :

$$\theta(f) = f(r(\theta)) \qquad (\forall f \in A_c(X)).$$

PROOF.

By restriction, a probability measure defines a state of $A_c(X)$. Conclusion follows from 2.2.

3. CHOQUET'S THEOREM OF INTEGRAL REPRESENTATION.

3.0. NOTATIONS.

In the sequel X will denote a fixed compact convex set. We shall identify X with the state space of $A_c(X)$ (see 2.2.). In particular, X is the base of the cone $\tilde{X} = \{y \in A_c(X)' \ ; \ y \geq 0\}$.

Consequently, to each $\theta \in \mathcal{M}_+(X)$ we can associate a point $x \in \tilde{X}$, denoted $r(\theta)$ and called the resultant of θ, such that :

$$\theta(f) = r(\theta)(f) \qquad (\forall f \in A_c(X)).$$

3.1. FINITE DIMENSION.

Let X be of finite dimension. We then have the famous CARATHEODORY's theorem which says that : For every $x \in X$, there exist $x_1,\dots,x_n \in E(X)$ and $r_1,\dots,r_n \in]0,1[$ such that :

$$\sum_i r_i = 1 \qquad \text{and} \qquad x = \sum_i r_i x_i .$$

In other words, if $d(y)$ denotes the Dirac measure at the point y of X, x is the resultant of the discrete measure $\sum_i r_i d(x_i)$ which is concentrated on $E(X)$.

Does the same hold for general convexes with the word Radon probability measure instead of discrete measure? The answer is very difficult even in a as simple case as when $E(X)$ is denumbrable.

3.2. THEOREM.

Suppose that X is a metrizable compact convex set. Then $E(X)$ is a G_δ set i.e. a denumbrable intersection of open sets, and every $x \in X$ is the barycenter of a probability measure concentrated on $E(X)$.

PROOF.(sketch of).

First step. The key idea is to introduce a relation on $\mathcal{M}_+(X)$ which says that a measure is "closer" to the extreme points than another. This is the following :

$$\theta \prec \theta' \Longleftrightarrow \theta(f) \leq \theta(f') \text{ for any } f \in Q_c(X).$$

As $Q_c(X)$ is total in $C(X)$, this relation is an order and as

$$Q_c(X) \cap Q_c(X) = A_c(X)$$

two comparable measures have the same resultant in X. For each $x \in X$, denotes by M_x the set of those $\theta \in \mathfrak{M}_+^1(X)$ with $r(\theta) = x$. One shows easely that :

(i) $E(X) = \{x \in X \; ; \; M_x = \{d(x)\}\}$;

(ii) The order $\prec$ is inductive.

Second step. Express that the maximal measures are in fact close to E(X). The key notion is the following : to each $f \in Q_c(X)$ associate :

$$\hat{f} = \inf \{g \in -Q_c(X) \; ; \; g \geq f\}.$$

The function $\hat{f}$ is concave and upper semi-continuous. And $\hat{f} = f$ on E(X). The main résult of that step is :

3.3. LEMMA.

A measure θ is maximal if, and only if, $\theta(f) = \theta(\hat{f})$ for any f $Q_c(X)$. Conséquently, θ is maximal if, and only if, θ is concentrated on each of the G_δ sets $B_f = \{\hat{f} = f\}$ for any $f \in Q_c(X)$.

Third step. The last step is the remark that if X is metrizable, there exists a convex continuous function f such that $B_f = E(X)$. It suffices to consider a sequence (f_n) of affines functions total in $A_c(X)$ s.t. $0 \leq f_n \leq 2^{-n}$ and to consider $f = \sum_n f_n^2$.

The third step proves that E(X) is a G_δ , thus a Borel set. And, if $x \in X$, consider a maximal measure θ which majorizes d(x), then θ has resultant x and is concentrated on E(X) by steps 2 and 3.

3.4. EXAMPLE.

Let Δ be the closed unit disk and f a continuous function on Δ harmonic in the interior. The classical Poisson formula :

$$f(z) = \frac{1}{2\pi} \int_0^{2\pi} f(\cos(\theta), \sin(\theta)) \frac{1-|z|^2}{|e^{i\theta}-z|^2} d\theta$$

can be interpreted in terms of maximal measures on a convenient compact convex set. It is the same for the Bernstein's theorem which says that if f is a completely monotone function on $\underline{R}$, i.e. a C^∞ function s.t. $(-1)^k f^k \geq 0$ for all $k \in \underline{N}$, there exists a Radon measure θ on $[0,+\infty[$, positive, such that :

$$f(x) = \int_0^\infty e^{-kx} d\theta(k)$$

4. UNICITY.

The problem we are interested in now is the question of the unicity of the maximal measure associated to a point of the compact convex set X.

If we look at some X in $\underline{R}^2$ we easily see that the only convex sets bearing this unicity property are the triangles. This unicity property has many different, at first surprising, expressions which I shall state in the following theorem.

In the sequel E will denote an order unit space, X the state space of E and $\tilde{X}$ the cone generated by X. We say that E is a simplex space if E satisfies the Riesz's interpolation property :

$$\forall u_1, u_2, v_1, v_2 \quad \text{such that} \quad u_i \leqslant v_j \ (i,j = 1,2),$$

there exists a $w \in E$ such that : $\quad u_i \leqslant w \leqslant v_j \ (i,j = 1,2).$

We shall say that X is a simplex if every x in X is the barycenter of a unique maximal measure.

4.1. EXAMPLE.

Let ω be any open set in $\underline{R}^2$ with compact closure. Let $A(\bar{\omega})$ be the space of continuous functions on $\bar{\omega}$, harmonic in ω , then $A(\bar{\omega})$ is a simplex space.

4.2. THEOREM.

The following statements are equivalent :

(a) X is a simplex ;

(b) E is a simplex space ;

(c) E' is lattice (when ordered by the positive cone $\tilde{X}$) ;

(d) Edward's separation property : For any f, -g in $Q_s(X)$ with $f \leqslant g$, there exists $h \in A_c(X)$ with $f \leqslant h \leqslant g$.

If X is metrizable, these four statements are equivalent to the following :

(e) The weak Dirichlet's problem : For every compact subset K of E(X) and every $f \in C(K)$ there exists an $\bar{f} \in A_c(X)$ (= E) such that :

$$\bar{f}_{|K} = f \qquad \text{and} \qquad \|f\|_K = \|\bar{f}\|$$

PROOF.

(a) $\Longrightarrow$ (c).

By (ii) of 3.2, the set of maximal measures M(X) is a cone hereditary in $\mathfrak{M}_+(X)$, thus lattice for its own order. By (a), the resultant map is a linear bijection from M(X) onto $\tilde{X}$. The latter is thus lattice for its own order and so E' is lattice.

(c) $\Longrightarrow$ (b).

Take $u_1, u_2, v_1, v_2 \in E$ (= $A_c(X)$) such that $u_i < v_j$ (i,j = 1,2), and consider

$f = \sup(u_1,u_2)$ and $g = \inf(v_1,v_2)$. By the Riesz decomposition property, it is easy to see that $\hat{f} \subset \check{g}$ (where $\check{g} = -(-g)\hat{}$). Thus, by Hahn-Banach, there exists $w \in A_c(X)$ s.t. $\hat{f} \leqslant w \leqslant \check{g}$. To replace $<$ by $\leqslant$ is a standard "passage à la limite" we shall admit. The same argument holds to prove (b) $\Longrightarrow$ (a).

(d) $\Longrightarrow$ (e). (Even in the non metrizable case).

Consider K and f as in (e). For sake of simplicity assume $0 \leqslant f \leqslant 1$. Define f_1, f_2 on X by the following conditions : $f = f_1 = f_2$ on K ; $f_1 = 1 = 1-f_2$ elsewhere. We have $f_1, -f_2 \in Q_s(X)$ and $f_2 \leqslant f_1$ thus, by (d) there exists $h \in A_c(X)$ such that $f_2 \leqslant h \leqslant f_1$. In particular, $0 \leqslant h \leqslant 1$ and $h = f$ on K.

(e) $\Longrightarrow$ (a). (In the metrizable case).

Suppose θ and θ' are two maximal probability measures with same barycenter. Sustracting $\theta \wedge \theta'$ and normalizing we can suppose θ and θ' disjoint. Thus, for a given $\varepsilon > 0$, there exists two disjoint compact subsets of E(X), say K_1 and K_2 such that : $\theta(K_1) \geqslant 1-\varepsilon$ and $\theta'(K_2) \leqslant \varepsilon$. Take $f \in A_c(X)$ with $0 \leqslant f \leqslant 1$, $f = 1$ on K_1 and $f = 0$ on K_2. By the barycenter formula :

$$\theta(f) = f(r(\theta)) \geqslant 1-\varepsilon \ ;$$

$$\theta'(f) = f(r(\theta')) \leqslant \varepsilon ,$$

but $f(r(\theta)) = f(r(\theta'))$, a contradiction if $\varepsilon < \frac{1}{2}$.

5. BAUER SIMPLEXES.

We are now going to characterize the lattice spaces among simplex spaces.

5.1. THEOREM.

The following are equivalent :

(a) X is a Bauer simplex i.e. a simplex with E(X) closed ;

(b) Solution of the Dirichlet's problem : Any bounded continuous function on E(X) extends to an affine continuous function on X ;

(c) E is lattice ;

(d) There exists a compact space T such that X is affinely homeomorphic to the compact convex set $\mathfrak{M}_+^1(T)$;

(e) There exists a compact space T and a bipositive linear isometry from E onto C(T).

PROOF.

Note that (e) $\Longrightarrow$ (d) as (d) is the dual statement of (e).

(a) $\Longrightarrow$ (b).

comes form 4.2.(e).

(b) $\Longrightarrow$ (c).

Denote $C_b(E(X))$ the space of bounded continuous functions on E(X). By Bauer's maximum principle, the map $f \mapsto f_{|E(X)}$ is a linear bipositive isometry form $A_c(X)$ onto $C_b(E(X))$. As the latter is lattice, so is $A_c(X)$ and E.

(c) $\Longrightarrow$ (a).

If E is lattice, E satisfies the Riesz interpolation property and X is a simplex by 4.2. Furthermore, one has :

$$E(X) = \{x \in X : x(a \vee b) = \max(x(a), x(b)), (\forall a, b \in E)\}$$

so, E(X) is closed.

(a) and (b) clearly give (e) with T = E(X) and (d) $\Longrightarrow$ (a) comes from the fact that $\mathfrak{M}^1_+(T)$ is a Bauer simplex.

6. FUNCTION SPACES.

It happens very often that an order unit space is given under the form of a function space : A function space on a compact space T is a closed separating subspace F of C(T) containing the constants. The following notions are considered :

- A Šilov set of F is a closed subset S of T such that :

$$\sup_{s \in S} |f(z)| = \sup_{t \in T} |f(t)| \qquad (\forall f \in F)$$

- If F admits a smallest Šilov set, this set is called the Šilov boundary of F.

- The Choquet boundary of F is the set of all $t \in T$ such that the following is true :

If $\theta \in \mathfrak{M}^1_+(T)$ is such that $\theta(f) = f(t)$ for any $f \in F$, then $\theta = d(t)$.

It is easy to interpret these notions in terms of compact convex sets. Consider the map $t \mapsto e(t)$ from T into the state space X of F, where e(t) is the evaluation at t. By 1.2, $e(T) \supset E(X)$ and by (i) of 3.3, one has that :

6.1. PROPOSITION.

The Choquet boundary is the inverse image of E(X) by the evaluation map.

Furthermore, by 1.2. :

6.2. THEOREM.

(a) A closed subset S of T is a Šilov set for F if and only if S contains the Choquet boundary.

(b) F admits a Šilov boundary which is the closure of the Choquet boundary.

By 3.3, we have :

6.3. THEOREM.

If T is metrizable, the Choquet boundary of F is a G_δ set of T and, for any continuous linear positive functional L on F there is at least one probability measure θ on T concentrated on the Choquet boundary such that :

$$L(f) = \theta(f) \qquad \text{for all} \quad f \in F.$$

This measure θ is unique if F satisfies the Riesz interpolation property. In that case, one can solve the weak Dirichlet's problem :

- For eveny compact K in the Choquet boundary and every $f \in C(K)$, there exists a norm preserving extension of f in F.

7. NOTES.

The book of ALFSEN [I] is the most comprehensive work on compact convex sets and its scholarly notes give precise references for the research of sources as well as for further reading. The following numbers correspond to the sections in the text :

1. For 1.1., 1.2. and 1.3., see [2], II,§7.
2. 2.1. is due to Kadison, see p. 74-75 of [1]. For 2.3., see [1], I,§2.
3. The first proof of 3.2. is due to Choquet. The present proof is due to Choquet and Meyer, see [1]. The idea of the order in the first step goes back to Bishop de Leeuw and in the present form to Mokobodzki,see [1], I,§4. For the examples 3.4., see [3], §31.
4. Theorem 4.2. is due to Choquet for the equivalences (a), (c) and (e) ; the others are due to Edwards, see [1], II,§3.
5. Due to Bauer, see [1] , II,§4.
6. See [1], I,§5.

BIBLIOGRAPHY

[1] E.M. ALFSEN. Compact convex sets and boundary integrals. Springer-Verlag, Berlin 1971

[2] N. BOURBAKI. Espaces vectoriels topologiques. Chap. I et II. Hermann, Paris 1966, 2ème éd. (ASI 1189)

[3] G. CHOQUET. Lectures on Analysis. Vol. II, W.A. Bengamin Inc, New-York, 1969

[3] G. CHOQUET et P.A. MEYER. Existence et unicité des représentations intégrales dans les convexes compacts quelconques. Ann. Inst. Fourier (Grenoble) 13, p. 139-154, 1963.

REPRESENTATION OF BANACH LATTICES

Alain GOULLET de RUGY

Equipe d'Analyse , Université de Paris VI

0. INTRODUCTION.

Recall that a Banach lattice is a couple (V,V_+) where V is a Banach space and V_+ a cone in V defining the order of V and for which V is a lattice space. The norm and the order are related by the following axiom :

$\forall x,y \in V : |x| \leq |y| \Longrightarrow \|x\| \leq \|y\|.$

This axiom implies that the lattice operations : $x,y \longmapsto x \vee y$ or $x \wedge y$ are continuous. Consequently, the cone V_+ is closed.

The problem we are concerned with is the following : Represent V as a concrete space : C(T), T compact ; $C_o(T)$,T locally compact ; $L^p(T)$; l^p,... . This is a very rich theory, where a lot has been done in the past twenty years, and I shall restrict myself to some of the most significant results.

Two kinds of representation will appear : Representation by means of continuous functions on some topological space with or without infinite values.

The kind of representation we shall obtain will depend upon the abundance of extreme generators in the cone P(V) of positive functionals on V i.e. on the abundance of real lattice homomorphisms. Without any restriction on V, there won't exist any real lattice homomorphism and we shall only have Davies's representation theorem (cf. 2.4.) of V by real continuous functions on some compact space, with possible infinite values on some rare subset. On the contrary, in particular cases, such as when V is an M-space, we shall have representation theorem by real, finite-valued, continuous functions on some non-compact topological space, due to the abundance of real lattice homomorphisms. The non-compactness of the representation space is not a handicap. On the contrary, its structure expresses precise features of the Banach lattice V.

References to sources and complements are relagated to the end of this paper.

1. CASE OF FINITE VALUED FUNCTIONS.

1.1. NOTATIONS.

V will be a fixed Banach lattice ; V_1 denotes its unit ball ; V' is topological dual ; V'_1 its dual unit ball ; P(V) the positive elements in V' and $P_1(V)=P(V)\cap V'_1$.

An extreme generator of $P(V)$ is, by definition, a generator D of $P(V)$ such that $(P(V) \setminus D)$ is convex. If $P(V)$ has a base B, D is extreme if, and only if, $D \cap B$ is an extreme point of B. $P(V)_g$ will denote the union of the extreme generators of $P(V)$ $P_1(V)_g = P(V)_g \cap P_1(V)$.

Recall that $L \in P(V)_g$ if, and only if, L is a lattice homomorphism i.e. :

$$L(a \vee b) = \max(L(a), L(b)), \ (\forall a, b \in V).$$

Thus $P(V)_g$ is closed in $P(V)$. In particular $P_1(V)_g$ is compact.

1.2. EXAMPLES.

If $V = C(T)$, the space of continuous real functions on a compact topological space T, $P(V) = \mathcal{M}_+(V)$ the cone of positive Radon measures on T ; if $V = C_o(T)$, the space of continuous real functions on some locally compact topological space T vanishing at infinity, $P(V) = \mathcal{M}_+^b(T)$ the cone of positive bounded Radon measures on T. In both cases, $P(V)_g$ consists of the ponctual measures $rd(t)$ where $r \in \underline{R}_+$ and $d(t)$ is the Dirac measure at the point $t \in T$. If $V = L^p(X,\theta)$, where $1 \leq p < +\infty$, and θ a positive Radon measure on some locally compact topological space, $P(V) = L^q(X,\theta)_+$ where q is the conjugate number of p and $P(V)_g$ is made of the functions with support reduced to a point of X of θ-measure non null.

The first theorem we state is simply a restatement of Bauer's theorem :

1.3. THEOREM.

If V is an order unit Banach lattice space, there exists a compact topological space T and a bipositive linear isometry of V onto $C(T)$.

Let us now consider a more general case :

1.4. DEFINITION.

We say that a Banach lattice V is an M-space if the following is true :

$$\|a \vee b\| = \max(\|a\|, \|b\|) \quad \text{for all} \quad a, b \in V_+.$$

The main interest of such spaces V is given by the following result which expresses the abundance of extreme generators :

1.5. LEMMA.

If V is an M-space, then $P_1(V)$ is a cap of $P(V)$ i.e. the complement of $P_1(V)$ in $P(V)$ is convex. In particular, $E(P_1(V)) \subset P(V)_g$.

From this, one gets Kakutani's theorem in a slightly modified version :

1.6. THEOREM.

Let V be an M-space. To each $v \in V$ associate the homogeneous function $\bar{v}$ on $P(V)_g$ defined by :

$$\bar{v} : L \mapsto L(v) \quad \text{for all} \quad L \in P(V)_g.$$

Then the map $v \longmapsto \bar{v}$ is a bipositive linear isometry of V onto the space $H_c(P(V)_g)$ of continuous homogeneous real functions on $P(V)_g$ endowed with the topology of uniform convergence on $P_1(V)_g$.

The representation theorem bear some interesting properties :

P1. To the supremum in V correspond the upper envelope in the function space.

P2. Every element of P(V) is represented by some measure on the underlying topological space (here $P(V)_g$).

In order to understand the next property let us recall that an <u>ideal</u> in an ordered linear space V is a positively generated subspace J of V such that : $\forall x \in V,\ \forall y \in J : 0 \leqslant x \leqslant y \Longrightarrow x \in J$. An ideal J is said to be <u>dense</u> if every positive element of V is the supremum of a net of positive elements of J.

P3. To every closed ideal J of V correspond a closed set S_J of the underlying topological space such that :

$$J = \left\{v : \bar{v} = 0 \text{ on } S_J\right\}.$$

All "good" representations must possess these properties. A fourth one can be added, also very important, which is not verified in theorem 1.6.

P4. The image of V in the function space is an ideal of continuous functions.

This last property is extremely strong : it implies that the representation space is "small". It will be possible to get a representation satisfying P1 to P4 for certain M-spaces by "cutting off" $P(V)_g$ in the following sense :

<u>1.7. DEFINITION.</u>

A positive element e of a Banach lattice V is a <u>topological unit</u> if the closed ideal generated by e is the whole space.

It is not difficult to prove that if e is a topological unit, then, the set $B_e = \left\{L \in P(V)\ ;\ l(e) = 1\right\}$, is a (non-compact) base of P(V), i.e. B_e meets each generator of P(V) at a point different from 0. Every separable Banach lattice has a topological unit.

<u>1.8. THEOREM.</u>

Let V be an M-space with topological unit e. Let us denote $T_e = B_e \cap P(V)_g$ and φ the restriction of the dual norm to T_e. To each $v \in V$ associate the continuous function $\bar{\bar{v}}$ on T_e defined by :

$$\bar{\bar{v}} : L \longmapsto L(v) \quad \text{for all} \quad L \in T_e.$$

Then, the map $v \longmapsto \bar{\bar{v}}$ is a bipositive linear isometry of V onto the space $\mathcal{D}_\varphi(T_e)$ of φ-<u>dominated</u> continuous functions that is the continuous functions f on T_e satisfying :

$$\forall \varepsilon > 0,\quad \exists K \text{ compact} \subset T_e \text{ s.t. } |f| \leqslant \varepsilon\varphi \quad \text{out of } K$$

endowed with the norm :

$$\|f\|_{\varphi} = \inf \{ r \, ; \, |f| \leq r\varphi \} .$$

Furthermore, this representation satisfies property P1 to P4.

The proof is too long to be summarized in a few lines. It rests mainly on the fact that the cone P(V) is a biréticulé cone, the theory of which I have developed in [1]. Note that property P4 is clear. P1 is a consequence of the fact that elements of T_e commute with the supremum. Property P2 can be made more precise by the following : P(V) can be identified with the positive Radon measures on the Cech-compactification $\check{T}_e$ of T_e concentrated on T_e (which is a K_{σ}), which integrate φ. Also, a stronger version of P3 holds : Closed ideals of V are in bijection, in a natural way, with the closed sets of T_e.

To end this section, let us caracterize the spaces of the form $C_o(T)$ among M-spaces :

1.9. PROPOSITION.

Let V be an M-space. Then, there exists a bipositive linear isometry of V onto some space $C_o(T)$ where T is a locally compact topological space if, and only if, the dual norm, when restricted to $(P(V)_g \div \{0\})$ is continuous.

2. INFINITE VALUED FUNCTIONS.

We shall first treat the case when V is order complete where results of algebraic nature are available. Then we shall treat the case when V is not order complete where results are known only when V has a topological unit.

2.1. THEOREM.

Suppose that F is an order complete vector space. Then, there exists a stonian compact topological space T and a linear bipositive isomorphism from F onto a dense ideal of the space $C_{\infty}(T)$ of continuous functions from T into $[-\infty, +\infty]$, finite on a dense subset of T.

Recall that a stonian compact topological space T is, by definition, a space such that the closure of every open set is open. We shall apply this theorem to represent L-spaces.

2.2. DEFINITION.

We say a Banach lattice V is an L-space if the given norm is additive on V_+.

2.3. THEOREM.

Suppose that V is an L-space. Then, there exists a locally compact topological space T, a positive Radon measure θ on T and a bipositive linear isometry from V onto the space $L^1(T,\theta)$.

PROOF (sketch of).

First note that V is order complete and by 2.1. can be represented by a dense ideal $J \subset C_{\infty}(T)$ for some stonian T. Using the abundance of projections in V, there exists a dense open set T' of T such that J contains the space $\mathcal{K}(T')$ of continuous real functions in T' with compact support as a dense ideal. Denote by L the positive linear functional on J which coincide with the L-norm on J_+. The restriction of L to $\mathcal{K}(T')$ defines a positive Radon measure θ on T' and some convergence arguments show that L and θ still coincide on J which almost ends the proof.

Let us now consider the second case. The best result is the following :

2.4. THEOREM.

Let F be a Banach lattice with topological unit e. Then, there exists a compact topological space T and a bipositive linear bijection from F onto an ideal C_F of continuous functions on T with values in $[-\infty,+\infty[$, finite on a dense subset of T, such that C_F contains C(T) as a dense ideal. Furthermore, F' can be identified with the set of Radon measures on T which integrate every function in C_F.

PROOF (idea of).

Consider the ideal J_e generated by e. With e as order unit, it is an order unit Banach space. Thus there exists a bipositive linear isometry from J_e onto some C(T) with T compact. In fact, this isometry extends to a bijection from F onto some space of continuous functions on T as described in the theorem.

3. NOTES.

3.1. I have said nothing about the uniqueness of the associated topological space in each of the representations studied. Simply note that, except for theorem 2.3., it is unique within an homeomorphism. Furthermore, this space is invariant by a change of norm respecting the locally convex space and the additional properties required to the initial norm in every statement.

3.2. All the theorems about representation of Banach lattices stated here generalize to locally convex lattices. For example let us restate theorem 2.3. :

3.3. THEOREM.

Let F be a locally convex L-space, that is a locally convex Hausdorff complete topological vector space together with a closed cone F_+ defining the order, for which F is lattice, such that the topology of F can be defined by a family P of semi-norms additive on F_+ and such that : $p(|x|) = p(x)$ for all $x \in F$ and $p \in P$.

Then, there exists a locally compact topological space T, a family $\Theta = (\theta_i)_{(i \in I)}$ of positive Radon measures on T and a bipositive linear isomorphism from F onto the space $L^1(\Theta)$ of the θ_i-integrable functions on T, for all $i \in I$, endowed with the topology associated with the semi-norms : $f \longmapsto \theta_i(|f|)$.

3.4. SOURCES AND REFERENCES.

The numbers below refer to sections in the paper.

1. Theorem 1.3. goes back to Kakutani ([4]). Theorems 1.6., 1.8. and proposition 1.9. are found in Goullet de Rugy [2], corollaires 1.31., 3.18. and proposition 2.31.
2. For theorem 2.1. and sources, see the Chapter 7 of [6]. This book of Luxemburg and Zaanen is so complete that it becomes confusing. So it might be better to look at Vulikh [8], theorem V.4.2. . Theorem 2.3. is due to Kakutani, see [5]. Theorem 2.4. is due to E.B. Davies. See theorem 10 of [3]. For another proof, see [7], theorem 1.

BIBLIOGRAPHIE

[1] A. GOULLET de RUGY. La théorie des cônes biréticulés. Ann. Inst. Fourier (Grenoble), 21 (4), 1-64, 1971

[2] A. GOULLET de RUGY. La structure idéale des M-espaces. J. Math. Pures et Appl. 51, 331-373, 1972

[3] E.B. DAVIES. The Choquet theory and representation of ordered Banach spaces. Illinois J. Math., 13, 176-187, 1969

[4] S. KAKUTANI. Concrete representation of abstract M-spaces. Ann. of Math. 42, 994-1024, 1941

[5] S. KAKUTANI. Concrete representation of abstract L-spaces and the mean ergodic theorem. Ann. of Math. 42, 523-537, 1941

[6] W.A.J. LUXEMBURG and A.C. ZAANEN. Riesz spaces. Vol. I, North Holland, Amsterdam, London 1971

[7] H.H. SCHAEFER. On the representation of Banach lattices by continuous numerical functions. Math. Z. 125, 215-232, 1972

[8] B.Z. VULIKH. Introduction to the theory of partially ordered spaces. Moscow 1961 (English translation, Groningen 1967)

ORDER IDEALS IN ORDERED BANACH SPACES

A.J. Ellis

Department of Pure Mathematics, University College of Swansea, Wales.

Let E be an ordered Banach space, i.e. the positive cone E_+ of E is closed, normal and generating. A linear subspace I of E is called an order ideal if $x \in I$, $y \in E$ and $0 \leq y \leq x$ implies that $y \in I$, i.e. if $I_+ = I \cap E_+$ is an extremal subset of E_+. If I_+ generates I then I is called an ideal.

The Banach dual space E^*, with the dual ordering, is also an ordered Banach space and so it is natural, and important, to study the relationship between order ideals I and their annihilators $I^o = \{f \in E^* : f(x) = 0, \forall x \in I\}$.

It is easy to verify that I^o is an order ideal whenever I is an ideal. However I^o may be an order ideal without I being an ideal, for example when $E = R^3$, $E_+ = \{(x,y,z): z \geq 0, x^2 + y^2 \leq z^2\}$, and I is any two-dimensional subspace of E which intersects E_+ in an extreme ray. The precise property which I must satisfy, a kind of approximate positive-generation, is described in the following result [12].

Theorem 1. Let I be an order ideal in E. Then I^o is an order ideal in E^* if and only if I is perfect, i.e. for each $x \in I$ $\exists$ sequences $w_n \in I$, $y_n, z_n \in E$ such that $\|y_n\| \leq 1$, $\|z_n\| \leq 1$ and $-w_n + \frac{1}{n}y_n \leq x \leq w_n + \frac{1}{n}z_n$, for each n.

For extensive generalizations of this result see Jameson [16] and Nagel [17].

For the remainder of these notes let E be a base normed Banach space, with base B and closed unit ball $co(B \cup -B)$. Then I is an ideal in E if and only if $I = \operatorname{lin} F$ for some face F of B; if I is closed then so is F, but the converse is much more subtle.

In fact if F is closed then $\operatorname{lin} F$ is closed if and only if each $f \in A^b(F)$ has an extension belonging to $A^b(B)$. Here we denote by $A^b(B)$ the Banach space of all bounded affine real-valued functions on B; this space is readily identified with E^*.

If K is a compact convex set then $A^b(K)$ is the second dual space of the ordered Banach space $A(K)$. If F is a closed face of K we write $F_\perp = \{f \in A(K): f(x) = 0, \forall x \in F\}$, and $(F_\perp)^\perp = \{x \in K: f(x) = 0, \forall f \in F_\perp\}$. Similarly if F is a norm-closed face of B we write $F^\perp = \{f \in A^b(B): f(x) = 0, \forall x \in F\}$, and $(F^\perp)_\perp = \{x \in B: f(x) = 0, \forall f \in F^\perp\}$. It is often of importance to know that $F = (F_\perp)^\perp$ or $F = (F^\perp)_\perp$; this is always the case if F is finite-dimensional. However, we have the following example due to J.D. Pryce.

Example 1. Let $E = L_2[0,1]$, $F = \{f \in E: 0 \leq f \leq 1\}$, $G = \{f \in E: f \geqslant 0, \|f\| \leq 1\}$ and let $h \in E_+$ be essentially unbounded on $[0,1]$. Then, if $K = \overline{co}(F \cup (G+h))$, K is weakly compact and F is a closed face of K, since all elements of F are essentially bounded. If $\phi \in F^\perp$ then, since $G-G$ is a neighbourhood of 0 in E and since $\operatorname{lin} F$ is dense in E, it follows that $\phi = 0$. Therefore $(F^\perp)_\perp = K \neq F$, and a fortiori $(F_\perp)^\perp = K$.

The bipolar theorem shows that if F is a closed face of K (or a norm-closed face of B) then $F = (F_\perp)^\perp$ $(F = (F^\perp)_\perp)$ if and only if $F = K \cap L$ $(F = B \cap L)$ where L is the w^*-closed (norm-closed) linear hull of F; these conditions are certainly satisfied if L is w^*-closed (norm-closed). The following result is due to Alfsen [2] and D.A. Edwards [10].

Theorem 2. If F is a closed face of K, then the following statements are equivalent: (i) $\operatorname{lin} F$ is norm-closed; (ii) $\operatorname{lin} F$ is w^*-closed; (iii) $\exists$ a constant M such that each $f \in A(F)$ has an extension $g \in A(K)$ with $\|g\| \leq M\|f\|$.

If these statements hold then $A(K)^+|F = A(F)^+$ if and only if $A(K)/F_\perp$ is Archimedean ordered.

Precisely analogous results hold for the space $A^b(B)$ (with the exception of (ii)).

An ideal I in $A(K)$ such that $A(K)/I$ is Archimedean ordered is called an Archimedean ideal; if, in addition, I^0 is positively generated then I is called a strongly Archimedean ideal. Since an

Archimedean ideal I satisfies $I = (I^{\perp})_{\perp}$ the bipolar theorem shows that $I^{\perp}$ is strongly Archimedean if and only if the conditions (i) - (iii) of Theorem 2 hold for $F = I^{\perp}$. Analogous definitions and results apply in the case of ideals in $A^b(B)$.

A closed face F of K (or of B) is called <u>semi-exposed</u> if for each $x \in K \setminus F$ $(B \setminus F)$ $\exists$ $f \in A(K)^+$ $(A^b(B)^+)$ with $f(x) > 0$ while $f(y) = 0$ for all $y \in F$. If f can be chosen independently of x then F is called <u>exposed</u>. Clearly a face is semi-exposed if and only if it is the intersection of a family of exposed faces. It is not difficult to show that a semi-exposed face F of K is exposed if and only if it is a G_δ-set, which is always the case when K is metrizable. Of course every norm-closed face of B is a G_δ-set; however not every semi-exposed face is exposed, as the following example shows.

<u>Example 2</u>. Let $Y = \{f: [0,1] \to R: \|f\| = \sup\{|f(t)| : 0 \leq t \leq 1\} < \infty\}$, with the natural ordering, and let $B = \{f \in Y: f \geq 0, \|f\| \leq 1\}$. Then $A^b(B)$ is isomorphic to $Y^* \times R$ and each positive linear functional in Y^* supports B at 0. Since $Y^* = Y^*_+ - Y^*_+$ it follows that 0 is an $A^b(B)$-semi-exposed face of B. However, no strictly positive linear functional in Y^* exists (cf. [10]) and hence 0 is not an $A^b(B)$-exposed face of B.

The following result gives dual characterizations of exposed and semi-exposed faces of K and B. For a proof of parts (i) and (ii) see [13].

<u>Theorem 3</u>. (i) For a closed face F of K, F is exposed if and only if $F = (F_{\perp})^{\perp}$ and $\exists$ an element $h \in F_{\perp}^+$ such that for each $f \in F_{\perp}$ and $\varepsilon > 0$ $\exists$ $\lambda(f,\varepsilon) > 0$ with $f \leq \lambda h + \varepsilon$.

(ii) The closed face F of K is semi-exposed if and only if $F = (F_{\perp})^{\perp}$ and, given $f \in F_{\perp}$, $\varepsilon > 0$ $\exists$ $g \in F_{\perp}^+$ with $f \leq g + \varepsilon$.

(iii) The closed face F of B is $A^b(B)$-exposed if and only if $F = (F^{\perp})_{\perp}$ and $\exists$ $g \in F^{\perp +}$ such that $\forall$ $f \in F^{\perp}$, $x \in B$, $\varepsilon > 0$ $\exists$ $\lambda > 0$ and $h \in A^b(B)$ with $|h(x)| \leq 1$ and $f \leq \lambda g + \varepsilon h$.

(iv) The closed face F of B is $A^b(B)$-semi-exposed if and only if $F = (F^{\perp})_{\perp}$ and given $f \in F^{\perp}$, $x \in B$, $\varepsilon > 0$ $\exists$ $g \in F^{\perp +}$, $h \in A^b(B)$ with $|h(x)| \leq 1$ and $f \leq g + \varepsilon h$.

It is possible for F to be an exposed face of K without $F^{\perp}$ being even a perfect ideal in $A^b(K)$, as the following example shows.

Example 3. Let $K = \{\underline{x} = \{x_n\} \in \ell_1 : \|\underline{x}\| \leq 1, \sum_{n=1}^{N} x_n \geq 0, \forall N\}$. Then K is $\sigma(\ell_1, c_0)$-compact and $A(K)$, $A^b(K)$ are norm-isomorphic to $c_0 \times R$ and $\ell^\infty \times R$ respectively. O is $A(K)$-exposed in K by the support functional $\{\frac{1}{2^{n-1}}\} \in c_0$. However $\{0\}^\perp$ is not a perfect ideal in $A^b(K)$.

In order to get a duality between faces of B and order ideals in $A^b(B)$ we need to define an order ideal I in $A^b(B)$ to be w^*-perfect if given $f \in I$, $x \in B$, $\varepsilon > 0$ $\exists g \in I$, $h_1, h_2 \in A^b(B)$ such that $|h_i(x)| \leq 1$ and $-g + \varepsilon h_1 \leq f \leq g + \varepsilon h_2$. It is then true that $I_\perp$ is a face of B if and only if I is w^*-perfect.

A face F of a convex set C is said to be split if there exists a disjoint face F' of C such that every point x of C has a unique decomposition $x = \lambda y + (1-\lambda)z$, $y \in F$, $z \in F'$, $0 \leq \lambda \leq 1$. If K is a simplex then [1] every closed face of K is split. If B is a simplex then [7] every norm-closed face of B is split. This latter result requires the completeness of the base normed space E, as the following example shows.

Example 4. Let $E = C[0,1]$ with $B = \{f \geq 0 : \int_0^1 f(x)dx = 1\}$, $F = \{f \in B : \int_0^{\frac{1}{2}} f(x)dx = 0\}$. The base norm for E induced by B is the relative $L^1[0,1]$-norm, and F is a closed face of B which is not split.

If F is a split face of B with complementary face F' then there exists an $f \in A^b(B)$ such that $0 \leq f \leq 1$ while $F = f^{-1}(0)$, $F' = f^{-1}(1)$. Moreover, if $g \in A^b(F)$ and $h \in A^b(F')$ then there exists an $f \in A^b(B)$ such that $f = g$ on F, $f = h$ on F'. Therefore F and F' are norm-closed and it is easy to check that F and F' are strongly Archimedean, with norm-preserving extensions existing. A face F of B is split if and only if $\operatorname{lin} F$ is an L-ideal in E in the sense of Alfsen and Effros [5], and it follows [5, II, 1.13] that the intersection of an arbitrary family of split faces of B is split, and that the closed convex hull of an arbitrary family of split faces of B is split.

The situation for closed split faces of K is rather different. For example, if K denotes the probability measures on $[0,1]$ then each extreme point δ_x is split but its complementary face is dense in K; in this example there are, of course, far more norm-closed split faces of K than closed split faces. In the next theorem we sum up some of the results of Alfsen and Andersen [4, 6] concerning closed split faces of K.

Theorem 4. Every closed split face of K is strongly Archimedean, and norm-preserving extensions exist. The family $\mathfrak{F}$ of closed split faces of K is closed under arbitrary intersections and the convex hull of finite unions. The sets $F \cap \partial K$, for $F \in \mathfrak{F}$, are the closed sets for a facial topology on ∂K; this topology is Hausdorff if and only if K is a Bauer simplex.

It is not generally true that the closed convex hull of an arbitrary family of closed split faces of K is split, as the following example of A. Gleit shows.

Example 5. Let $A(K) = \{f \in C[-1,1] : f(0) = \frac{1}{2}(f(-1) + f(1))\}$, $I_n = \{f \in A(K) : f(\frac{1}{n}) = 0\}$. Each I_n is an ideal in the simplex space $A(K)$ so that $I_n^{\perp}$ is a split face of K. However $(\overline{co} \bigcup_{n=1}^{\infty} I_n^{\perp}) = \bigcap_{n=1}^{\infty} I_n = \{f \in A(K) : f(\frac{1}{n}) = 0, \forall n\}$, which is not positively generated, so $\overline{co} \bigcup_{n=1}^{\infty} I_n^{\perp}$ is not split (cf. [11]).

K is said to satisfy Størmer's axiom if $\overline{co} \bigcup_{\alpha} F_{\alpha}$ is a closed split face of K whenever each F_{α} is a closed split face of K. A simplex satisfies Størmer's axiom if and only if it is a Bauer simplex. However the state space K of any unital B^*-algebra satisfies Størmer's axiom. Alfsen and Andersen [4] have shown that for any K which satisfies Størmer's axiom a hull-kernel topology may be defined, and this topology gives, in the case of a unital B^*-algebra, precisely the Jacobson topology of the primitive ideal space. Some other relevant results (cf. [3,4,18]) are contained in the following theorem.

Theorem 5. Let K be the state space of a unital B^*-algebra A, and let F be a closed face of K. Then the following statements are equivalent: (i) F is a split face of K; (ii) $F_{\perp}$ is an Archimedean ideal in $A(K)$; (iii) $F_{\perp}$ is the self-adjoint part of a closed two-sided ideal in A.

Chu has shown that, for K as in Theorem 5, every closed face of K is semi-exposed. Moreover he has proved the following result [8, 9].

Theorem 6. If K is the state-space of a unital B^*-algebra then the following statements are equivalent: (i) A is a prime algebra; (ii) $A(K)$ is an anti-lattice, i.e. $f \wedge g$ only exists in $A(K)$ if either $f \leq g$ or $g \leq f$; (iii) K is prime, i.e. $K = co(F \cup G)$ with F, G semi-exposed faces implies either $F = K$ or $G = K$.

In connection with Theorem 6 we recall that a unital B^*-algebra is commutative if and only if $A(K)$ is a lattice.

Let A be a function algebra on a compact Hausdorff space Ω, let K be the state space of A and let $Z = co(K \cup -iK)$. The split faces of Z are also connected with the algebraic structure of A, as the following result shows (cf. [14] and [15]).

Theorem 7. Let F be a closed face of K. Then: (i) $co(F \cup -iF)$ is a split face of Z if and only if $F \cap X$ is a generalized peak set for A; (ii) F is a split face of Z if and only if $F \cap X$ is a generalized peak interpolation set for A.

References

1. E.M. ALFSEN, 'On the decomposition of a Choquet simplex into a direct convex sum of complementary faces', Math. Scand. 17(1965) 169-176.
2. E.M. ALFSEN, 'Facial structure of compact convex sets', Proc. London Math. Soc. 18 (1968) 385-404.
3. E.M. ALFSEN, 'Compact convex sets and boundary integrals', Springer-Verlag, Berlin, 1971.
4. E.M. ALFSEN and T.B. ANDERSEN, 'Split faces of compact convex sets', Proc. London Math. Soc. 21 (1970) 415-442.
5. E.M. ALFSEN and E.G. EFFROS, 'Structure in real Banach spaces I, II', Ann. Math. 96 (1972) 98-173.
6. T.B. ANDERSEN, 'On dominated extensions of continuous affine functions on split faces', Math. Scand. 29 (1971) 298-306.
7. L. ASIMOW and A.J. ELLIS, 'Facial decomposition of linearly compact simplexes and separation of functions on cones', Pac. J. Math. 34 (1970) 301-310.
8. CHU CHO-HO, 'Anti-lattices and prime sets', Math. Scand. 31 (1972) 151-165.
9. CHU CHO-HO, 'Prime faces in C^*-algebras', J. London Math. Soc. (to appear).

10. D.A. EDWARDS, 'On the homeomorphic affine embedding of a locally compact cone into a Banach dual space endowed with the vague topology', Proc. London Math. Soc. 14 (1964) 399-414.

11. E.G. EFFROS, 'Structure in simplexes', Acta Math. 117 (1967) 103-121.

12. A.J. ELLIS, 'Perfect order ideals', J. London Math. Soc. 40 (1965) 288-294.

13. A.J. ELLIS, 'On faces of compact convex sets and their annihilators', Math. Ann. 184 (1969) 19-24.

14. A.J. ELLIS, 'On split faces and function algebras', Math. Ann. 195 (1972) 159-166.

15. B. HIRSBERG, 'M-ideals in complex function spaces and algebras', Israel J. Math. 12(1972) 133-146.

16. GRAHAM JAMESON, '<u>Ordered linear spaces</u>', Lecture Notes in Mathematics, No.141, Springer-Verlag, Berlin, 1970.

17. R.J. NAGEL, 'Ideals in ordered locally convex spaces', Math. Scand. 29 (1971) 259-271.

18. E. STØRMER, 'On partially ordered vector spaces and their duals with applications to simplexes and C^*-algebras', Proc. London Math. Soc. 18 (1968) 245-265.

ORDER BOUNDED OPERATORS AND CENTRAL MEASURES

W. Wils

1. Introduction

Attempts to use the setting of Partially Ordered Vectors Spaces (P.O.S.) in the theory of the mathematical foundation of Quantum Mechanics are very old. The general theory of these spaces, however, seemed too weak to prove the kind of results which one wanted. Therefore it became necessary to consider special classes of P.O.S.

Considerable development took place on the one hand within the field of C* -algebras, and on the other hand for the class of Riesz-spaces. In due time it became apparent that many analogies could be made between the two fields. Attempts were made to build a unified theory which would cover a substantial class of P.O.S., which included both C* -algebras and Riesz-spaces. In these lectures I want to show how some of the powerful notions of C* -algebras, as two sided ideals, quotient-algebras, central decomposition theory and factors carry over to a more general setting.

2. Decomposition

Mathematicians often attempt to simplify the study of complicated objects by writing them as sums of simple components. Thus:

Definition: A splitting of a P.O.S. (E,E^+), (with $E = E^+ - E^+$), is a family of subspaces $\{E_i\}_i$ of E such that $E = \bigoplus_i E_i$ and $E^+ = \Sigma_i (E_i \cap E^+)$.

Hence every element $k \in E$ can be written in a unique way as a sum of finitely many k_i, $k_i \in E_i$, and moreover if $k \in E^+$ then every $k_i \in E_i \cap E^+$.

The subspaces E_i which appear in a splitting of E are called *split-subspaces* of E and the $E_i \cap E^+$ are called *split-faces* of E^+.

Proposition 1: There is a one-one correspondence between split-subspaces of E and the range spaces of linear maps $P: E \longrightarrow E$ with the property $0 \leq Pk \leq k$ for all $k \in E^+$ and $P^2 = P$. Any two such maps P commute and hence the set of split-faces of E^+ is a Boolean algebra.

The above proposition means firstly that there is an operator-characterization of split-subspaces and secondly that sums and intersections of split-subspaces (faces) are again split-subspaces (faces). Thus it is always possible to find a refinement of any two splittings. Later on, we shall ask: Does there exist a finest splitting of E ?

In this context we introduce a localization of the notion of splitting as follows. Let $k \in E^+$ and $F_k = \{h \in E^+ \mid 0 \leq h \leq k\}$,
$C_p = \bigcup_{\lambda \geq 0} \lambda F_k$ and $V_p = C_p - C_p$.

Definition: Two subspaces E_1 and E_2 are called *disjoint*, notation $E_1 \delta E_2$, if $\{E_1, E_2\}$ is a splitting of $E_1 + E_2$. Two split-faces G and H of E^+ are called *disjoint*, notation $G \delta H$, if G-G and H-H are *disjoint*. Two elements $p, p' \in E^+$ are said to be *disjoint*, notation $p \delta p'$, if $C_p \delta C_{p'}$. And $p = \Sigma_i p_i$ is a splitting of p if $\{V_{p_i}\}_i$ is a splitting of V_p.

As long as we consider disjoint elements it is as if we are dealing with a lattice. We have:
Suppose $g, h \in E^+$, $g \delta h$ and $o \leq k \leq g + h$.

i) Then there are unique $k_1, k_2 \in E^+$, with $k = k_1 + k_2$, $k_1 \leq g$, $k_2 \leq h$ and if $k' \leq k$, $k' \leq g$ then $k' \leq k_1$.

ii) If in addition $k \leq g'$ with $g' \delta h$ then $k \leq g$.

In many cases a splitting in direct summands does not exist. The operator-characterization of split-subspaces allows a generalization which can be used to describe smooth decompositions.

Definition. The set of order-bounded operators L_{ob} (E), of a positively generated P.O.S. E is defined to be the set of linear maps $T:E \rightarrow E$ such that there existed a $\lambda > o$ with $-\lambda I \leq T \leq +\lambda I$. There I is the identity map on E and $T \leq \lambda I$ means: for all $k \in E^+$, $Tk \leq \lambda k$.

Then $L_{ob}(E)$ is an algebra of operators. The positive cone $L_{ob}(E)^+ = \{T \in L_{ob}(E) \mid T > o\}$ is closed for multiplication. The algebra-unit I for $L_{ob}(E)$ is also an order-unit for $L_{ob}(E)$. If the ordering on E is Archimedean, i.e., k, $f \in E$ and $k \leq \alpha f$ for all $\alpha \geq o$ implies $k \leq o$, then the ordering of $L_{ob}(E)$ is Archimedean too. It is well known that an ordered algebra which is Archimedean is isomorphic to an algebra of functions and hence, in particular, is commutative.

If all the spaces V_p with $p \in E^+$ as their order unit and equipped with the corresponding norm are complete, then $L_{ob}(E)$ is complete in its order-unit norm. Every $T \in L_{ob}(E)$ leaves the spaces V_p invariant.

The use of $L_{ob}(E)$ in discoverning remnants of lattice structure in E is illustrated in the next proposition.

Proposition 2:

i. Let k be an order unit for E. Then the map $L_{ob}(E) \ni T \rightarrow Tk \in E$ is bipositive.

ii. Suppose S, $T \in L_{ob}(E)$ are such that sup $(S,T) \in L_{ob}(E)$. If k, $f \in E$ are such that Sk, $Tk \leq f$ then sup $(S,T)\, k \leq f$.

This closes the elementary theory of sets of order bounded operators. There are two main directions for further development. In both cases E is given the additional structure of an ordered Banach space and duality theory is used. The first development is a further elaboration on closed split faces, ideals in A, quotient spaces and ration extension theorems. Several aspect of this theory have been treated in A. Ellis's lectures. Here I shall add only a few remarks. The second line of thought concerns the construction of decompositions of a partially ordered space. The main tools are representing measures on compact convex sets. This topic will be covered in these lectures. [For a more extensive treatment of the topic treated in § 2, Alfsen [7], Ch. II, or Wils [5]].

3. Order bounded operators

Although the theory exposed in this section has been extended considerably by Alfsen and Effros [2] we shall consider only the now almost classical case of special classes of partially ordered Banach spaces. An order-unit space, (A,e) denotes an Archimedean ordered vector space A with distinguished order unit e considered as a normal vector space in the order-unit norm, $||a|| = \inf\{\lambda > o \mid -\lambda e \leq a \leq +\lambda e\}$ for $a \in A$.

A base-norm space (E,K) denotes a directed vector space E for which E^+ has a base K such that $B = \text{conv}(K \cup -K)$ is radially compact, considered in the norm $||x|| = \inf\{\lambda \geq o \mid x \in \lambda B\}$.

Let us suppose first that E is a norm-complete base norm space and $A = E^*$ is the Banach dual of E. Let e be the linear functional on E which assumes the value 1 on K. Then $e \in E^*$ and Ellis has shown that (A,e) is an order-unit space. [See [7], II, 1.15]. In this case Alfsen and Effros [2] proved the following theorem.

Theorem I: The set of order bounded operators $L_{ob}(E)$ is a complete commutative normed operator algebra and the map $L_{ob}(E) \ni T \to T^* \in L_{ob}(A)$ is an isometric isomorphism onto. Moreover the representation space Ω for $L_{ob}(A)$ is hyper stonean and for each $x \in E^+$ and $f \in A^+$. The map $L_{ob}(E) \ni S \to \langle S x, f \rangle \in \mathbb{R}$ is a normal positive linear functional.

Moreover Edwards and Gerzon [3] have shown that all linear combinations of normal positive linear functionals on A are of the form $L_{ob}(A) \ni T \to \langle Te, x \rangle$ for some $x \in E$.

The converse case where (A,e) is a given order-complete order-unit space and $E = A^*$ with $K = \{f \in E \mid \langle f,e \rangle = ||f|| = 1\}$ is a base-norm space is also of considerable interest. This is the setting for the rest of this paper. E.g. A is a C* -algebra with unit e and E is its dual. We provide K with the relative weak*-topology. Then A can be identified with the set of continuous affine functions on K. This identification will be used throughout the paper. [See Alfsen 73, II, § 1].

We want to find different representations for $L_{ob}(A)$. (See Alfsen [1] II, §7 or Wils [5]).

Theorem 2: The map $L_{ob}(A) \ni T \rightarrow Te \in A$ is an isometric isomorphism of the ordered space $L_{ob}(A)$ into A.

Hence $L_{ob}(A)$ can be identified with a subspace of A. This subspace is called the center of A. For a C* -algebra A with unit e, $L_{ob}(A)e$ coincides with the algebraic center of A. The second representation theorem requires more work.

The center $L_{ob}(E)$ of E is order complete, and the set of weak* -closed split-faces of E^+ is closed under arbitrary intersection and finite sums. The intersections of the closed split-faces of E^+ with the extreme boundary $\partial_e K$ of K, defines a topology on $\partial_e K$ the so called facial topology.

Theorem 3: If $x \in \partial_e K$ and $T \in L_{ob}(A)$, then there exists a unique constant $\lambda_T(x)$ such that $\lambda_T(x)x = T^*x$ Let $\lambda_T : x \rightarrow \lambda_T(x)$. The map $L_{ob}(A) \ni T \rightarrow \lambda_T \in C_f(\partial_e K)$ is a bipositive algebra isomorphism of $L_{ob}(A)$ onto $C_f(\partial_e K)$, the algebra of bounded, facially continuous functions on $\partial_e K$.

Theorem 4: For every $g \in C_f(\partial_e K)$ there is a unique $\bar{g} \in A$ such that $\bar{g}|\partial_e K = g$. Moreover $\bar{g} \in L_{ob}(A)$ e and for every $a \in A$ there exists $b \in A$ with $b|\partial_e K = \bar{g}|\partial_e K \cdot a|\partial_e K$.

Suppose we restrict A, viewed as the space of continuous affine functions on K to $\partial_e K$. Then the last theorem tells us that $L_{ob}(A)$ e consists of those elements in $A|\partial_e K$ with which one can multiply other arbitrary elements in $A|\partial_e K$ and still stay in $A|\partial_e K$. Hence there exist three representations of $L_{ob}(A)$.

1. as the set of order bounded operators on A.
2. as the set of facially continuous functions on $\partial_e K$.
3. as the set of multipliers within $A|\partial_e K$.

Further developments of this part of the theory leads to the consideration of the restriction of A to a closed split face and the subspace of A considering of those elements which vanish on such a closed split-face. The quotient of A with respect to the last subspace is in a natural way isomorphic with the first space. The properties of these spaces are being studied and sharp extension theorems for continuous affine functions on closed split-faces of K can be given. The theory does not depend on the existence of an order unit $e \in A$. Extensions to the case of weakly complete cones have been given by W. Habre [4]. Alfsen and Effros [2] considered similar theories for general real Banach spaces. Several results have been obtained, especially in the context of function algebras, for complex Banach spaces. For more references see Alfsen [1].

4. Central decomposition

We return to the ideas on decomposition, which were developed in §2. The setting with (A,e) an order-unit space, and $E = A^*$ a base-norm space, stays unaltered.

Suppose $k \in E^+$ and $\{E_i\}_i$ a splitting of E, then $k = \Sigma_i k_i$ with $k_i \in E_i^+$. When the splittings of E become finer and finer, there are, in general, more and more k_i, and, because $||k|| = \Sigma_i ||k_i||$, since E is a base-norm, k_i has a tendency to go to zero so that it becomes impossible to take a limit.

Thus either one has to develop a theory of direct integral decompositions for partially ordered spaces or one has to introduce representing measures as is done in the Choquet-theory for elements of a compact convex set, where similar difficulties as here are encountered. Both approaches have been worked out for the case of C* -algebras. The direct integral set up, involves a great deal of measure theory, and gives good results in the separable case. The use of representing measures offers less technical proofs and applies also to non-separable spaces. The results however are less detailed. Here the second approach is presented for partially ordered spaces.

Thus let $k \in \Sigma E^+$. With every splitting $k = \Sigma_i k_i$ of k we associate a measure, $\Sigma_i ||k_i|| \delta_{(k_i/||k_i||)}$, on K, which represents k. Here δ_g, $g \in K$, of course, is the point evaluation at g. As the splittings become finer, the measures increase in the order of Choquet and have a limit, which is the central measure corresponding to k. The hope is that this measure is concentrated in some sense on the set of points, which can no longer be splitted. The idea in the rest of the paper is to make the above heuristic approach more precise, to find properties of the central measure, and to indicate further possibilities for research.

As before, for $g \varepsilon E^+$, C_g denotes the smallest face E^+ which contains g and $V_g = C_g - C_g$. The ordering of E induces an ordering on V_g. If μ is a positive measure on K, we let $\Phi\mu : L^\infty(K,\mu) \rightarrow E$ be the map $< a, \Phi\mu(\varphi) > = \int \varphi \, a \, d\mu$, $a \varepsilon A$, $\varphi \varepsilon L^\infty(K,\mu)$. $\Phi\mu(\varphi)$ is defined as an element in $A^* = E$ by the above formula.

The following theorem is more general than is necessary for just central measures, but it possible to find other applications.

<u>Theorem 5</u>. Let $g \varepsilon K$, $||g|| = 1$, and $g \varepsilon W \varepsilon V_g$, where W is a complete linear lattice in the induced ordering. Then, the set of discrete probability measures $\Sigma_i a_i \delta_{f_i}$ with $f_i \varepsilon W \cap K$, $\Sigma_i \alpha_i f_i = g$, is directed in the order of Choquet-Meyer. Let μ be the supremum of this net of measures then μ is the unique probability measure such that $\Phi\mu$ is a lattice isomorphism from $L^\infty(K,\mu)$ onto W.

Various choices for W can be made. Let me indicate two which for the case of C* -algebra coincide, but in general are different. If $W = L_{ob}(E)\, g$ then W is a complete linear sublattice of V_g [Proposition 2]. This means that only splittings of all of E occur, and one obtains a kind of central measures which has not been studied yet. Another choice is to take $W = L_{ob}(V_g)\, g$. It follows once more from proposition 2 that W is a lattice, and it is not difficult to verify that W is complete.

<u>Definition</u>: For $h \varepsilon K$, we denote $L_{ob}(V_g)$ by Z_h. A probability measure μ on K, which represent $h \varepsilon K$, is said to be central iff $\Phi\mu$ maps $L^\infty(K,\mu)$ isomorphically onto the lattice $Z_h h \subseteq V_h$.

<u>Theorem 6</u>: For $g \varepsilon K$, there is a unique central measure μ which represents g.

The proofs of both theorems 5 and 6 do not contain many new ideas. The next result is much harder to obtain. It concerns the support of the central measure.

<u>Definition</u>: A point $k \varepsilon K$ is called <u>primary</u> when $L_{ob}(V_g)$ consists

only of multiples of the identity map on V_k. The union of all primary points in K is denoted by ∂_{pr} K.

In other words: A point is primary when it can not be split in two disjoint elements.

<u>Theorem 7</u>: Every $g \in K$, can be represented by a unique central measure μ and μ (O) = O for every Baire-set $O \subseteq K$ with $O \cap \partial_{pr} K = \phi$.

The Baire sets and the Borel sets coincide in the separable case and it has been shown by J. R. Christensen (Kopenhagen) that ∂_{pr} K in that case is universally measurable. His proof uses the Effros-Borel structure on the set of closed subsets of K. In the appendix another simpler proof is given.

<u>Theorem 8</u>: (J. P. Reus Christensen). Let K be a metrizable compact convex subset of a locally convex space E. Then the set, ∂_{pr} K, of primary points in K, is co-analytic.

Co-analytic means that the complement of the set is analytic and it implies that for every Radon-measures the set differs at most by a nul-set from a Borel set. Hence co-analytic sets are measurable for all Radon-measures. It is unknown under what conditions ∂_{pr} K is a Borel set.

In view of theorem 8 it would be nice to have a simpler proof of theorem 7 than is available, especially in the separable case.

When (A,e) is a C* algebra with unit and K is the state space of A, then $L_{ob}(A)$ e coincides with the center of A, and a state $f \in K$ is primary iff if the representation π_f, of A, constructed via the G.N.S. construction, is a primary representation, i.e. the weak closure of $\pi_f(A)$ is a factor. Two states f and g are disjoint iff the representations π_f and π_g are disjoint. To every closed split face F of K there corresponds an ideal $I \subseteq A$ such that $F = \{ f \in K | f(I) = \{0\} \}$ and vice versa.

For lattices, disjointness as introduced here coincides with the usual notion of disjointness. The interpretations of split-faces and order bounded operations are self-evident in this case.

5. Areas for further research

The following lines of development have been started in [5].

1. Do there exist intrinsic characterizations for central measures?
2. As remarked earlier, for a C* -algebra (A,e) with unit e, the set $L_{ob}(A)$ e coincides with the algebraic center of A. Therefore, in order to interpret the results on $L_{ob}(A)$ in a context of C* -algebras one has to study the centers of C* -algebras. Various sequential closures for C* -algebras have been considered and E. B. Davies has shown that the center of these enlarged C* -algebras, in the separable case, separates disjoint primary points. It is easy to see that these sequential closures also exist when A is an order-unit space. How big is the center now ? Answers to these questions can be used in the formation of a theory of direct integrals of partially ordered spaces.
3. Because closed split-faces have so many important properties, it is desirable to develop techniques to handle more general kinds of split-faces. What is the behaviour of central measures with respect to split-faces?
4. Consider the map $K \ni g \to \mu_g$ where μ_g is the central measure corresponding to g. In the case of simplices K, where the central measure coincides with the unique maximal measure, the above map is weak*-measurable. What are the properties of this map in general?

ad 1: A probability measure μ on K, is said to be sub-central if for every Borel set $B \subseteq K$, with $0 < \mu(B) < 1$, the resultants of the restricted measures $\mu|_B$ and $\mu_{K/B}$ are disjoint.

$[\mu_B(A) = \mu(B \cap A)$ for a Borel set $A \subseteq K]$

Subcentral measures have many-nice properties.

i) The subcentral measures representing a given point $g \in K$, form a complete lattice for the Choquet-Meyer ordering of measures. This lattice is isomorphic with the sublattices of $Z_g \cdot g$ and the central measure of g is the unique maximal subcentral measure representing g.

ii) Suppose μ and ν represent a point $g \in K$, and ν is subcentral. Then there exists a smallest measure $\rho_{\mu,\nu}$ with respect to the ordering of Choquet-Meyer, which majorizes μ and ν. If also μ is subcentral, $\mathcal{S}\mu,\nu$ is subcentral.

iii) Let $g \varepsilon K$, and μ the central measure of g. Then μ is majorized by every maximal measure on K which represents g. For C* -algebras, the central measure is the largest element in the set of measures which are majorized by all maximal measures, which representing a given point. This property does not hold in general as can be seen from simple examples.

Another characterization of central measures, in the case of separable C* -algebras, is that they are minimal, in the order of Choquet-Meyer again, among all measures representing a given point and with support in the set of primary points.

The proof of this last fact is not very difficult but its generalization hinges on questions, touched upon in the next section, which are unsolved.

ad 2: The set of all bounded affine functions on K, can be naturally identified with the second Banach dual A** of A. Let A^m be the smallest set in A**, which contains A and is closed with respect to the supremum norm over K. Then $A^m = (A^m)^+ - (A^m)^+$. It is not obvious that elements in the center of A^m are restrictions of elements in the center of A**. But, for every $T^* \varepsilon L_{ob}(A^m)$ there exists a $\tilde{T} \varepsilon L_{ob}(E)$ and then $\tilde{T}^{**} \varepsilon L_{ob}(A^{**})$ such that for all $a \varepsilon A^m$, $Ta = a \circ \tilde{T}$. A conjecture in this connection is that $L_{ob}(A^m)$ consists of the restructions to A^m of those elements $T \varepsilon L_{ob}(A^{**})$ such that $Te \varepsilon A^m$.

The most important question is, whether, at least in the separable case. $L_{ob}(A^m)$ e is big enough to separate disjoint states in K. For C* -algebras the answer is yes.

ad 3: For every split face F of K there is a complimentary split-face F' of K. F' is the biggest split face of K, which is disjoint of F. Then K = Conv(F,F'). Every affine function on F has an affine extension to K, which vanishes on the complementary face F'. In particular if G is a split face let P_G be the affine function which equals 1 on G and 0 on the complement G' of G. We call G admissable if P_G satisfies the barycentric calculus for central measures, that is, if $a \varepsilon A$ and $g \varepsilon K$, with associated central measure μ, then $\int P_G \, d\mu = \langle P_G, g \rangle$.

Admissable faces have many appealing properties. The set of admissible faces is closed for relative complementation, that is, if G and H are admissable and $G \subset H$, then $G' \cap H$, with G' the complement of G, is admissible. The set is closed for monotone sequential limits and contains the closed split-faces, their complements and intersections of those. But it is not know whether the intersections of admissible faces is again admissible. Are there enough admissible faces to separate disjoint states? The answer to these questions can help in building a theory of direct integrals.

ad 4: No comments.

Several other problems have been mentioned in [5]. Can one recover the results of the von-Neuman-Murray theory of direct integrals of operator algebras, using just central measures ? What are the geometric characterizations of the types of a state for C* -algebras (Type I, II, III)? Do such types or others exist in a setting of just compact convex sets?

Appendix

Theorem 8: (J.P. Reus Christensen). Let K be a compact convex set in a locally convex space E, such that the relative topology of K is metrizable. Then the set of primary points of K is the complement of an analytic set.

The notion of primary point can be defined for every convex set, but without loss of generality we may assume that K is the state space of A(K), where A(K) is the Banach space of affine continuous functions on K with the supremum norm. Let the function in A(K) which is identically equal to 1 on K be denoted by e and the Banach dual A(K)* of A(K) by E. Then $K = \{f \varepsilon E \mid \|f\| = \langle f, e \rangle = 1\}$. The set K is endowed with the relative $\sigma(E, A(K))$ topology [1, chapter 2, §2]. Finally, for $x \varepsilon E^+$, let

$$F_x = \{h \varepsilon E^+ \mid h \leq x\}, \quad C_x = \bigcup_{\lambda \geq 0} \lambda F_x \text{ and } V_x = C_x - C_x$$

Definition 1: Two elements $x, y \varepsilon E^+ \setminus 0$ are said to be disjoint, notation $x \,\delta\, y$, if 1. $V_{x+y} = V_y \oplus V_y$ (i.e., V_{x+y} is the algebraic direct sum of V_x and V_y) 2. $C_{x+y} = C_x + C_y$.

2. A point $x \in K$ is said to be primary if it cannot be written as $x = y + z$ with $y, x \in E^+ \setminus 0$ and $x \,\delta\, y$.

<u>Lemma 1</u>: Two points $x, y \in E^+$ are disjoint iff
1. $F_{x+y} = F_x + F_y$ and 2. $F_x \cap F_y = \{0\}$ [[5] prop.1.].

<u>Lemma 2</u>: Let T be a Hansdorff space and K a separable metric space. If R_1 and R_2 are compact subsets of $T \times K$ and $\pi : (t,k) \ni T \times K \to t \in T$, then $\pi\,(R_1 \cup R_2 \setminus R_1 \cap R_2)$ is a Borel subset of T.

<u>Proof</u>: We may assume $R_2 \subseteq R_1$. Let $\{0_n\}_n$ be a basis for the neighborhoods in K, consisting of closed sets. Put $\bar{0}_n = T \times 0_n$. Then $\pi\,(R_1 \setminus R_2) = \bigcup_n \pi(R_1 \cap \bar{0}_n) \setminus \pi(R_2 \cap \bar{0}_n)$. Clearly, the sets $R_1 \cap \bar{0}_n$ and $\pi\,(R_1 \cap 0_n)$ are closed and hence the lemma follows.

<u>Theorem 9</u>: Let $K = \{h \in E \mid h \geq 0, ||h|| \leq 1\}$. The set $\{(x,y) \mid x,y \in K, x \,\delta\, y\}$ is a Borel subset of $K \times K$.

<u>Proof</u>. The following four sets are all closed in $K \times K \times 2K$.

$$A_1 = \{(x,y,z) \mid x, y \in K;\ z \in F_{x+y}\}.$$

$$A_2 = \{(x,y,z) \mid x, y \in K;\ z \in F_x + F_y\}.$$

$$B_1 = \{(x,y,z) \mid x, y \in K;\ z \in F_x + F_y\}.$$

$$B_2 = \{(x,y,0) \mid x, y \in K\}.$$

Let $\pi : (x,y,z) \to (x,y)$, then it follows from lemma 2 that $\pi\,(A_1 \setminus A_2)$ and $\pi\,(B_1 \setminus B_2)$ are Borel subsets of $K \times K$ and consequently $[\pi\,(A_1 \setminus A_2) \cup \pi\,(B_1 \setminus B_2) \cup \{K \times 0\} \cup \{0 \times K\}]^c = \{(x,y) \mid x \,\delta\, y\}$ is Borel too. The last equality follows from lemma 1.

Proof of theorem 8: The continuous map $(x,y) \to x + y$ maps $\{(x,y) \mid x \,\delta\, y\}$ onto the complement of the set of primary points in K, which therefore is analytic.

References

[1] Alfsen, E.M., Compact convex sets and boundary integrals
Berlin, Heidelberg, New York, Springer 1971

[2] Alfsen E.M., Effros, E., Structure in real Banach spaces
Annals of Math. 96: 98-173, (1972).

[3] Edwards D.A., The ideal centre of certain partial ordered
Banach spaces (preprint)

[4] Habre W., Thèse du troisieme cycle, Paris 1972

[5] Wils, W., The ideal center of partially ordered Banach
spaces
Acta.Math. 127: 41-77 (1971)

ORDERED NORMED TENSOR PRODUCTS

Gerd Wittstock

Fachbereich Mathematik der Universität des Saarlandes

Saarbrücken, Germany

Recently ordered topological tensor products have been studied in the literature. In general there are several possible order structures and topologies on the tensor product. Much more is known if the factors are spaces of a special type. The aim of this paper is to build up a more general theory. On the other hand, to obtain nice results, we restrict our self to the class of normed spaces ordered by a normal strict B-cone, respectively regular ordered spaces in the sense of E.B. Davies. This paper should be viewed as a short report on the subject.

1 Notation and preliminary results

As usual we define the tensor product of two real linear spaces E,F by an universal mapping property:

1.1 Definition. The pair $(E \otimes F, \omega)$, $E \otimes F$ a real linear space and $\omega: E\times F \to E \otimes F$ a bilinear mapping, is called the tensor product of E and F, if for every bilinear mapping $\varphi: E\times F \to G$ there exists an unique linear mapping $\hat{\varphi}: E \otimes F \to G$ such that $\varphi = \hat{\varphi} \circ \omega$. We denote $\omega(x,y) = x \otimes y$ $(x \in E, y \in F)$.

By this definition the tensor product is unique up to an isomorphism. If $\hat{\omega}$ is injective, it is possible and often convenient to identify $E \otimes F$ as a subspace of G. Then we have a special model of the tensor product.

1.2 Example. Denote by $\mathfrak{F}(M)$ the vector space of all real functions on a set M. If $E \subset \mathfrak{F}(M)$, $F \subset \mathfrak{F}(N)$ are subspaces, define a bilinear mapping $E\times F \to \mathfrak{F}(M\times N)$, where (x,y) is mapped onto the function $\{(m,n) \mapsto x(m)y(n)\}$, for all $x \in E, y \in F, m \in M, n \in N$. The induced linear mapping $E \otimes F \to \mathfrak{F}(M \times N)$ is injective. We embed $E \otimes F \subset \mathfrak{F}(M \times N)$ and interpret $x \otimes y$ as a function of two variables:

$$x \otimes y : (m,n) \mapsto x(m)y(n) \quad \text{for all } m \in M, n \in N.$$

A special case is the usual representation of tensorproducts of finite dimensional spaces as matrices.
We identify in a natural way the space $B(E,F)$ of all bilinear real forms on $E \times F$ with the dual $(E \otimes F)^*$ and denote this duality by $\langle E \otimes F, B(E,F) \rangle$. Thus we write

$$\varphi(x,y) = \langle x \otimes y, \varphi \rangle \quad \text{for } x \in E, y \in F, \varphi \in B(E,F).$$

Since the algebraic duals $E^* \subset \mathfrak{F}(E), F^* \subset \mathfrak{F}(F)$, the mapping

$$x^* \otimes y^* : (x,y) \mapsto \langle x,x \rangle \cdot \langle y,y \rangle \quad (x \in E, x^* \in E^*, y \in F, y^* \in F^*)$$

is defined and bilinear. We obtain an embedding $E^* \otimes F^* \subset (E \otimes F)^*$.
We denote the corresponding bilinear mapping by

$$\omega^* : E^* \times F^* \to (E \otimes F)^*.$$

1.3 <u>Example</u>: Let $\mathfrak{F}(M;F)$ be the space of all F-valued functions on M. If $E \subset \mathfrak{F}(M)$ is a linear subspace, we have a bilinear mapping $E \times F \to \mathfrak{F}(M;F)$, where (x,y) is mapped onto the function $\{m \mapsto x(m)y\}$. The induced linear mapping $E \otimes F \to \mathfrak{F}(M;F)$ is injective. We embed $E \otimes F \subset \mathfrak{F}(M;F)$ and interpret $x \otimes y$ as a F-valued function

$$x \otimes y : m \mapsto x(m)y.$$

As a special case we obtain from $E^* \subset \mathfrak{F}(E)$ the embedding $E^* \otimes F \subset L(E,F)$ into the space of all linear operators from E into F.

1.4 <u>Example</u>. If $f \in L(E_1,E_2)$, $g \in L(F_1,F_2)$, then $\omega_2 \cdot (f \times g) : E_1 \times F_1 \to E_2 \times F_2$ is bilinear. There exists an unique linear mapping, denoted by $f \otimes g$, such that the diagramm

$$\begin{array}{ccc} E_1 \times F_1 & \xrightarrow{f \times g} & E_2 \times F_2 \\ \downarrow \omega_1 & & \downarrow \omega_2 \\ E_1 \otimes F_1 & \xrightarrow{f \otimes g} & E_2 \otimes F_2 \end{array}$$

commutes. The bilinear mapping $(f,g) \mapsto f \otimes g$ induces an embedding $L(E_1,E_2) \otimes L(F_1,F_2) \subset L(E_1 \otimes F_1, E_2 \otimes F_2)$.

1.5 <u>Remark</u>. Let H_1, H_2 be Hilbert spaces, $L_b(H_1)$ the space of all bounded linear operators on H_1, $K(H_1)$ the space of all compact linear operators and $N(H_1)$ the space of all nuclear operators (the trace class). The trace form induces a natural identification $K(H_1)' = N(H_1)$, and $N(H_1)' = L_b(H_1)$, more over we have $H_1' = H_1$. If now $f:H_1 \to H_1, g:H_2 \to H_2$ are bounded respectively compact or nuclear operators, $x, x_1 x_2 \in H_1, y \in H_2$ there are a lot of possible meanings of symbols like $f \otimes g$, $x \otimes y$, $x \otimes f$, $x_1 \otimes x_2$ a.s.o.. But all the identifications behave in a natural manner, there is no danger of confusion.

1.6 If E,F have an additional norm or order structure, we may introduce norms or orderings on $E \otimes F$ by a restricted universal mapping property - take only bounded or positive mappings $\varphi, \hat{\varphi}$ in definition 1.1 - or we may introduce them by special representations as 1.2 to 1.5.

1.7 Definition. A norm $\|.\|_\alpha$ is called a tensor norm (cross-norm) if the bilinear mappings

$$\omega : E \times F \to E \otimes_\alpha F, \omega' : E' \times F' \to (E \otimes_\alpha F)'$$

are defined and bounded with norm less then one. We denote $E \otimes_\alpha F = (E \otimes F, \|.\|_\alpha)$. This is equivalent to

$$\|x \otimes y\|_\alpha = \|x\| \; \|y\| \qquad \text{for all } x \in E, y \in F$$

$$\|x' \otimes y'\|_{\alpha'} = \|x'\| \; \|y'\| \qquad \text{for all } x' \in E', y' \in F',$$

where $\|.\|_{\alpha'}$ denotes the dual norm.

1.8 The solution of the corresponding mapping problem for bounded bilinear operators is the projective tensor norm $\|.\|_\pi$. It is the distance functional of the convex hull of the tensorproduct of the unit balls S_E, S_F.

$$\|t\|_\pi = \inf\{\sum_{\nu=1}^{n} \|x_\nu\| \; \|y_\nu\| : n \in N, t = \sum_{\nu=1}^{n} x_\nu \otimes y_\nu\}.$$

The dual of $E \otimes_\pi F$ is the space $B_b(E,F)$ of all bounded bilinear forms.

1.9 By a function space representation $E \subset \mathfrak{F}_b(M), F \subset \mathfrak{F}_b(N)$, isometrically embedded into spaces of bounded functions, we obtain the injective tensornorm $\|.\|_\varepsilon$; $F \otimes_\varepsilon F \subset \mathfrak{F}_b(M,N)$. The induced injective norm is independent of the representation. By $E \subset \mathfrak{F}_b(E'), F \subset \mathfrak{F}_b(F')$ we obtain

$$\|t\|_{\varepsilon} = \sup\{|\langle t, x'\otimes y'\rangle| : x'\in E', \|x'\| \le 1, y'\in F', \|y'\| \le 1\}.$$

1.10 <u>Proposition</u>. If $\|,\|_{\alpha}$ is a tensor norm then

$$\|.\|_{\varepsilon} \le \|.\|_{\alpha} \le \|.\|_{\pi} .$$

1.11 <u>Definition</u>. Let E,F be ordered linear spaces with proper cones E_+, F_+. We call a cone $C_{\alpha}\subset E\otimes F$ a <u>tensor cone</u> and write $E\otimes_{\alpha}F = (E\otimes F, C_{\alpha})$, if the canonical bilinear mappings $\omega: E\times F\to E\otimes_{\alpha}F$ and $\omega^*: E^*\times F^*\to(E\otimes_{\alpha}F)^*$ are positive:

$$x\otimes y\in C_{\alpha} \qquad \text{for all } x\in E_+, y\in F_+$$

and

$$x^*\otimes y^*\in C_{\alpha}^* \qquad \text{for all } x\in E^*_+, y\in F^*_+,$$

where E^*_+, F^*_+ and C_{α}^* denote the dual cones.

1.12 The solution of the corresponding universal problem for positive bilinear mappings φ, $\varphi(E_+,F_+)\subset G_+$, is the <u>projective cone</u>

$$C_p = \mathrm{co}(E_+\otimes F_+) = \{\sum_{\nu=1}^{n} x_{\nu}\otimes y_{\nu} : x_{\nu}\in E_+, y_{\nu}\in F_+, n\in\mathbb{N}\}.$$

If E_+, F_+ are proper cones then C_p is a proper cone.

1.13 <u>Definition</u>. Let E,F be ordered linear spaces, such that

$$E_+ = \{x\in E: \langle x, E^*_+\rangle \ge 0\}.$$

Let $E\in\mathfrak{F}(M)$, $F\subset\mathfrak{F}(N)$ be order embeddings, then the embedding $E\otimes F\subset\mathfrak{F}(M\times N)$ induces the <u>injective tensor cone</u> C_i. It is independent of the special representation:

$$C_i = \{t\in E\otimes F: \langle t, E^*_+\otimes F^*_+\rangle \ge 0\}.$$

1.14 <u>Proposition</u>. If C_{α} is a tensor cone then $C_p\subset C_{\alpha}\subset C_i$.

1.15 <u>Example</u>. We show an example of a tensor cone, which is different from the projective and the injective cone.

Let $A_n\subset L(\mathbb{C}^n)$ be the real linear space of complex hermitian matrices, ordered by the cone A_{n+} of positive semidefinite matrices. Then $A_n\otimes A_m = A_{nm}$. An easy computation shows, that the closure $\overline{C}_p\subset A_{nm+}$. The trace form induces a duality $\langle A_n, A_n\rangle$ with

$$\langle x,y\rangle = \mathrm{Tr}(xy) \qquad x,y\in A_n.$$

By this duality we obtain $(A_n)^*_+ = A_{n+}$ and $C_p{}^* = C_i$. It follows that

$$\overline{C}_p \subset A_{nm+} = (A_{nm})^*_+ \subset C_p{}^* = C_i.$$

We show now, that $\overline{C}_p \neq A_{nm+}$ if $n,m \geq 2$. then $(A_{nm})^*_+ \neq C_p{}^*$. Let $e_1, \dots, e_n$ respectively $e_1, \dots, e_m$ be the canonical basis of $\mathbb{C}^n(\mathbb{C}^m)$. Define linear operators $P_{\nu\mu}:\mathbb{C}^n\to\mathbb{C}^n$ respectively $P_{\nu\mu}:\mathbb{C}^m\to\mathbb{C}^m$ $(\nu,\mu=1,2)$ by

$$P_{\nu\mu}: e_\nu \mapsto e_\mu, P_{\nu\mu}: e_\lambda \mapsto 0 (\lambda \neq \nu).$$

Then

$$\sum_{\nu,\mu=1}^{2} P_{\nu\mu}\otimes P_{\nu\mu} \in A_{nm+}$$

since it is hermitian and

$$\langle \sum_{\nu,\mu=1}^{2} P_{\nu\mu}\otimes P_{\nu\mu} a,a\rangle = \sum_{\nu,\mu=1}^{2} \alpha_{\nu\nu}\overline{\alpha}_{\mu\mu} \geq 0 \quad \text{for } a = \sum_{\varkappa=1}^{n}\sum_{\lambda=1}^{m} \alpha_{\varkappa\lambda} e_\varkappa \otimes e_\lambda.$$

The mapping $\tau: L(\mathbb{C}^n)\to L(\mathbb{C}^n)$ (transposed matrix) maps A_{n+} into A_{n+} and $(\tau\otimes id)A_{nm}\subset A_{nm}$, $(\tau\otimes id)\overline{C}_p\subset\overline{C}_p$. But

$$(\tau\otimes id)\sum_{\nu,\mu=1}^{2} P_{\nu\mu}\otimes P_{\nu\mu} = \sum_{\nu,\mu=1}^{2} P_{\mu\nu}\otimes P_{\nu\mu} \notin A_{nm+}$$

since

$$\langle \sum_{\nu,\mu=1}^{2} P_{\mu\nu}\otimes P_{\nu\mu} a,a\rangle = \sum_{\nu,\mu=1}^{2} \alpha_{\mu\nu}\overline{\alpha}_{\nu\mu}$$

is not positive semidefinite. This example is due to W.F. Stinespring [15].

2 Regular ordered normed tensor products

Let $(E,E_+,\|.\|)$, $(F,F_+,\|.\|)$ be ordered normed spaces, such that E_+ and F_+ are closed normal strict B-cones (see Schaefer [13]).
Later on we will assume, that E and F are regular ordered (see definiton 2.2 below)

2.1 Definition. We call $(E\otimes F, C_\alpha, \|.\|_\alpha)$ an ordered normed tensor product if

(i) C_α is a closed normal strict B-cone

(ii) C_α is a tensor cone: $E_+ \otimes F_+ \subset C_\alpha$ and $E'_+ \otimes F'_+ \subset C_\alpha{}'$

(iii) $\|x \otimes y\|_\alpha \leq \|x\| \, \|y\|$ for all $x \in E_+, y \in F_+$ and

$\|x' \otimes y'\|_{\alpha'} \leq \|x'\| \, \|y'\|$ for all $x' \in E'_+, y' \in F'_+$.

(where $C_\alpha{}'$ is the cone of bounded positive linear forms and $\|.\|_{\alpha'}$ is the dual norm.)

An ordered normed space has a normal strict B-cone if and only if it posesses an equivalent regular norm $\|.\|_1$ in the sense of E.B.Davies [4]. Take

$$\|x\|_1 = \inf \{ \|y\| : y \pm x \in E_+ \}.$$

2.2 <u>Definition</u>. We call a normed ordered space E <u>regular</u> ordered if the cone is closed and

(i) if $y \pm x \in E_+$ then $\|x\| \leq \|y\|$

(ii) if $\|x\| < 1$ then there exists a $y \in E_+$ such that $\|y\| < 1$ and $y \pm x \in E_+$.

2.3 <u>Definition</u>. Let E,F be regular ordered normed spaces. $E \otimes_\alpha F = (E \otimes F, C_\alpha, \|.\|_\alpha)$ is called a <u>regular ordered normed tensor product</u>, if it is an ordered normed tensorproduct and the norm is regular.

2.4 <u>Remark</u>. Regular ordered normed spaces have the following properties

(i) If E is regular ordered, then the completion $\tilde{E}$ is regular ordered.

(ii) The dual E' is regular ordered. if and only if $\tilde{E}$ is regular ordered.

(iii) If E is ordered by a cone E_+, and the norm fulfills the properties (i) and (ii) of definition 2.2 then $(E, \overline{E}_+, \|.\|)$ is a regular ordered normed space.

<u>Now we consider always regular ordered normed spaces</u>.

2.5 <u>Lemma</u>. If E,F are regular ordered normed spaces, then

(i) $\|x\| = \sup \{ \langle x,x' \rangle : x' \in E'_+, \|x'\| \leq 1 \}$ for all $x \in E_+$

(ii) $\|x'\| = \sup \{ \langle x,x' \rangle : x \in E_+, \|x\| \leq 1 \}$ for all $x \in E'_+$

(iii) $\|\varphi\| = \sup \{\|\varphi(x,y)\| : x \in E_+, \|x\| \leq 1,\ y \in F_+, \|y\| \leq 1\}$
for all bounded positive bilinear mappings $\varphi : E \times F \to G$ into a regular ordered normed space G.

(iv) If $\varphi \pm \psi \in B_b(E,F)_+$ then $\|\varphi\| \geq \|\psi\|$.

We denote by $B_b(E,F)_+$ the cone of all bounded positive bilinear forms on $E \times F$.

2.6 Proposition. If $(E \otimes F, C_\alpha, \|.\|_\alpha)$ is a regular ordered normed tensor product of regular ordered normed spaces E,F then

$$\|.\|_\varepsilon \leq \|.\|_\alpha \leq \|.\|_\pi$$

and $\|.\|_\alpha$ is a tensor norm (see definition 1.7)

Proof. If $\varphi \in C_\alpha{}' = (E \otimes_\alpha F)'_+$ then

$$\|\varphi\|_{\alpha'} = \sup \{\langle \varphi, t\rangle : t \in C_\alpha, \|t\|_\alpha \leq 1\}$$

$$\geq \sup \{\langle \varphi, x \otimes y\rangle : x \in E_+, \|x\| \leq 1,\ y \in F_+, \|y\| \leq 1\} = \|\varphi\|$$

If $\varphi \in C_\alpha{}'$ then $\varphi \in B_b(E,F)_+$. Since $\|.\|_{\alpha'}$ is a regular norm

$$\|\psi\|_{\alpha'} = \inf \{\|\varphi\|_{\alpha'} : \varphi \pm \psi \in C_\alpha{}'\} \quad \text{for all } \psi \in (E \otimes_\alpha F)'.$$

By Lemma 5(iv) we obtain

$$\|\psi\|_{\alpha'} \geq \inf \{\|\varphi\| : \varphi \pm \psi \in B_b(E,F)_+\} = \|\psi\| = \|\psi\|_{\pi'} .$$

Therefore $\|.\|_\alpha \leq \|.\|_\pi$.

If $t \in C_\alpha$ then by 2.5(iii)

$$\|t\|_\alpha = \sup \{\langle t, \varphi\rangle : \varphi \in C_\alpha{}', \|\varphi\|_{\alpha'} \leq 1\}$$

$$\geq \sup \{\langle t, x' \otimes y'\rangle : x' \in E'_+, \|x'\| \leq 1,\ y' \in F'_+, \|y'\| \leq 1\}$$

$$= \|t\|_\varepsilon .$$

2.7 The projective ordered normed tensor product. Le E,F be regular ordered normed spaces. The projective cone $C_p = co(E_+ \otimes F_+)$ is a proper cone. The functional

$$\|t\|_p = \sup \{\langle t, \varphi\rangle : \varphi \in B_b(E,F)_+, \|\varphi\| \leq 1\} \quad \text{for all } t \in C_p$$

is monoton, subadditiv and positiv homogeneous on C_p. Thus the functional

$$\|s\|_p = \inf\{\|t\|_p : t \pm s \in C_p\}$$

is a norm on $E \otimes F$. It follows from the remark 4(iii) that $E \otimes_p F = (E \otimes F, \overline{C}_p, \|.\|_p)$ is a regular ordered normed tensor product. The dual space is $B_b(E,F)_+ - B_b(E,F)_+$ with the norm

$$\|\psi\|_{p'} = \inf\{\|\varphi\| : \varphi \pm \psi \in B_b(E,F)_+\} .$$

$E \otimes_p F$ has the following <u>universal property</u>: if $\varphi : E \times F \to G$ is a positive bounded bilinear mapping into a regular ordered normed space G, then the induced linear mapping $\hat{\varphi} : E \otimes_p F \to G$ is positive bounded and $\|\hat{\varphi}\| = \|\varphi\|$. It has a <u>functorial property</u>: if $f : E_1 \to E_2$, $g : F_1 \to F_2$ are positive and bounded then $f \otimes g : E_1 \otimes_p F_1 \to E_2 \otimes_p F_2$ is positive bounded and $\|f \otimes g\| = \|f\|\ \|g\|$.

2.8 <u>The injective ordered normed tensor product</u>. We take the injective cone (biprojective cone)

$$C_i = \{t : \langle t, x' \otimes y' \rangle \geq 0 \quad \text{for all } x' \in E'_+,\ y' \in F'_+\}.$$

The injective tensor norm $\|.\|_\varepsilon$ is monoton on C_i. We define the ordered injective norm

$$\|s\|_i = \inf\{\|t\|_\varepsilon : t \pm s \in C_i\}.$$

$E \otimes_i F = (E \otimes F, C_i, \|.\|_i)$ is a regular ordered normed tensor product. It has a <u>functoriel property</u>: if f,g are positive bounded linear operators, then $f \otimes g$ is positive bounded and $\|f \otimes g\| = \|f\|\ \|g\|$.

2.9 <u>Proposition</u>. If $E \otimes_\alpha F$ is a regular ordered normed tensorproduct, then $\overline{C}_p \subset C_\alpha \subset C_i$ and $\|.\|_p \geq \|.\|_\alpha \geq \|.\|_i$.

2.10 <u>Example</u>. In the case of the Hilbert space l_2 we have the sequence

$$l_2 \otimes_\pi l_2 \to l_2 \otimes_p l_2 \to l_2 \otimes_2 l_2 \to l_2 \otimes_i l_2 \to l_2 \otimes_\varepsilon l_2 .$$

The cones $\overline{C}_p$ and C_i coincide but the corresponding normes are all different and not equivalent.

2.11 <u>Proposition</u>. The natural isometric order embedding $F \subset F''$ induces

(i) an isometric order embedding $E \otimes_p F \subset E \otimes_p F''$ and

(ii) an order embedding $E \otimes_i F \subset E \otimes_i F''$.

<u>Proof</u>. (i) If $\varphi \in (E \otimes_p F)'_+ = B_b(E,F)_+$, $x \in E_+$, then the positive bounded linear form $\varphi(x,.) : y \mapsto \varphi(x,y)$ has a natural extension to F'', induced by the embedding $F' \subset F'''$. Thus we get a natural bilinear extension $\varphi_1 : E \times F'' \to R$. Of cours $\varphi_1 \in B_b(E,F'')_+$. By duality we obtain

$$E \otimes F \cap (E \otimes_p (F'')_+ = (E \otimes_p F)_+ .$$

The linear extension mapping $\varphi \mapsto \varphi_1$ has norm less than 1, thus $E \otimes_p F \subset E \otimes_p F''$ is an isometry.

(ii) F'_+ is $\sigma(F''', F'')$ dense in F'''. Therefore

$$\langle t, x' \otimes y' \rangle \geq 0 \quad \text{for all } x' \in E'_+,\ y' \in F'_+$$

if and only if

$$\langle t, x' \otimes y''' \rangle \geq 0 \text{ for all } y''' \in F'''_+ .$$

It follows that $E \otimes_i F \subset E \otimes_i F''$ is an order embedding.

3. A tensoriel characterisation of the Riesz decomposition property

E posesses the Riesz decomposition property, if

$$[0,x] + [0,y] = [0,x+y] \qquad \text{for all } x,y \in E_+ .$$

This is equivalent to the Riesz interpolation property:

If $y_\nu - x_\mu \in E_+ (\nu,\mu = 1,2)$ then there exists a z, such that $y_\nu - z \in E_+$ $z - x_\mu \in E_+ (\nu,\mu = 1,2)$. A regular ordered normed space $\tilde{E}$ has the Riesz decomposition property if and only if the dual E' is a lattice.

3.1 Theorem. (i) If $\tilde{E}$ has the Riesz decomposition property, then for every F the closure of C_p in the projective tensor norm $\|.\|_\pi$ is the injective cone $C_i \subset E \otimes F$.

(ii) If C_p is dense in C_i in $E \otimes A(Q)$, Q a square, then E has the Riesz decomposition property.

If K is a compact convex set then A(K) is the space of all affine continous functions on K with the pointwise order and the sup-norm.

We need the following lemma for the proof of the theorem.

3.2 Lemma. Let E be a Banach lattice, $x_1, \dots, x_n, u \in E$, such that $u \pm x \in E_+$ and $\eta > 0$. Then there exist $u_\mu \in E_{u+}$, and $u' \in (E_u)'_+, \mu = 1, \dots, m$, such that $\langle u_\lambda, u'_\mu \rangle = \delta_{\lambda\mu}$ and

$$\|x_\nu - \sum_{\mu=1}^{n} \langle x_\nu, u'_\mu \rangle u_\mu\| < \eta .$$

Proof. The linear space $E_u = \{x\colon \text{there exists a } \lambda > 0, \lambda u \pm x \in E_+\}$ is a Banach lattice with the norm

$$\|x\|_u = \inf \{\lambda > 0,\ \lambda u \pm x \in E_+\} \qquad \text{for all } x \in E_u .$$

E_u is an (AM)-space. By the Kakutani representation theorem, there exists a compact set T such that $E_u \cong C(T)$. Define a continous mapping $f:T \to R^n$, $f:t \mapsto (x_\nu(t))_{\nu=1}^n$. In R^n we find a partition of unity $\gamma_\mu(\mu=1,\dots,m)$ and points $\xi_\mu \in f(T)$ such that

$$0 \le \gamma_\mu, \sum_{\mu=1}^m \gamma_\mu = 1, \ \gamma_\lambda(\xi_\mu) = \delta_{\lambda\mu}, \ \|\xi - \sum_{\mu=1}^m \xi_\mu \gamma_\mu(\xi)\|_{R^n} < \eta$$

for all $\xi \in f(T)$. Then $u_\mu = \gamma_\mu \circ f \in C(T)_+ = E_{u+}$, $u'_\mu \in f^{-1}(\xi_\mu) \in (E_u)'_+$ and they have the desired properties.

Proof of theorem 3.1. (i) Assume first that E is a Banach lattice. Let

$$t = \sum_{\nu=1}^n x_\nu \otimes y_\nu$$

be an element of $C_i \subset E \otimes F$. Choose $u \in E_+$, such that $u \pm x \in E_+$. For $\eta > 0$ there exist by lemma 3.2 elements $u_\mu \in E_{u+}$, $u'_\mu \in (E_u)'_+$ such that $\langle u_\lambda, u'_\mu \rangle = \delta_{\lambda\mu}$ and

$$\|x_\nu - \sum_{\mu=1}^m \langle x_\nu, u'_\mu \rangle u_\mu\| < \eta \ .$$

Define

$$t_\eta = \sum_{\nu=1}^n \sum_{\mu=1}^m \langle x_\nu, u'_\mu \rangle u_\mu \otimes y_\nu = \sum_{\mu=1}^m u_\mu \otimes z_\mu \ .$$

Since $E_{u+} = E_+ \cap E_u$ it follows that the restriction $E'_+|E_u$ is $\sigma\langle (E_u)', E_u\rangle$ dense in $(E_u)'_+$. Since $t \in E_u \otimes F$ we obtain

$$\langle t, u' \otimes y' \rangle \ge 0 \quad \text{for all } u' \in (E_u)'_+, y' \in F'_+ .$$

Now

$$0 \le \langle t, u'_\mu \otimes y' \rangle = \sum_{\nu=1}^n \langle x_\mu, u'_\mu \rangle \langle y_\mu, y' \rangle = \langle t_\eta, u'_\mu \otimes y' \rangle = \langle z_\mu, y' \rangle \text{for all } y' \in F'_+ .$$

Hence $z_\mu \in F_+$ and $t_\eta \in C_p \subset E \otimes F$. Now

$$\|t - t_\eta\|_\pi \le \sum_{\nu=1}^n \|x_\nu - \sum_{\mu=1}^m \langle x_\nu, u'_\mu \rangle u_\mu\| \cdot \|y_\nu\| \le \eta \sum_{\nu=1}^n \|y_\nu\| .$$

Assume now that E has the Riesz decomposition property. Then E" is a Banach lattice. By proposition 2.11 we have the relations

$$\overline{C}_p = (E \otimes_p F)_+ = (E'' \otimes_p F)_+ \cap (E \otimes F) = (E'' \otimes_i F)_+ \cap (E \otimes F) = (E \otimes_i F)_+ = C_i .$$

Here $\overline{C}_p$ is the closure of C_p in the ordered projective tensornorm $\|.\|_p$. But consider the duality $\langle E \otimes_\pi F, B_b(E,F)\rangle = \langle E \otimes_\pi F, (E \otimes_\pi F)'\rangle$. The polar of C_p is $B_b(E,F)_+$, and the bipolar is $\overline{C}_p$. By the bipolar theorem we obtain, that $\overline{C}_p$ is the $\|.\|_\pi$-closure of C_p.

(ii) Since $A(Q)$ is a 3-dimensional space all possible tensor norms $\|.\|_\alpha$ coinside : $E \otimes_\alpha A(Q) = E{\times}E{\times}E$. Let $q_1, \dots, q_4$ be the extremal points of Q, and $v'_\nu : v \to v(q_\nu)$. The functional $v' \in A(Q)'$ generate extremal rays of $A(Q)'_+$ and these extremal rays generate the cone $A(Q)'_+$. We have the relation

$$v'_1 - v'_2 + v'_3 - v'_4 = 0.$$

The affine functions $v_1, \dots, v_4 \in A(Q)$ defined by

$$v_\nu(q_\nu)=1,\ v_\nu(q_{\nu+1})=1,\ v_\nu(q_{\nu+2})=0,\ v_\nu(q_{\nu+3})=0$$

where $q_5 = q_1$ a.s.o, generate extremal rays of $A(Q)_+$ and these extremal rays generate the cone $A(Q)_+$.

Now choose $x, y_1, y_2 \in \widetilde{E}$, such that $0, x \le y_1, y_2$. We define a bilinear form $t \in \widetilde{E} \otimes A(Q) \subset B(E', A(Q)')$ by the formulae

$$t(.,v'_1) = y_1$$
$$t(.,v'_2) = y_2$$
$$t(.,v'_3) = y_2 - x.$$

Then $t(.,v'_4) = t(.,v'_1 - v'_2 + v'_3) = y_1 - x$. The bilinear form t is positive, $t \in C_i$.

Assume for a moment, that $t \in C_p$, then

$$t = \sum_{\nu=1}^{4} z_\nu \otimes v_\nu$$

where $z_\nu \in \widetilde{E}_+$, because the v_ν generate the cone $A(Q)_+$. Then we obtain $y_1 = t(.v_1') = z_1 + z_2, y_2 = z_2 + z_3, y_2 - x = z_3 + z_4$. Thus $0, x \le z_2 \le y_1, y_2$. In general we only have $t \in C_i = \overline{C}_p$, but by an iteration process we obtain the same result.

For $\eta > 0$ there exists

$$s = \sum_{\nu=1}^{4} z_\nu \otimes v_\nu \in C_p,\ z_\nu \in \tilde{E}_+,\ \text{such that } \|t-s\| < \frac{\eta}{12}.$$

We obtain

$$\|y_1-(z_1+z_2)\|<\frac{\eta}{12}, \|y_2-(z_2+z_3)\|<\frac{\eta}{12}, \|(y_2-x)-(z_3+z_4)\|<\frac{\eta}{12}.$$

Thus there exists an $u \in \tilde{E}_+$, $\|u\| < \frac{\eta}{3}$ such that

$$-u \le y_1-(z_1+z_2) \le u$$
$$-u \le y_2-(z_2+z_3) \le u$$
$$-u \le (y_2-x)-(z_3+z_4) \le u.$$

We obtain $0,x \le z_2 \le y_1+w, y_2+w$ where $w = 3u \in \tilde{E}_+$, $\|w\| < \eta$. By iteration we find sequences $a_n, w_n \in \tilde{E}_+$ such that $\|w_n\| \le 2^{-n}$,

$$0,x \le a_1 \le y_1+w_1, y_2+w_1$$

and

$$0,x,a_{n-1}-w_{n-1} \le a_n \le y_1+w_n, y_2+w_n, a_{n-1}+w_n.$$

It follows that

$$-(w_{n-1}+w_n) \le a_n-a_{n-1} \le w_{n-1}+w_n$$

and therefore

$$\|a_n-a_{n-1}\| < 2^{-n+2}$$

The limit $z = \lim a_n \in \tilde{E}$ exists and $0,x \le z \le y_1, y_2$. The space $\tilde{E}$ has the Riesz interpolation property.

3.3 _Theorem_. If $\tilde{E},\tilde{F}$, have the Riesz interpolation property, then every complete regular ordered tensor product $E \tilde{\otimes}_\alpha F$ has the Riesz interpolation property.

Proof. If E,F, have the Riesz interpolation property, then $B_b(E,F)_+ = L_b(E,F')_+$ is a lattice. Then $(E \otimes_p F)' = B_b(E,F)_+ - B_b(E,F)_+$ is a Banach lattice and the predual $E \tilde{\otimes}_p F$ has the Riesz interpolation property. (see Ng [9] and Peressini [10])

We show now that $(E \tilde{\otimes}_\alpha F)'$ is a lattice. If $\varphi_1,\varphi_2 \in (E \tilde{\otimes}_\alpha F)'_+ \subset B_b(E,F)_+$, then there exists $\varphi = \inf(\varphi_1,\varphi_2) \in B_b(E,F)_+$. Since

$$\|\varphi\|_{\alpha'} = \sup\{\langle t,\varphi\rangle : \|t\|_\alpha \le 1,\ t \in C_\alpha\} \le$$
$$\le \inf_{\nu=1,2} \sup\{\langle t,\varphi_\nu\rangle : \|t\|_\alpha \le 1, t \in C_\alpha\} = \inf\{\|\varphi_1\|\ \|\varphi_2\|\}$$

we obtain that $\varphi \in (E \tilde{\otimes}_\alpha F)'$ and $(E \tilde{\otimes}_\alpha F)'$ is a lattice. Therefore the predual $E \tilde{\otimes}_\alpha F$ has the Riesz interpolation property.

4. Ordered tensorproducts of base normed and order unit normed spaces.

As special examples of the preceding results we consider now base normed and order unit normed spaces.

4.1 Definition. A regular ordered normed space is a base normed space, if the norm is additive on the positive cone.

4.2 Definition. A regular ordered normed space is an order unit normed space, if there is a greatest element u in the unit ball. u is called the norm determing order unit.
The norm is determined by the formula

$$\|x\| = \inf \{\lambda>0 : \lambda u \pm x \in E_+\} \qquad \text{for all } x \in E.$$

4.3 Example. (i) If K is a compact convex set, then the space A(K) of all continous affine functions on K with the pointwise order and the sup-nom is an order unit normed space.
(ii) The dual A(K)' is a base normed space.

4.4 Remark. E is a base normed space, if and only if E' is an order unit normed space. Then

$$\|x\| = \langle x, u' \rangle \qquad \text{for all } x \in E_+,$$

where u' is the norm determing order unit in E'.

4.5 Theorem. Let E,F be base normed spaces. Then
(i) $E \otimes_p F$ is a base normed space and $\|.\|_p = \|.\|_\pi$.
(ii) $E \otimes_i F$ is base normed.

Proof. (i) Let $u' \in E'$, $v' \in F'$ be the norm determing order units.
Since $\|x\| = \langle x, u' \rangle$ for all $x \in E_+$ we obtain that

$$\|\varphi\| \leq 1 \text{ if and only if } u' \otimes v' \pm \varphi \in B_b(E,F)_+$$

for all $\varphi \in B_b(E,F)_+$. It follows that

$$\|t\|_p \leq \langle t, u' \otimes v' \rangle \leq \|t\|_p \qquad \text{for all } t \in \overline{C}_p.$$

The norm is additive on $\overline{C}_p$.

If $\varphi \in B_b(E,F)$, $\|\varphi\| \leq 1$ then $u' \otimes v' \pm \varphi \in B_b(E,F)_+$. Thus $B_b(E,F) = (E \otimes_p F)'$ and $\|.\|_p = \|.\|_\pi$.

(ii) If $t \in C_i$ then $\|t\|_i = \|t\|_\varepsilon = \sup\{\langle t, x' \otimes y' \rangle : x' \in E'_+, y' \in F'_+\} = \langle t, u' \otimes v' \rangle$. The norm is additive on the positive cone C_i.

4.6 Theorem. Let E,F be order unit normed spaces with norm determing order units u,v. Then

(i) $E \otimes_p F$ is an order unit normed space with norm determing order unit $u \otimes v$ and

(ii) $E \otimes_i F$ is an order unit normed space with norm determing order unit $u \otimes v$ and $\|.\|_i = \|.\|_\varepsilon$.

Proof. (i) If $\varphi \in B_b(E,F)_+$ then by lemma 2.5(iii) we get

$$\|\varphi\| = \varphi(u,v). \text{ If } \|t\|_p \leq 1, \text{ then } \varphi(u,v) - \langle t, \varphi \rangle \geq 0.$$

It follows that $u \otimes v - t \in \overline{C}_p$. So $u \otimes v$ is the greatest element of the unit ball.

(ii) If $\|t\| \leq 1$, $x' \in E'_+$, $y' \in F'_+$ then

$$0 \leq \|x'\| \|y'\| - \langle t, x' \otimes y' \rangle = \langle u \otimes v - t, x' \otimes y' \rangle.$$

It follows that $u \otimes v - t \in C_i$ and $\|t\|_i \leq 1$. Therefore $u \otimes v$ is the norm determing order unit and $\|.\|_i = \|.\|_\varepsilon$.

4.7 Theorem. Let E,F be base normed spaces (or both order unit normed spaces). If $\widetilde{E}$ has the Riesz decomposition property, then $E \otimes_p F = E \otimes_i F$ norm and order isomorphic.

Proof. (i) Let E,F are base normed, $u' \in E'_+$, $v' \in F'_+$ the norm determing order units. Since $\overline{C}_p = C_i$ we obtain

$$\|t\|_p = \langle t, u' \otimes v' \rangle = \|t\|_i \qquad \text{for all } t \in \overline{C}_p = C_i.$$

Thus $\|.\|_p = \|.\|_i$.

(ii) Let E,F are order unit normed spaces, $u \in E_+$, $v \in F_+$ the norm determing order units. Since $\overline{C}_p = C_i$, we obtain

$$\|t\|_p = \inf\{\lambda > 0 : \lambda u \otimes v \pm t \in \overline{C}_p = C_i\} = \|t\|_i \quad \text{for all } t \in E \otimes F.$$

4.8 Example. Consider the space

$$\lambda_1 = \{(\xi_\nu)_{\nu=0}^\infty : \xi_\nu \in \mathbb{R}, \sum_{\nu=0}^\infty |\xi_\nu| < \infty\}$$

with the cone

$$\lambda_{1+} = \{(\xi_\nu)_{\nu=0}^{\infty} : \xi_o - \sum_{\nu=1}^{\infty} |\xi_\nu| > 0\}.$$

As a Banach space we have $\lambda_1 = l_1$, but we consider another ordering. With the norm

$$\|(\xi_\nu)_{\nu=0}^{\infty}\| = \sum_{\nu=0}^{\infty} |\xi_\nu|$$

it is an order unit normed space, where $(1,0,0,\ldots)$ is the norm determing order unit. With the equivalent norm

$$\|(\xi_\nu)_{\nu=0}^{\infty}\| = \max(|\xi_0|, \sum_{\nu=1}^{\infty} |\xi_\nu|)$$

it is a base normed space. We need this example for the next proposition.

4.9 <u>Proposition</u>. Let $f: E \to F$ be a positive bounded linear operator, E an order unit normed space and F a base normed space. If (i) $\tilde{E}$ or (ii) $\tilde{F}$ has the Riesz decomposition property, then f and f' are absolutely summing.

<u>Proof</u>. A linear mapping f is absolutely summing if $f \otimes id: E \otimes_\varepsilon l_1 \to F \otimes_\pi l_1$ is bounded. We have $\lambda_1 \cong l_1$.

(i) The mapping $f \otimes id: E \otimes_p \lambda_1 \to F \otimes_p \lambda_1$ is positive and bounded. By theorem 4.7 and the properties of λ_1 we obtain that $E \otimes_p \lambda_1 = E \otimes_i \lambda_1 = E \otimes_\varepsilon \lambda_1$ and $F \otimes_p \lambda_1 = F \otimes_\pi \lambda_1$.

(ii) The mapping $f \otimes id: E \otimes_i \lambda_1 \to F \otimes_i \lambda_1$ is positive and bounded. By theorem 4.7 and the properties of λ_1 we obtain that $E \otimes_i \lambda_1 = E \otimes_\varepsilon \lambda_1$ and $F \otimes_i \lambda_1 = F \otimes_p \lambda_1 = F \otimes_\pi \lambda_1$.

5. <u>An approximation problem</u>

We construct another regular ordered product $E \iota F$. In some cases $E \iota F = E \otimes_i F$.

5.1 <u>The ι-product</u>. Let E, F be regular ordered normed spaces. The norm $\|\cdot\|_\varepsilon$ is monotone on the cone

$$C_\iota = \{t \in E \tilde{\otimes}_\varepsilon F : \langle t, E'_+ \otimes F'_+ \rangle \geq 0\}.$$

The linear space $E \iota F = C_\iota - C_\iota$ is ordered by the cone C_ι and normed by the regular norm

$$\|t\|_\iota = \inf\{\|s\|_\varepsilon : s \pm t \in C_\iota\} \quad \text{for all } t \in E \iota F.$$

It is $\|.\|_\varepsilon \le 2\|.\|_\iota$. Let $(t_n)_{n=1}^\infty, \|t_n\|_\iota < 2^{-n}$, be a sequence in $E\iota F$. Then there exists a sequence $(s_n)_{n=1}^\infty, s_n \in C_\iota, \|s_n\| < 2^{-n}$ such that $s_n \pm t_n \subset C_\iota$.

Then

$$t = \sum_{n=1}^\infty t_n, s = \sum_{n=1}^\infty s_n \in E \tilde{\otimes}_\varepsilon F.$$

Since $s \pm t \in C_\iota$, is t an element of $E\iota F$ and the sum converges in the norm $\|.\|_\iota$. The ι-product $E\iota F$ is a Banach space.

5.2 The <u>approximation</u> <u>problem</u> is whether $E \tilde{\otimes}_i F = E\iota F$. In the case of two (AL)-spaces, the same problem was raised by H.Schaefer ([14] page 410): Is every positive compact operator of an (AM)-space E to an (AL)-space F already nuklear? In this case we have $E' \tilde{\otimes}_\pi F = E' \tilde{\otimes}_p F = E' \tilde{\otimes}_i F$, where $E' \tilde{\otimes}_\pi F$ is the space of all nuklear operators from E into F. The problem is whether $C_\iota = \tilde{C}_i$.

In the case of two order unit spaces E,F we have by theorem 4.6 $E \tilde{\otimes}_i F = E\iota F = E \tilde{\otimes}_\varepsilon F$. We will show a more general result in the case of a simplex space.

5.3 <u>Definition</u>. An regular ordered normed space is <u>approximately order unit normed</u>, if the open unit ball is directed upwards.

5.4 <u>Remark</u>. E is approximately order unit normed, if and only if the dual E' is base normed (see K.F.Ng[9])

5.5 <u>Example</u>. The space of compact selfadjoint operators on a Hilbert space is an approximately order unit normed space. The dual is the base normed space of all selfadjoint nuklear operators.

5.6 <u>Definition</u>. A <u>simplex</u> <u>space</u> is an approximately order unit normed space with the Riesz decomposition property.

E is a simplex space, if and only if the dual E' is an (AL)-space.

5.7 <u>Theorem</u>. Let $\tilde{E}, \tilde{F}$ be approximately order unit normed spaces. Then

(i) $E \tilde{\otimes}_p F$ is an approximately order unit normed space and

(ii) $E\iota F$ is an approximately order unit normed space and $E\iota F = E \tilde{\otimes}_\varepsilon F$.

<u>Proof</u>. (i) We show that $(E \otimes_p F)'$ is a base normed space. If $\varphi_1, \varphi_2 \in B_b(E,F)_+$, $\eta > 0$ then there exist $x_\nu \in \tilde{E}_+$, $\|x_\nu\| < 1, y_\nu \in \tilde{F}_+, \|y_\nu\| < 1$

$(\nu=1,2)$ such that

$$(1-\eta)\|\varphi_\nu\| \le \varphi_\nu(x_\nu,y_\nu) \le \|\varphi_\nu\| \qquad (\nu=1,2)$$

There exists $x_3 \in E_+$, $\|x_3\| < 1$, $y_3 \in F_+$, $\|y_3\| < 1$ with $x_3 - x_\nu \in E_+$, $y_3 - y_\nu \in F_+$. Then

$$(1-\eta)(\|\varphi_1\| + \|\varphi_2\|) \le (\varphi_1+\varphi_2)(x_3,y_3) \le \|\varphi_1\| + \|\varphi_2\|.$$

(ii) Denote by $K = \{x' \in E'_+ : \|x'\| \le 1\}$, $L = \{y' \in F'_+ : \|y'\| \le 1\}$. K and L are w*-compact. If $t_1, t_2 \in E \tilde{\otimes}_\varepsilon F$, $\|t_\nu\| < 1$ then

$$|\langle t_\nu, x'\otimes y'\rangle| < \frac{1}{2} + \frac{1}{2}\|x'\|\|y'\| \quad \text{for all } x' \in K, y' \in L.$$

Since $\|x'\| = \sup\{\langle x,x'\rangle : x \in E_+, \|x\| < 1\}$, K is w*-compact and $\tilde{E}$ approximately order unit normed, there exist $u_1 \in \tilde{E}_+$, $\|u_1\| < 2^{-\frac{1}{2}}$, $v_1 \in \tilde{F}_+$, $\|v_1\| < 2^{-\frac{1}{2}}$ such that

$$|\langle t_\nu, x'\otimes y'\rangle| < \frac{1}{2} + \langle u_1\otimes v_1, x'\otimes y'\rangle \quad \text{for all } x' \in K, y' \in L.$$

Inductively we find $u_n \in E_+$, $\|u_n\| < 2^{-\frac{n}{2}}$, $v_n \in F_+$, $\|v_n\| < 2^{-\frac{n}{2}}$ such that

$$|\langle t_\nu, x'\otimes y'\rangle| < 2^{-n} + \langle \sum_{\nu=1}^{n} u_\nu\otimes v_\nu, x'\otimes y'\rangle \quad \text{for all } x' \in K, y' \in L.$$

The series converge

$$s = \sum_{n=1}^{\infty} u_n\otimes v_n, \quad \|s\|_\varepsilon < 1, \text{ and } s \pm t_\nu \subset C_\iota.$$

Thus $E \tilde{\otimes}_\varepsilon F$ is an approximate order unit normed space and $E \tilde{\otimes}_\varepsilon F = C_\iota - C_\iota = E \iota F$.

5.8 Theorem. If $\tilde{E}$ is a simplex space, $\tilde{F}$ an approximate order unit normed space then

$$E \tilde{\otimes}_p F = E \tilde{\otimes}_i F = E \iota F = E \tilde{\otimes}_\varepsilon F = BA_o(K,L)$$

Proof. Since $E \tilde{\otimes}_p F$ and $E \tilde{\otimes}_\varepsilon F$ are approximately order unit normed spaces the dual cones have bases and are generated by there extreme rays. In the same manner as in [2] we can show, that these extreme rays coincide. Thus $E \tilde{\otimes}_p F = E \tilde{\otimes}_i F = E \iota F = E \tilde{\otimes}_\varepsilon F$. Since E has the approximation property the latter space coincides with the space $BA_o(K,L)$ of all biaffine continous functions on $K\times L$, which vanish at the origin.

References

1. Behrends E. und Wittstock G.: Tensorprodukte kompakter konvexer Mengen. Inventiones math. 10,251-266 (1970)

2. Behrends E. und Wittstock G.: Tensorprodukte und Simplexe. Inventiones math. 11,188-198 (1970)

3. Davies E.B., Vincent-Smith G.F.: Tensor products, infinite products and projective limits of Choquet simplexes. Math. Scand. 22,145-164 (1968)

4. Davies E.B.: The structure and ideal theory of the predual of a Banach lattice. Trans. Amer. Math. Soc. 131, 544-555 (1968)

5. Fremlin D.H.: Tensor products of archimedian vector lattices. Amer. Journal Math. 94, 777-798 (1972)

6. Hulanicki A. and Phelps R.R.: Some applications of tensor products of partially ordered spaces. J. Functional Analysis 2, 177-201 (1968)

7. Lazar, A.J.: Affine products of simplexes. Math. Scand. 22, 165-175 (1968)

8. Namioka. F., Phelps R.R.: Tensor products of compact convex sets. Pacific J. Math. 31,469-480 (1969)

9. Ng. Kung-Fu: On a computation rule for polars. Math. Scand. 26, 14-16 (1970)

10. Peressini A.L.: Ordered topological vector spaces. New York, Evanston and London: Harper & Row:1967

11. Peressini A.L and Sherbert D.R.: Ordered topological tensor products. Proc. London Math. Soc. 19,177-190 (1969)

12. Popa N.: Produit tensoriels ordonnés. Rev. Roumani Math. Pures et Appl. 13, 235-246 (1968)

13. Schaefer H.H.: Topological vector spaces. Berlin-Heidelberg-New York, Springer:1971

14 Schaefer H.H.: Normed tensor products of Banach lattices. Israel J. Math. 13, 400-415 (1973)

15. W.F. Stinespring: Positive functions on C*-algebras. Proc. Amer. Math. Soc. 6, 211-216 (1955)

16. Wittstock G.: Choquet Simplexe und nukleare Räume. Inventiones math. 15,251-258 (1972).

POSITIVE LINEAR MAPS OF C*-ALGEBRAS

Erling Størmer

University of Oslo,

Oslo, Norway

I. INTRODUCTION

The theory of positive linear maps of C*-algebras appears in the literature in a rather scattered and usually special form. The best known examples are states and *-representations, but they have so many special properties that they are usually not considered in the general theory. With this in mind the theory falls roughly into five parts, namely the study of completely positive maps, inequalities, projection maps (also called expectations), Jordan homomorphisms, and extremal maps. These five topics will be discussed in separate sections in the present notes. There is still another field waiting to be developed, namely spectral theory for positive linear maps. Since this subject has only been studied in the case of automorphisms [4, 14, 22, 33] we shall not be concerned with it here.

Let now φ be a positive linear map of one C*-algebra $\mathfrak{A}$ into another $\mathfrak{B}$, and by this we mean a linear map which carries positive operators into positive operators. Then clearly φ is self-adjoint, i.e. $\varphi(A^*) = \varphi(A)^*$ for all A in $\mathfrak{A}$. If $\mathfrak{A}$ and $\mathfrak{B}$ do not have identities, φ can be extended to the algebras with identities adjoined, so we shall always assume our C*-algebras have identities, denoted by I. If $\varphi(I)$ is invertible, we may replace φ by $\varphi(I)^{-\frac{1}{2}}\, \varphi\, \varphi(I)^{-\frac{1}{2}}$, and thus assume $\varphi(I) = I$. If $\varphi(I)$ is not invertible, approximation arguments reduce consideration to maps which carry I into an invertible operator. We shall say φ is normalized when $\varphi(I) = I$ and restrict attention to such maps.

We refer the reader to the two books of Dixmier [17, 18] for the general theory of von Neumann and C*-algebras.

These notes are extensions of lecture notes [53] written in connection with a symposium on C*-algebras at Louisiana State University, Baton Rouge, Louisiana, in March 1967.

2. COMPLETELY POSITIVE MAPS

In the study of states of C*-algebras the key result is the decomposition of a state into the composition of a vector state and a *-representation. A similar result is desirable for positive maps, and can be obtained adding stronger positivety assumptions. Let $\mathfrak{A}$ and $\mathfrak{B}$ be C*-algebras and φ a positive map of $\mathfrak{A}$ into $\mathfrak{B}$. Let M_n denote the complex $n \times n$ matrices and 1_n the identity map of M_n onto itself. We say φ is <u>n-positive</u> if $\varphi \otimes 1_n$ is a positive map of $\mathfrak{A} \otimes M_n$ into $\mathfrak{B} \otimes M_n$ [11].

Since we can consider $\mathfrak{A} \otimes M_m$ as a subspace of $\mathfrak{A} \otimes M_n$ for $m \leq n$, it is almost immediate that if φ is n-positive, then it is also m-positive whenever $1 \leq m \leq n$. An example of an (n-1)-positive map which is not n-positive is given by $\varphi : M_n \to M_n$ defined by

$$\varphi(A) = \frac{1}{n^2-n-1}((n-1)\operatorname{tr}(A)I - A),$$

where tr is the usual trace on M_n [12]. We thus have a hierarchy of maps with different positivety properties.

We shall mainly be concerned with <u>completely positive</u> maps. They are the ones which are n-positive for all positive integers n. These maps are also studied in the article of Kraus. The following result is very useful and gives us the desired decomposition theorem [36, 46].

<u>Theorem 2.1</u> Let φ be a normalized positive linear map of a C*-algebra $\mathfrak{A}$ into another $\mathfrak{B}$ acting on a Hilbert space $\mathcal{H}$. Then φ is completely positive if and only if there exist a Hilbert space $\mathcal{K}$, a linear isometry V of $\mathcal{H}$ into $\mathcal{K}$, and a *-representation π of $\mathfrak{A}$ on $\mathcal{K}$ such that

$$\varphi(A) = V^*\pi(A)V$$

for all $A \in \mathfrak{A}$. Moreover, if $\mathfrak{A}$ is abelian, then every positive linear map of $\mathfrak{A}$ into $\mathfrak{B}$ is completely positive.

It is not difficult to show that if $\varphi = V^*\pi V$ then φ is completely positive. For the converse assume φ is completely positive and consider the vector space $\mathfrak{A} \otimes \mathcal{H}$, the algebraic tensor product of $\mathfrak{A}$ and $\mathcal{H}$ endowed with the inner product

$$(\Sigma_i A_i \otimes x_i, \Sigma_j B_j \otimes y_i) = \Sigma_{ij}(\varphi(B_j^* A_i)x_i, y_j) .$$

Since φ is completely positive this inner product is a positive Hermitian form. Let $\mathcal{N}$ denote the set of null vectors with respect to the form, and let $\mathcal{K}$ denote the closure of $\mathfrak{A} \otimes \mathcal{H}/\mathcal{N}$. Then $\mathcal{K}$ is a Hilbert space, and V defined by $Vx = I \otimes x + \mathcal{N}$ is a linear isometry of $\mathcal{H}$ into $\mathcal{K}$. Letting π be the representation induced on $\mathcal{K}$ by the representation π' of $\mathfrak{A}$ on $\mathfrak{A} \otimes \mathcal{H}$ defined by

$$\pi'(A)\Sigma_j B_j \otimes y_j = \Sigma_j AB_j \otimes y_j$$

it is easily checked that $\varphi = V^*\pi V$. For the last statement see [46].

If π is an irreducible representation of $\mathfrak{B} \otimes M_n$ then $\pi = \pi_1 \otimes \pi_2$ with π_1 an irreducible representation of $\mathfrak{B}$ and π_2 one of M_n. Let φ be a positive map of $\mathfrak{A}$ into $\mathfrak{B}$. Then φ is completely positive if and only if $\pi \circ (\varphi \otimes 1_n)$ is positive for all irreducible representations π of $\mathfrak{B} \otimes M_n$, $n = 1,2,\ldots,$ hence by the remark above if and only if $\pi_1 \circ \varphi$ is completely positive for all irreducible representations π_1 of $\mathfrak{B}$. In particular, if $\mathfrak{B}$ is abelian, then φ is completely positive [47] (note: states are completely positive by Theorem 2.1).

We have thus obtained large classes of maps which are completely positive. There is yet another important class of maps which are completely positive, namely the projection maps to be studied in chapter 4. We follow [12] and first show a closely related result. Call φ <u>pseudo-multiplicative</u> if for all $A \in \mathfrak{A}$ there is $A' \in \mathfrak{A}$ such that $\varphi(BA') = \varphi(B)\varphi(A)$ for all $B \in \mathfrak{A}$.

<u>Theorem 2.2</u> Every positive linear pseudo-multiplicative map φ of C*-algebra $\mathfrak{A}$ into another $\mathfrak{B}$ is completely positive.

<u>Proof</u>. Say $\mathfrak{B}$ acts on a Hilbert space $\mathcal{H}$ and let $x \in \mathcal{H}$. Since φ in pseudo-multiplicative its image $\varphi(\mathfrak{A})$ is a *-algebra, hence we may assume $\varphi(\mathfrak{A})$ is dense in $\mathfrak{B}$. Since the projection $[\mathfrak{B}x]$ onto the closure of the linear manifold $\mathfrak{B}x$ in $\mathcal{H}$ belongs to the commutant

of $\mathcal{B}$, it suffices to show $A \to \varphi(A)[\mathcal{B}x]$ is completely positive. By density and continuity it suffices to show the restriction map $A \to \varphi(A)|\varphi(\mathfrak{A})x$ is completely positive i.e. given an integer n, a positive operator $(A_{jk})_{1\leq j,k\leq n}$ in $\mathfrak{A} \otimes M_n$ (which equals the $n \times n$ matrices over $\mathfrak{A}$) and $x_i \in \varphi(\mathfrak{A})x$, we have $\Sigma_{jk}(\varphi(A_{jk})x_k, x_j) \geq 0$. Now $x_i = \varphi(B_i)x$, where B_i has the property that $\varphi(BB_i) = \varphi(B)\varphi(B_i)$ for all $B \in \mathfrak{A}$. Thus

$$\begin{aligned}\Sigma_{jk}(\varphi(A_{jk})x_k, x_j) &= \Sigma_{jk}(\varphi(A_{jk})\varphi(B_k)x, \varphi(B_j)x)\\ &= \Sigma_{jk}(\varphi(B_j^*)\varphi(A_{jk})\varphi(B_k)x, x)\\ &= \Sigma_{jk}(\varphi(B_j^*A_{jk}B_k)x, x)\end{aligned}$$

which is positive since $\Sigma_{jk}B_j^*A_{jk}B_k$ is positive. QED.

In the definition of pseudo-multiplicative φ the operator A' had very nice multiplicative properties relative to φ. We can generalize this slightly and say a positive linear map φ is <u>definite</u> on a self-adjoint operator A if $\varphi(A^2) = \varphi(A)^2$. Then the following identities follow from inequality 3.1 below on self-adjoint operators [Broise, unpublished].

$$\varphi(AB+BA) = \varphi(A)\varphi(B) + \varphi(B)\varphi(A)$$

$$\varphi(ABA) = \varphi(A)\varphi(B)\varphi(A)$$

for all B in $\mathfrak{A}$.

In order to define positive and completely positive linear maps it is not necessary to assume $\mathfrak{A}$ and $\mathcal{B}$ are C*-algebras; it suffices to assume they are linear subspaces of C*-algebras. This situation was studied extensively by Arveson [2, 3]. He could show the following Hahn-Banach theorem for completely positive maps.

<u>Theorem 2.3</u> Let $\mathfrak{A}$ be a norm-closed self-adjoint linear subspace of a C*-algebra $\mathcal{B}$ containing the identity of $\mathcal{B}$. Let φ be a completely positive linear map of $\mathfrak{A}$ into the bounded operators $\mathcal{B}(\mathcal{H})$ on a Hilbert space $\mathcal{H}$. Then there is a completely linear map $\varphi_1 : \mathcal{B} \to \mathcal{B}(\mathcal{H})$ such that $\varphi_1|\mathfrak{A} = \varphi$.

The proof [2] or [38] is too complicated for us to get into here. In the special case when $\mathfrak{A}$ is a C*-algebra a simple proof is available. Let by Theorem 2.1 $\varphi = V^*\pi V$, where V is a linear isometry of $\mathcal{H}$ into a Hilbert space $\mathcal{K}$ and π a representation of $\mathfrak{A}$ on $\mathcal{K}$. Extend π to a representation ρ of $\mathcal{B}$ on a Hilbert space $\mathcal{K}'$ containing $\mathcal{K}$. Then

$V^*\rho V$ is by Theorem 2.1 completely positive, extends φ , and maps $\mathcal{B}$ into $\mathcal{B}(\mathcal{H})$.

3. INEQUALITIES

The most famous inequality in analysis is probably the Cauchy-Schwarz inequality, which for states of a C*-algebra takes the form

$$|\omega(A^*B)|^2 \leq \omega(A^*A)\omega(B^*B) .$$

This inequality has been generalized in different directions to positive linear maps. The by now classical inequality is the one of Kadison [27], which states that if $\|\varphi\| \leq 1$ then $\varphi(A^2) \geq \varphi(A)^2$ for all self-adjoint A . This inequality follows from

Theorem 3.1 Let φ be a normalized completely positive linear map from a C*-algebra $\mathfrak{A}$ into another $\mathcal{B}$. Then φ satisfies the Cauchy-Schwarz inequality

3.1 $$\varphi(A^*A) \geq \varphi(A)^*\varphi(A)$$

for all $A \in \mathfrak{A}$. In particular, whenever φ is positive and A normal in $\mathfrak{A}$ then 3.1 holds for A .

Proof. If φ is completely positive and $\mathcal{B}$ acts on $\mathcal{H}$, $\varphi = V^*\pi V$ with V and π as in Theorem 2.1. Therefore, with $A \in \mathfrak{A}$, $x \in \mathcal{H}$ we have

$$(\varphi(A^*A)x,x) = (V^*\pi(A)^*\pi(A)Vx,x) = \|\pi(A)Vx\|^2$$
$$\geq \|V^*\pi(A)Vx\|^2 = (\varphi(A)^*\varphi(A)x,x) ,$$

and 3.1 follows. In the general case with A normal, then the C*-algebra C*(A) generated by A and the identity is abelian. By Theorem 2.1 φ restricted to C*(A) is completely positive, hence 3.1 holds for A .

Choi [12] has shown that 3.1 holds for all 2-positive maps. Another generalization of Kadison's inequality can be obtained as follows. Say a real valued measureable function f on an interval (-a,a) is operator-convex [5, 15] if $\frac{1}{2}(f(A)+f(B)) \geq f(\frac{1}{2}(A+B))$ for all self-adjoint operators A and B of norm less than a , e.g. $f(x) = x^2$ is operator convex. Then we have [12, 15].

Theorem 3.2 Let φ be a normalized positive linear map from a C*-algebra $\mathfrak{A}$ into another $\mathfrak{B}$. Let f be an operator convex function on (-a,a). Then $\varphi(f(A)) \geq f(\varphi(A))$ for all self-adjoint A in $\mathfrak{A}$ of norm less than a.

Again we restrict to the abelian subalgebra C*(A) on which φ is completely positive. For a completely positive map φ we may by Theorem 2.1 assume $\varphi = P\pi P$ with P a projection. By [15] f is operator convex if and only if for all operators A with spectrum in (-a,a) and all projections P we have $Pf(PAP)P \leq Pf(A)P$. Thus the theorem follows.

Since the operator convex functions are of the form [5, 16]

$$f(t) = \int_{-a}^{a} \frac{t^2}{a^2 - tx}\, dm(x) + bt + c$$

with m a finite positive regular Borel measure on [-a,a] one can also show, putting $a = b = c = 1$ and m the point measure with support {1}, that φ satisfies the inequality

3.2 $$\varphi(A^{-1}) \geq \varphi(A)^{-1}$$

for A positive and invertible in $\mathfrak{A}$.

We conclude this section with an application of 3.1 to the determination of the norm of a positive linear map. Note that it is clear that the norm of a normalized positive linear map φ is 1 when restricted to the self-adjoint operators. However, it is not clear what the norm is on all operators. The answer is [42],

Theorem 3.3 Let φ be a normalized self-adjoint linear map from a C*-algebra $\mathfrak{A}$ into another $\mathfrak{B}$. Then φ is positive if and only if $\|\varphi\| = 1$.

If $\|\varphi\| = 1$ then from the same theorem for linear functionals, $\rho \circ \varphi$ is a state of $\mathfrak{A}$ for each state ρ of $\mathfrak{B}$, hence φ is positive. Conversely, if φ is positive and U is a unitary operator in $\mathfrak{A}$ then by the inequality 3.1

$$\|\varphi(U)\|^2 = \|\varphi(U)^*\varphi(U)\| \leq \|\varphi(U^*U)\| = \|\varphi(I)\| = 1 .$$

But for a self-adjoint linear map φ, $\|\varphi\| = \sup\|\varphi(U)\|$, the sup being taken over all unitary operators in $\mathfrak{A}$ [42], hence $\|\varphi\| = 1$.

4. PROJECTION MAPS

The most useful maps studied so far fall into the following situation: Let $\mathfrak{A}$ be a C*-algebra, $\mathfrak{C}$ a C*-subalgebra of $\mathfrak{A}$ containing the identity I in $\mathfrak{A}$, and $\mathfrak{B}$ a von Neumann algebra. If φ is a normalized positive map of $\mathfrak{A}$ into $\mathfrak{B}$ such that φ restricted to $\mathfrak{C}$ is a homomorphism onto a weakly dense subalgebra of $\mathfrak{B}$, then φ is said to be a projection of $\mathfrak{A}$ on $\mathfrak{B}$ relative to $\mathfrak{C}$. Often $\mathfrak{B} = \mathfrak{C}$ and $\varphi | \mathfrak{C}$ is the identity map, in which case φ is said to be a projection (or expectation, or conditional expectation, or projection of norm 1) of $\mathfrak{A}$ onto $\mathfrak{B}$. It should be remarked that most authors make the extra assumption that $\varphi(ABC) = \varphi(A)\varphi(B)\varphi(C)$ for $A, C \in \mathfrak{C}$, $B \in \mathfrak{A}$. But this assumption is redundant as the following result shows [12, 53, 56, 57] and [Broise, unpublished].

Theorem 4.1 Let $\mathfrak{A}$, $\mathfrak{B}$ and $\mathfrak{C}$ be as above and φ a projection of $\mathfrak{A}$ on $\mathfrak{B}$ relative to $\mathfrak{C}$. Then φ is completely positive and satisfies the identity

$$\varphi(ABC) = \varphi(A)\varphi(B)\varphi(C) \tag{4.1}$$

for $A, C \in \mathfrak{C}$, $B \in \mathfrak{A}$.

If we can show 4.1, then since $\varphi(\mathfrak{C})$ is strongly dense in $\mathfrak{B}$ then the same argument that was used to prove Theorem 2.2, shows φ is completely positive. Note that we don't assume $\varphi(\mathfrak{C}) = \varphi(\mathfrak{A})$, hence we cannot conclude that φ is pseudo-multiplicative. But 4.1 follows from [53] or [57]. We won't go into the proof since 4.1 is usually trivially satisfied in concrete cases.

In Sakai's proof [43] of the equivalence of GCR and type I for C*-algebras the key lemma was about projection maps. He showed that if in Theorem 4.1 $\mathfrak{B}$ is a factor of type III on a separable Hilbert space then $\mathfrak{A}$ has a type III factor representation.

If $\mathfrak{B} = \mathfrak{C}$, and φ is a projection of $\mathfrak{A}$ onto $\mathfrak{B}$ more information is known, see e.g. [1]. We shall be concerned with the situation when $\mathfrak{B} = \mathfrak{C}$ is the fixed point set of a group G of *-automorphisms of the von Neumann algebra $\mathfrak{A}$, viz. $\mathfrak{B} = \{A \in \mathfrak{A} : g(A) = A \text{ for all } g \in G\}$. The following theorem results from a slight modification of an argument in [44].

Theorem 4.2 Let $\mathfrak{A}$ be a von Neumann algebra acting on a separable Hilbert space $\mathcal{H}$. Let G be a group of *-automorphisms of $\mathfrak{A}$, and let $\mathcal{B}$ be its fixed point algebra. Assume

$$4.2 \qquad \operatorname{conv}(g(A) : g \in G)^{-} \cap \mathcal{B} \neq \emptyset$$

for all A in $\mathfrak{A}$, where conv denotes the convex hull. Then there exists a projection φ of $\mathfrak{A}$ onto $\mathcal{B}$ such that for all A in $\mathfrak{A}$, $\varphi(A) \in \operatorname{conv}(g(A) : g \in G)^{-}$. Hence, if this set intersects $\mathcal{B}$ in exactly one point, then $\varphi(A) = \varphi(g(A))$ for all $g \in G$.

The assumption 4.2 is a generalization of Schwartz's property P. Indeed, let $\mathfrak{A} = \mathcal{B}(\mathcal{H})$, $\mathcal{C}$ a von Neumann algebra acting on $\mathcal{H}$, and G the group of *-automorphisms $A \to UAU^*$ of $\mathcal{B}(\mathcal{H})$ where U runs through the unitary operators in $\mathcal{C}$. Then $\mathcal{C}$ has property P if and only if 4.2 holds with $\mathcal{B} = \mathcal{C}'$. If furthermore $\mathcal{C}$ is abelian and generated by projections $\{E_n\}_{n=1,2,\ldots,}$ then 4.2 is satisfied, and $\varphi : \mathcal{B}(\mathcal{H}) \to \mathcal{B} = \mathcal{C}'$ is the diagonal process studied in [34]. For a discussion of closely related projection maps, see the article of Poole.

If $\mathfrak{A}$ is a von Neumann algebra with center $\mathcal{C}$, and G is the group of inner *-automorphisms of $\mathfrak{A}$, then by [17, p.253],

$$\operatorname{conv}(g(A) : g \in G)^{-} \cap \mathcal{C} \neq \emptyset \quad \text{for } A \in \mathfrak{A},$$

hence 4.2 holds with $\mathcal{B} = \mathcal{C}$. In this example, when $\mathfrak{A}$ is finite, the map φ is the center trace of $\mathfrak{A}$ [17], and is both normal and G-<u>invariant</u>, i.e. $\varphi \circ g = \varphi$ for $g \in G$. In order to conclude this in general, if $\mathfrak{A}$ is a von Neumann algebra and G a group of *-automorphisms of $\mathfrak{A}$, we say $\mathfrak{A}$ is G-<u>finite</u> if for all nonzero positive A in $\mathcal{B}$ - the fixed point algebra - there exists a normal G-invariant state ρ of $\mathfrak{A}$ such that $\rho(A) \neq 0$ [35]. Then we can draw the desired conclusions [35].

Theorem 4.3 With the notation above, if $\mathfrak{A}$ is G-finite, then $\operatorname{conv}(g(A) : g \in G)^{-} \cap \mathcal{B}$ consists of exactly one point $\varphi(A)$ for each A in $\mathfrak{A}$. The map φ is the unique faithful G-invariant normal projection of $\mathfrak{A}$ onto $\mathcal{B}$. Conversely, the existence of such a map implies $\mathfrak{A}$ is G-finite.

The proof of Theorem 4.3 is quite different from the known proofs in the finite von Neumann algebra case [17, 29, 31]. The main idea is the application of an ergodic result of Alaoglu-Birkhoff [40, § 146], which implies that if $\mathcal{U}$ is a group of unitary operators on a Hilbert space $\mathcal{H}$, and P is the projection on the subspace $\{x \in \mathcal{H} : Ux = x$ for all $U \in \mathcal{U}\}$ then $P \in \operatorname{conv}(U : U \in \mathcal{U})^-$. Let ρ be a normal G-invariant state of $\mathfrak{A}$. $\rho = \omega_{x_\rho} \circ \pi_\rho$ with π_ρ a normal *-representation of $\mathfrak{A}$ on a Hilbert space $\mathcal{H}_\rho$, and x_ρ a unit vector in $\mathcal{H}_\rho$ cyclic under $\pi_\rho(\mathfrak{A})$. There exists a unitary representation $g \to U_\rho(g)$ of G on $\mathcal{H}_\rho$ such that

4.3
$$\pi_\rho(g(A)) = U_\rho(g)\pi_\rho(A)U_\rho(g)^{-1}$$
$$U_\rho(g)x_\rho = x_\rho$$

for all $g \in G$, $A \in \mathfrak{A}$. Let $P_\rho = \{x \in \mathcal{H}_\rho : U_\rho(g)x = x \text{ for all } g \in G\}$. Then $x_\rho \in P_\rho$, so $P_\rho \neq 0$.

Using that there is a net $\{\Sigma_i \lambda_i^\alpha U_\rho(g_i^\alpha)\}_{\alpha \in J}$ in $\operatorname{conv}(U_\rho(g) : g \in G)$ which converges strongly to P_ρ one exhibits a map Φ_ρ like the one in the theorem, of $\pi_\rho(\mathfrak{A})$ onto the fixed points in $\pi_\rho(\mathfrak{A})$ such that $\Phi_\rho(\pi_\rho(A))P_\rho = P_\rho\pi_\rho(A)P_\rho$ for all A in $\mathfrak{A}$. Then one considers the direct sum $\oplus\pi_\rho$ of the π_ρ and $\oplus\Phi_\rho$ of the Φ_ρ with ρ G-invariant and normal to conclude the proof. For details see [20].

Results of the same type as above have been obtained in the C*-algebra setting of quantum physics. Then one is given a C*-algebra $\mathfrak{A}$ and a group G of *-automorphisms of $\mathfrak{A}$ such that $\mathfrak{A}$ has a certain asymptotical abelian property with respect to G, e.g. if G is the translation group R^n then one assumes $\lim_{g \to \infty} \|[g(A),B]\| = 0$ for all $A, B \in \mathfrak{A}$. One is then interested in the G-invariant states of $\mathfrak{A}$. Via the representation 4.3 above one is in the situation of spatial automorphisms of von Neumann algebras. Due to the asymptotical abelian property one obtains a G-invariant normal projection of the von Neumann algebra $\pi_\rho(\mathfrak{A})^-$ onto an abelian von Neumann algebra contained in its center [6, 19, 20, 52] the results being quite similar to Theorem 4.3 (note that the proof in [6] is incomplete). In this way valuable information is obtained on the G-invariant states of $\mathfrak{A}$.

One of the conclusions in Theorem 4.3 was that φ is unique. This type of result is quite common when $\mathfrak{B}$ is in some sense large in $\mathfrak{A}$. Connes [14] has shown a result like this in a case when there is no natural group of automorphisms to be used.

Theorem 4.4 Let $\mathfrak{A}$ be a von Neumann algebra, and $\mathfrak{B}$ a von Neumann subalgebra of $\mathfrak{A}$ such that $\mathfrak{B}' \cap \mathfrak{A} \subset \mathfrak{B}$. Then there is atmost one normal projection map of $\mathfrak{A}$ onto $\mathfrak{B}$.

The condition $\mathfrak{B}' \cap \mathfrak{A} \subset \mathfrak{B}$ holds e.g. when $\mathfrak{B}$ is a maximal abelian subalgebra of $\mathfrak{A}$. Note that since the relative double commutant theorem does not necessarily hold for $\mathfrak{B}$ [30], we cannot use the natural approach of considering the group of inner automorphisms of $\mathfrak{A}$ defined by unitaries in $\mathfrak{B}' \cap \mathfrak{A}$ to prove the theorem.

The existence problem for projection maps can also be viewed as an extension problem for positive maps. Except for the examples mentioned little is known even for the identity map, cf. [59]. Modifying the proof that a state of a C*-algebra can be extended to a state of a larger C*-algebra, one can show the analogue for positive maps into abelian von Neumann algebras [34].

Except for Theorem 2.3, which holds for completely positive maps into $\mathfrak{B}(\mathcal{H})$, the extension results obtained have been to mappings of the weak closures. For example, if φ is a positive map of the C*-algebra $\mathfrak{A}$ into the C*-algebra $\mathfrak{B}$ acting on a Hilbert space $\mathcal{H}$ then φ has a positive extension φ^{**} of the second dual $\mathfrak{A}^{**}$ of $\mathfrak{A}$ into $\mathfrak{B}^-$. If $\mathfrak{A}$ acts on a Hilbert space, one can show that φ has a normal positive extension $\bar{\varphi}$ of $\mathfrak{A}^-$ into $\mathfrak{B}^-$ if and only if the functionals $\omega_x \circ \varphi$, x a unit vector in $\mathcal{H}$. are weakly continuous on the unit ball in $\mathfrak{A}$ at 0 [28]. Kadison [28] has also shown a much deeper result concerning the same problem.

We shall finally in this section discuss another aspect of projection maps. Let $\mathfrak{A}$ be a von Neumann algebra and $\mathfrak{B}$ a von Neumann subalgebra of $\mathfrak{A}$ (containing the identity of $\mathfrak{A}$). Suppose φ is a normal projection map of $\mathfrak{A}$ onto $\mathfrak{B}$. Then the problem is: which properties of $\mathfrak{A}$ are inherited by $\mathfrak{B}$? Typical example, what is the type of $\mathfrak{B}$ when the type of $\mathfrak{A}$ is known? Recall that $\mathfrak{A}$ is finite if and only if the *-operation is strongly continuous on bounded sets in $\mathfrak{A}$ [17, p.303]. One thus has a technique for deciding whether a von Neumann algebra is semi-finite or not. Using this we have [58],

Theorem 4.5 Let $\mathfrak{A}$ be a semi-finite von Neumann algebra and $\mathfrak{B}$ a von Neumann subalgebra of $\mathfrak{A}$. Suppose there is a normal projection map of $\mathfrak{A}$ onto $\mathfrak{B}$. Then $\mathfrak{B}$ is semi-finite.

In the type I case a stronger result is available. Let $\mathfrak{Z}$ be an abelian von Neumann algebra and $\mathfrak{C}$ a von Neumann subalgebra of $\mathfrak{Z}$.

$\mathcal{Z}$ is said to be totally atomic over $\mathcal{C}$ if every nonzero projection in $\mathcal{Z}$ majorizes a nonzero projection E in $\mathcal{Z}$ such that $E\mathcal{Z} = E\mathcal{C}$ [23, 30]. Then we have [55, 58], see also [54].

Theorem 4.6 Let $\mathcal{A}$ be a von Neumann algebra of type I and $\mathcal{B}$ a von Neumann subalgebra of $\mathcal{A}$ containing the center $\mathcal{C}$ of $\mathcal{A}$. Then the following two conditions are equivalent.

i) For each nonzero positive operator A in $\mathcal{A}$ there is a normal projection map φ of $\mathcal{A}$ onto $\mathcal{B}$ such that $\varphi(A) \neq 0$.

ii) $\mathcal{B}$ is of type I and its center is totally atomic over $\mathcal{C}$.

5. JORDAN HOMOMORPHISMS

When a C*-algebra is viewed from an order theoretic or a Jordan algebra point of view, the maps of key importance are the Jordan homomorphisms, also called C*-homomorphisms, which are defined to be those normalized positive maps which preserve squares of self-adjoint operators (viz $\varphi(A^2) = \varphi(A)^2$ for A self-adjoint). We shall in this paragraph dicuss several characterizations of Jordan homomorphisms, but first we show they are just sums of *-homomorphisms and *-anti-homomorphisms [25, 26, 32, 47, 48].

Theorem 5.1 Let φ be a Jordan homomorphism of the C*-algebra $\mathcal{A}$ into the bounded operators on a Hilbert space. Let $\mathcal{B}$ denote the C*-algebra generated by $\varphi(\mathcal{A})$. Then there exist two orthogonal central projections E and F in $\mathcal{B}^-$ such that the map $\varphi_1 : A \to \varphi(A)E$ is a homomorphism, $\varphi_2 : A \to \varphi(A)F$ is an anti-homomorphism, $E + F = I$, and $\varphi = \varphi_1 + \varphi_2$ as linear maps. In particular, if π is an irreducible representation of $\mathcal{B}$, then $\pi \circ \varphi$ is either a homomorphism or an anti-homomorphism.

This theorem was first proved algebraically [25] when $\mathcal{A}$ was an $n \times n$ matrix ring over a ring with identity. From the structure theory of von Neumann algebras, such an algebra is a direct sum of $n \times n$ matrix rings over rings with identities. Hence, if $\mathcal{A}$ is a von Neumann algebra φ is of the form $\varphi_1 + \varphi_2$ with φ_1 and φ_2 as in the theorem. A little argument is here necessary to show φ_1 and φ_2 are self-adjoint. Finally, with $\mathcal{A}$ and $\mathcal{B}$ as in the theorem φ has a normal extension

$\varphi^{**}: \mathfrak{A}^{**} \to \mathfrak{B}^{-}$, where $\mathfrak{A}^{**}$ denotes the second dual of $\mathfrak{A}$. An argument like that for *-homomorphisms shows φ^{**} is a Jordan homomorphism, and the result for von Neumann algebras is now applied to conclude the proof.

As immediate consequences of this theorem it follows that i) if U is a unitary operator in $\mathfrak{A}$ then $\varphi(U)$ is a unitary operator, ii) if A is self-adjoint and $|A| = (A^2)^{\frac{1}{2}}$ then $\varphi(|A|) = |\varphi(A)|$, iii) if A is invertible then so is $\varphi(A)$, and $\varphi(A^{-1}) = \varphi(A)^{-1}$.

The interesting thing is that the converse results also hold, [12, 26, 27, 41, 42]. The proof that i) implies φ is a Jordan homomorphism is a typical application of the Cauchy-Schwarz inequality 3.1 for self-adjoint operators. Indeed, with S and T self-adjoint and $T \geq 0$, $\|S\| \leq 1$, $S + iT$ is unitary if and only if $T = (I-S^2)^{\frac{1}{2}}$. Thus, if the positive map carries unitary operators into unitary operators and A is self-adjoint of norm less than or equal to 1 then $\varphi((I-A^2)^{\frac{1}{2}}) = (I-\varphi(A)^2)^{\frac{1}{2}}$, hence, using the Cauchy-Schwarz inequality twice, $I-\varphi(A)^2 = \varphi((I-A^2)^{\frac{1}{2}})^2 \leq \varphi(I-A^2) = I-\varphi(A^2) \leq I-\varphi(A)^2$, and $\varphi(A^2) = \varphi(A)^2$. Note that the assumption that φ be positive is redundant, cf. Theorem 3.3 and its proof. The proof that $\varphi(A^{-1}) = \varphi(A)^{-1}$ for A positive and invertible implies φ is a Jordan homomorphism follows from inequality 3.2 [12].

Broise [7, 8, 9, 10] has given other interesting characterizations of Jordan homomorphisms, but we shall rather concentrate on their order structure together with the isometric properties of Jordan isomorphisms. When one wants to generalize the theory of positive maps to such maps of partially ordered vector spaces, the maps that play the role of homomorphisms are the order-homomorphisms. Let $\mathfrak{A}$ and $\mathfrak{B}$ be C*-algebras and φ a positive normalized map of $\mathfrak{A}$ into $\mathfrak{B}$. φ is said to be an <u>order-homomorphism</u> if its null space $\mathfrak{N}$ is positively generated (linearly) and $\varphi^{-1}(\varphi(A))$ contains a positive operator for each A in $\mathfrak{A}$ for which $\varphi(A) \geq 0$. If φ is one-to-one, it is an <u>order-isomorphism</u>. Now, with φ an order-homomorphism $\mathfrak{N}$ is a two-sided ideal. Indeed, let $\mathfrak{J} = \{A \in \mathfrak{A} : A^*A \in \mathfrak{N}\}$. Since $\mathfrak{J}$ is the intersection of the left kernels of all the states $\rho \circ \varphi$ as ρ runs through the states of $\mathfrak{B}$, $\mathfrak{J}$ is a left ideal, hence if A is positive in $\mathfrak{N}$, $A = (A^{\frac{1}{2}})^2 \in \mathfrak{J}$, and $BA \in \mathfrak{N}$ for all $B \in \mathfrak{A}$.

Since $\mathfrak{N}$ is positively generated it is a left ideal and symmetrically a right ideal. Thus the study of order-homomorphisms is reduced to that of order-isomorphisms. The latter maps are closely related to

Jordan isomorphisms and isometries, as the following theorem shows, [26, 27, 47].

Theorem 5.2 Let $\mathfrak{A}$ and $\mathfrak{B}$ be C*-algebras and φ a normalized linear self-adjoint map of $\mathfrak{A}$ into $\mathfrak{B}$. Then the following three conditions are equivalent.

1) φ is an Jordan isomorphism.

2) φ is an isometry of $\mathfrak{A}$ onto a Jordan algebra.

3) φ is an order-isomorphism of $\mathfrak{A}$ onto a Jordan algebra.

It is an easy consequence of the last statement in Theorem 5.1 that 1) $\Longrightarrow$ 2), see [48]. If 2) holds, then from the trivial part of Theorem 3.3 φ is positive. Let A be positive in $\varphi(\mathfrak{A})$. Then the C*-algebra C*(A) generated by A and I is contained in $\varphi(\mathfrak{A})$, since the latter is a uniformly closed Jordan algebra. Since φ^{-1} is an isometry on C*(A) the above remarks show φ^{-1} restricted to C*(A) is positive, hence $\varphi^{-1}(A) \geq 0$, and φ is an order-isomorphism. Finally, the Cauchy-Schwarz inequality applied to φ and φ^{-1} shows 3) $\Longrightarrow$ 1).

A quite different order-theoretic approach to Jordan homomorphisms is motivated by the duality of the Heisenberg and Schrödinger pictures of quantum dynamics. Now the map $A \to \hat{A}$ is an order-isomorphism of the C*-algebra $\mathfrak{A}$ into the w*-continuous complex functions on the state space $S(\mathfrak{A})$ of $\mathfrak{A}$, where $\hat{A}$ is defined by $\hat{A}(\rho) = \rho(A)$. This map carries the self-adjoint part of $\mathfrak{A}$ onto the w*-continuous real affine functions on $S(\mathfrak{A})$ [32, 45]. Thus if φ is a normalized positive map from the C*-algebra $\mathfrak{A}$ into another $\mathfrak{B}$ then $\varphi^* : \rho \to \rho \circ \varphi$ is a w*-continuous affine map of $S(\mathfrak{B})$ into $S(\mathfrak{A})$, and conversely, given such a map ν of $S(\mathfrak{B})$ into $S(\mathfrak{A})$ then $\nu = \varphi^*$ with φ as above. In some applications [32, 43] one is only given ν on a dense subset of $S(\mathfrak{B})$, but then continuity assumptions extend it. For simplicity we consider only maps defined on all of $S(\mathfrak{B})$. We call a convex subset F of $S(\mathfrak{A})$ a <u>face</u> if $\rho \in F$, $\omega \in S(\mathfrak{A})$, $\omega \leq k\rho$ for some $k > 0$, implies $\omega \in F$, F is an <u>invariant face</u> if the states $A \to \rho_B(A) = \rho(B^*AB)\rho(B^*B)^{-1}$ are in F whenever ρ is and $\rho(B^*B) \neq 0$ with $B \in \mathfrak{A}$ (see also the article of Poole). Now a w*-closed convex subset F of $S(\mathfrak{A})$ is an invariant face if and only if it is the annihilator in $S(\mathfrak{A})$ of a norm closed two-sided ideal in $\mathfrak{A}$ [21, 39]. Using this, Theorem 5.2, and the remarks preceding it, one obtains the following result, which has immediate generalizations to partially ordered vector spaces [32, 51].

<u>Theorem 5.3</u> Let $\mathfrak{A}$ and $\mathfrak{B}$ be C*-algebras and φ a normalized positive map of $\mathfrak{A}$ into $\mathfrak{B}$. Then

1) φ is an order-homomorphism if and only if $\varphi^*(S(\mathfrak{B}))$ is an invariant face of $S(\mathfrak{A})$.
2) φ is an order-isomorphism if and only if φ^* is onto.
3) φ is a Jordan-isomorphism of $\mathfrak{A}$ onto $\mathfrak{B}$ if and only if φ^* is an affine isomorphism of $S(\mathfrak{B})$ onto $S(\mathfrak{A})$.

A related characterization of Jordan homomorphisms of $\mathfrak{A}$ onto $\mathfrak{B}$ is given in [13]. We thus obtain a condition for φ to be a Jordan homomorphism in terms of its image and φ^*.

In order to avoid the condition that $\varphi(\mathfrak{A})$ be a Jordan algebra, more delicate assumptions are needed on φ^*. We shall first need a characterization of those φ for which φ^* carries pure states into pure states [47].

<u>Theorem 5.4</u> Let $\mathfrak{A}$ and $\mathfrak{B}$ be C*-algebras and φ a normalized positive map of $\mathfrak{A}$ into $\mathfrak{B}$. Then φ^* carries pure states of $\mathfrak{B}$ into pure states of $\mathfrak{A}$ if and only if for each irreducible representation ψ of $\mathfrak{B}$ on a Hilbert space $\mathcal{H}_\psi$ either $\psi \circ \varphi$ is of the form $A \to \omega(A)I$ with ω a pure state of $\mathfrak{A}$, or $\psi \circ \varphi = V^*\rho V$, where V is a linear isometry of $\mathcal{H}_\psi$ into a Hilbert space $\mathcal{K}_\psi$, and ρ is an irreducible *-homomorphism or *-anti-homomorphism of $\mathfrak{A}$ into $\mathcal{B}(\mathcal{K}_\psi)$.

In the proof one replaces $\psi \circ \varphi$ by φ and thus assumes $\omega_x \circ \varphi$ is a pure state of $\mathfrak{A}$ for all unit vectors x in $\mathcal{H}_\psi$. Moreover these states are all unitarily equivalent, thus of the form $\omega_x \circ \varphi = \omega_y \circ \pi$, where π is an irreducible representation of $\mathfrak{A}$ on a Hilbert space $\mathcal{K}_\psi$. There is now induced a normal map η of $\pi(\mathfrak{A})$ into $\mathfrak{B}$ such that $\eta \circ \pi = \varphi$, and the problem is thus reduced to the case when φ is normal, $\mathfrak{A} = \mathcal{B}(\mathcal{K}_\psi)$, $\omega_x \circ \varphi = \omega_y$ for all unit vectors x in $\mathcal{H}_\psi$. One now shows φ is induced by a linear or conjugate linear isometry of $\mathcal{H}_\psi$ into $\mathcal{K}_\psi$, thus generalizing the well known result of the same type by Wigner [60]. A slight reformulation now completes the proof.

It follows from Theorem 5.1 that if φ is a Jordan homomorphism of $\mathfrak{A}$ into $\mathfrak{B}$, and $\mathfrak{B}$ equals the C*-algebra generated by $\varphi(\mathfrak{A})$, then φ^* carries pure states into pure states. Thus Theorem 5.4 is applicable, and from the largeness of its image, the case that $\psi \circ \varphi$ is a state may be excluded from considerations. We need then a condition that forces V to be a unitary map of $\mathcal{H}_\psi$ onto $\mathcal{K}_\psi$ for each ψ . For this, if ρ is a state of $\mathfrak{A}$ then $[\rho]$ shall denote the norm closure of the set of states $\sum_{j=1}^{n}\lambda_j\rho_{A_j}$, where $\Sigma_j\lambda_j = 1$, $\lambda_j > 0$, $A_j \in \mathfrak{A}$. Then $[\rho] = \{\omega\circ\pi_\rho : \omega$ a normal state of $\pi_\rho(\mathfrak{A})^-\}$, and $[\rho]$ is a norm closed invariant face of $S(\mathfrak{A})$, the smallest containing ρ [51]. Using Theorems 5.1, 5.3 and 5.4 we can show [51],

<u>Theorem 5.5</u> Let φ be a normalized positive map of a C*-algebra $\mathfrak{A}$ into another C*-algebra. Let $\mathfrak{B}$ be the C*-algebra generated by $\varphi(\mathfrak{A})$. Then φ is a Jordan homomorphism if and only if for each pure state ρ of $\mathfrak{B}$ $\varphi^*(\rho)$ is a pure state of $\mathfrak{A}$, and $\varphi^*([\rho]) = [\varphi^*(\rho)]$.

In other words, nice behaviour of φ^* on the norm closed facial structure of $S(\mathfrak{B})$ associated with the pure states is equivalent to φ being a Jordan homomorphism. As a corollary of Theorems 5.3 and 5.5 it follows that if ν is a w*-continuous affine isomorphism of $S(\mathfrak{B})$ onto $S(\mathfrak{A})$, then $\nu([\rho]) = [\nu(\rho)]$ for all pure states ρ of $\mathfrak{B}$.

From Theorem 5.1 it is clear that a better knowledge about *-anti-homomorphisms is necessary in order to understand Jordan homomorphisms. Very little is known on the subject, and even for von Neumann algebras little is known about their *-anti - automorphisms, e.g. the existence of *-anti-automorphisms in every von Neumann algebra is unknown, except for type I algebras. It seems that the *-anti-automorphisms of main importance are those of order 2. Say φ is one of the C*-algebra $\mathfrak{A}$. Let $\mathfrak{R} = \{A \in \mathfrak{A} : \varphi(A) = A^*\}$. Then $\mathfrak{R}$ is a self-adjoint real operator algebra (weakly closed if $\mathfrak{A}$ is) such that

i) $\mathfrak{R} \cap i\mathfrak{R} = \{0\}$, where $i\mathfrak{R} = \{iA : A \in \mathfrak{R}\}$.

ii) $\mathfrak{R} + i\mathfrak{R} = \mathfrak{A}$.

Conversely, the existence of such an $\mathfrak{R}$ implies that the map $A + iB \to A^* + iB^*$, $A,B \in \mathfrak{R}$, is a *-anti-automorphism of order 2 of $\mathfrak{A}$. Thus the study of *-anti-automorphisms of order 2 is reduced to that of real self-adjoint algebras satisfying i) and ii). Using the structure theory for weakly closed such algebras one can show [49],

<u>Theorem 5.6</u> Let $\mathfrak{A}$ be a von Neumann algebra with center $\mathfrak{C}$ acting on a Hilbert space $\mathcal{H}$. Let φ be a *-anti-automorphism of $\mathfrak{A}$ of order 2 such that $\varphi(A) = A$ for all $A \in \mathfrak{C}$. Then there exist two orthogonal projections P and Q in $\mathfrak{A}'$ with sum I and conjugate linear isometries J_1 and J_2 of the Hilbert spaces $P\mathcal{H}$ and $Q\mathcal{H}$ respectively

such that $J_1^2 = P$, $J_2^2 = -Q$, and

$$\varphi(A) = J_1 A^* J_1 - J_2 A^* J_2$$

for all A in $\mathfrak{A}$. Furthermore, there exists a conjugate linear isometry J of $\mathcal{H}$ such that $J^2 = I$, $J \mathfrak{A} J = \mathfrak{A}$, and $JA^*J = A$ for all $A \in \mathcal{C}$.

If $\mathfrak{A}$ is of type I one can always find a J like the one above (such operators are called conjugations), and if $\mathfrak{A}$ is of type III or is finite with a separating and cyclic vector one can in the theorem assume $Q = 0$. If one defines an inner *-anti-automorphism to be one which leaves the center elementwise fixed and which is of the form $A \to V^{-1}A^*V$ with V a conjugate linear isometry of $\mathcal{H}$ such that $V^2 \in \mathfrak{A}$, then φ is inner if and only if there exists a J like in Theorem 5.6 and a unitary operator U in $\mathfrak{A}$ such that $\varphi(A) = U^{-1}JA^*JU$ for all A in $\mathfrak{A}$. Moreover, if $\mathfrak{A}$ is of type I, then every *-anti-automorphism of $\mathfrak{A}$ leaving the center elementwise fixed, is inner, cf. the analogous result for *-automorphisms, see [49].

6. EXTREMAL MAPS

The normalized positive linear maps of a C*-algebra $\mathfrak{A}$ into another $\mathfrak{B}$ form a convex set. We say φ is extremal if it is an extreme point of this set. If $\mathfrak{B}$ is a von Neumann algebra, the convex set is compact in the point-open topology on the space of linear transformations of $\mathfrak{A}$ into $\mathfrak{B}$, where $\mathfrak{B}$ is taken in the weak topology [29]. Hence the study of extremal maps is a natural consequence of the Krein-Milman Theorem. Several well known maps are extremal [47].

1. With φ of the form $\varphi(A) = \omega(A)I$ with ω a pure state of $\mathfrak{A}$, then clearly φ is extremal.
2. Every Jordan homomorphism of $\mathfrak{A}$ into $\mathfrak{B}$ is extremal. This is a simple consequence of the Cauchy-Schwarz inequality 3.1 applied to self-adjoint operators.
3. If $\mathfrak{A}$ and $\mathfrak{B}$ act on Hilbert spaces $\mathcal{K}$ and $\mathcal{H}$ respectively and V is a linear isometry of $\mathcal{H}$ into $\mathcal{K}$ such that $VV^* \in \mathfrak{A}^-$, $V^*\mathfrak{A}V \subset \mathfrak{B}$, then the map $\varphi(A) = V^*AV$ is extremal.

A natural problem is to find all extremal maps. If $\mathfrak{A}$ and $\mathfrak{B}$ are both equal to M_2 - the complex 2×2 matrices - the extremal maps where computed in [47], the result being so complicated that it seems to be unfruitful to try to do the same for more general C*-algebras. For a related problem, see the article of Gorini. An essentially different case is when either $\mathfrak{A}$ or $\mathfrak{B}$ is abelian. In this case we saw in section 2 that every positive map is completely positive. More generally, notice that it is immediate from the definition that the normalized completely positive maps form a convex set. This set seems more promising to study in view of 3) above and the decomposition theorem 2.1, and indeed this is so, as shown in [2]. A clear cut result is the following [12].

Theorem 6.1 Let φ be a normalized map of M_n into M_m. Then φ is extreme among the completely positive maps if and only if $\varphi(A) = \Sigma_i V_i{}^*AV_i$, where $\Sigma_i V_i{}^*V_i = I$ and $\{V_i{}^*V_j\}_{i,j}$ is a linearly independent set in M_m.

In order to study the case when $\mathfrak{A}$ or $\mathfrak{B}$ is abelian it is at the present necessary to make restricted assumptions on $\mathfrak{B}$. If $\mathfrak{B}$ is abelian, the following result is known [50].

Theorem 6.2 Let $\mathfrak{A}$ be a GCR-algebra and $\mathfrak{B}$ an abelian von Neumann algebra. Then a normalized positive linear map φ of $\mathfrak{A}$ into $\mathfrak{B}$ is extremal if and only if $\rho \circ \varphi$ is a pure state of $\mathfrak{A}$ for all ρ in a w*-dense subset of the pure states of $\mathfrak{B}$.

In the case when $\mathfrak{A}$ is abelian the only complete results are obtained when $\mathfrak{B} = M_n$. We first show that if $\mathfrak{A}$ is abelian and φ is extremal then the nullspace of φ is a two-sided ideal in $\mathfrak{A}$. This is immediate from a more general result [47].

Theorem 6.3 Let $\mathfrak{A}$ and $\mathfrak{B}$ be C*-algebras and φ an extremal normalized positive map of $\mathfrak{A}$ into $\mathfrak{B}$. Let A be a central operator in $\mathfrak{A}$ such that $\varphi(A)$ is in the center of $\mathfrak{B}$. Then $\varphi(AB) = \varphi(A)\varphi(B)$ for all B in $\mathfrak{A}$.

Before we discuss the proof notice that if we combine Theorem 6.3 with 2. above, if $\mathfrak{A}$ and $\mathfrak{B}$ are abelian, then the extremal maps are exactly the *-homomorphisms of $\mathfrak{A}$ into $\mathfrak{B}$ [24, 37, 47].

For the proof of Theorem 6.3 we quickly reduce to the case when A is positive and $\|A\| < \frac{1}{2}$. Thus $\frac{1}{2}I - A$ and $\frac{1}{2}I - \varphi(A)$ are positive invertible central operators in $\mathfrak{A}$ and $\mathfrak{B}$ respectively. Define a normalized positive linear map ψ of $\mathfrak{A}$ into $\mathfrak{B}$ by

$$\psi(B) = \varphi(B(\tfrac{1}{2}I-A))(\tfrac{1}{2}I-\varphi(A))^{-1} .$$

Since $\frac{1}{2}I - \varphi(A)$ is positive and invertible there is $k > 0$ such that $kI \leq \frac{1}{2}I - \varphi(A)$. Thus it is easy to see that $k\psi \leq \varphi$. But φ is extremal so $\psi = \varphi$, and an easy computation completes the proof. Thus in the case when $\mathfrak{A}$ is abelian and $\mathfrak{B} = M_n$ in order to classify the extremal maps it suffices to study the case when $\mathfrak{A}$ is finite dimensional. For this we say a family $\{P_i\}$ of projections in a finite dimensional Hilbert space $\mathcal{H}$ is <u>weakly independent</u> [12] if for $T_i \in \mathcal{B}(\mathcal{H})$

$$\Sigma_i P_i T_i P_i = 0 \quad \text{implies} \quad P_i T_i P_i = 0$$

for all i. The theorem then states [2, 12, 47],

<u>Theorem 6.4</u> Let $\mathfrak{A}$ be an abelian C*-algebra generated by its minimal projections $\{E_1,\ldots,E_m\}$. Then a normalized positive linear map φ of $\mathfrak{A}$ into M_n is extremal if and only if $\{\text{range } \varphi(E_i)\}_{i=1,\ldots,m}$ is weakly independent.

REFERENCES

1. Arveson, W.: Analyticity in operator algebras, Amer.J.Math. 89, 578-642 (1967).

2. " : Subalgebras of C*-algebras, Acta math. 123, 141-224 (1969).

3. " : Subalgebras of C*-algebras, II, Acta math. 128, 271-308 (1972).

4. " : On groups of automorphisms of operator algebras, To appear.

5. Bendat, J. and Sherman, S.: Monotone and convex operator functions, Trans.Amer.Math.Soc. 79, 58-71 (1955).

6. Borchers, H.J.: On the vacuum state in quantum field theory, II, Commun.math.Phys. 1, 57-79 (1965).

7. Broise, M.: Une caractérisation des représentations unitaires, Bull.Sc.math. 2 ser., 59-64 (1964).

8. " : Une caractérisation des représentations unitaires de certains semi-groups, Bull.Sc.math. 2 ser., 69-79 (1966).

9. " : Sur certaines applications unitaires de l'espace des opérateurs de Hilbert-Schmidt, C.R.Acad.Sc.Paris, 263, 722-725 (1966).

10. " : Sur les isomorphismes de certaines algèbres de von Neumann, Ann.scient, Ec.Norm.Sup. 83, 91-111 (1966).

11. Choi, M.D.: Positive linear maps on C*-algebras, Canad.J.Math. 24, 520-529 (1972).

12. " : Positive linear maps on C*-algebras, Thesis, University of Toronto (1972).

13. Civin, P. and Yood, B.: Lie and Jordan structures in Banach algebras, Pacific J.Math. 15, 775-797 (1965).

14. Connes, A.: Thèse, To appear.

15. Davis, C.: A Schwarz inequality for convex operator functions, Proc.Amer.Math.Soc. 8, 42-44 (1957).

16. " : Notions generalizing convexity for functions defined on spaces of matrices, Amer.Math.Soc., Proc. of Symposia in Pure Math. Providence, vol.7, Convexity 187-201 (1962).

17. Dixmier, J.: Les algèbres d'opérateurs dans l'espaces hilbertien, Paris, Gauthier-Villars, 2. ed. (1969).

18. " : Les C*-algèbres et leurs représentations, Paris, Gauthier-Villars, (1964).

19. Doplicher, S., Kadison, R., Kastler, D. and Robinson, D.W.: Asymptotically abelian systems, Commun,math.Phys. 6, 101-120 (1967).

20. Doplicher, S., Kastler, D. and Størmer, E.: Invariant states and asymptotic abelianness, J.Fnal.Anal. 3, 419-434 (1969).

21. Effros, E.: Order ideals in a C*-algebra and its dual, Duke Math. J. 30, 391-412 (1963).

22. Gardner, L.T.: On isomorphisms of C*-algebras, Amer.J.Math. 87, 384-396 (1965).

23. Guichardet, A.: Une caractérisation des algèbres de von Neumann discrètes, Bull.Soc.Math.France 89, 77-101 (1961).

24. Ionescu-Tulcea, A. and C.: On the lifting property (1), J. of Math.Anal. and Applic. 3, 537-546 (1961).

25. Jacobson, N. and Rickart, C.: Homomorphisms of Jordan rings, Trans.Amer.Math.Soc. 69, 479-502 (1950).

26. Kadison, R.V.: Isometries of operator algebras, Ann.Math. 54, 325-338 (1951).

27. " : A generalized Schwarz inequality and algebraic invariants for operator algebras, Ann.Math. 56, 494-503 (1952).

28. " : Unitary invariants for representations of operator algebras, Ann.Math. 66, 304-379 (1957).

29. " : The trace in finite operator algebras., Proc.Amer. Math.Soc. 12, 973-977 (1961).

30. " : Normalcy in operator algebras, Duke Math.J. 29, 459-464 (1962).

31. " : The trace in finite operator algebras, Lecture notes, Columbia University (1963).

32. " : Transformations of states in operator theory and dynamics, Topology, suppl. 2, 177-198 (1965).

33. Kadison, R.V. and Ringrose, J.R.: Derivations and automorphisms of operator algebras, Commun.math.Phys. 4, 32-63 (1967).

34. Kadison, R.V. and Singer, I.M.: Extensions of pure states, Amer. J.Math. 81, 383-400 (1959).

35. Kovács, I. and Szücs, J.: Ergodic type theorems in von Neumann algebras, Acta.Sc.Math. 27, 233-246 (1966).

36. Neumark, M.A.: On a representation of additive operator set functions, C.R. (Doklady) Acad.Sci. URSS 41, 359-361 (1943).

37. Phelps, R.R.: Extreme positive operators and homomorphisms, Trans.Amer.Soc. 108, 265-274 (1963).

38. Powers, R.T.: Algebras of unbounded operators, II, To appear.

39. Prosser, R.: On the ideal structure of operator algebras, Mem, Amer.Math.Soc. 45, 1-28 (1963).

40. Riesz, F. and Nagy, B.Sz.: Functional Analysis, Ungar, New York (1955).

41. Russo, B.: Linear mappings of operator algebras, Proc.Amer.Math. Soc. 17, 1019-1022 (1966).

42. Russo, B. and Dye, H.A.: A note on unitary operators in C*-algebras, Duke Math.J. 33, 413-416 (1966).

43. Sakai, S.: On a characterization of type I C*-algebras, Bull. Amer.Math.Soc. 72, 508-511 (1966).

44. Schwartz, J.: Two finite, non hyperfinite, non isomorphic factors, Comm.Pure Appl.Math. 16, 19-26 (1963).

45. Semadeni, Z.: Free compact convex sets, Bull,Acad.Sc.Pol. 13, 141-146 (1965).

46. Stinespring, W.F.: Positive functions on C*-algebras, Proc.Amer. Math.Soc. 6, 211-216 (1955).

47. Størmer, E.: Positive linear maps of operator algebras, Acta math. 110, 233-278 (1963).

48. " : On the Jordan structure of C*-algebras, Trans.Amer. Math.Soc. 120, 438-447 (1965).

49. " : On anti-automorphisms of von Neumann algebras, Pacific J.Math. 21, 349-370 (1967).

50. " : On extremal maps of operator algebras, Aarhus University (1966).

51. " : On partially ordered vector spaces and their duals, with applications to simplexes and C*-algebras, Proc. London Math.Soc. 18, 245-265 (1968).

52. Størmer, E.: Large groups of automorphisms of C*-algebras, Commun.math.Phys. 5, 1-22 (1967).

53. " : Positive linear maps and Jordan homomorphisms, Oslo University (1969).

54. " : States and invariant maps of operator algebras, J.Fnal.Anal. 5, 44-65 (1970).

55. " : On projection maps of von Neumann algebras, Math. Scand. 30, 46-50 (1972).

56. Takesaki, M.: Conditional expectations in von Neumann algebras, J.Fnal.Anal. 9, 306-321 (1972).

57. Tomiyama, J.: On the projection of norm one in W*-algebras, Proc.Japan Acad. 33, 608-612 (1957).

58. " : On the projection of norm one in W*-algebras, III, Tôhoku Math.J. 11, 125-129 (1959).

59. " : The extension property of von Neumann algebras and a class of C*-algebras associated to them, To appear.

60. Wigner, E.P.: Gruppentheorie und ihre Anwendung auf die Quantenmechanik der Atomspektren, Friedr.Vieweg, Braunschweig (1931).

AXIOMATICS OF PREPARING AND MEASURING PROCEDURES

A. Hartkämper
Fachbereich Dynamische Systeme der Universität Osnabrück
Osnabrück, Germany

The aim of an axiomatic foundation of a physical theory is to get a better understanding of mathematical structures of an already existing theory whose formation is mostly based on a more intuitive comprehension of physical phenomena and to find out, what are the physical facts leading to certain mathematical structures. In Quantum mechanics the underlying mathematical structure is the Hilbert space structure. We will therefore try to deduce this structure from the description of quantum mechanical phenomena.

First we briefly explain what is understood by an axiomatic foundation of a physical theory in our concept. For details see (1). The scheme given here is of course not the only possible way to get an axiomatic foundation of quantum mechanics (2). We only sketch the general concept of a physical theory, to clarify the framework we use for our axiomatic foundation of quantum mechanics.

A physical theory consists of three main parts:

1. a mathematical theory $\mathcal{MT}$
2. mapping principles $\leftrightarrow$
3. a reality domain, $\mathcal{W}$ for short:

$$\mathcal{PT} = \mathcal{MT} \leftrightarrow \mathcal{W}$$

We do not want to talk about the axiomatic construction of a mathematical theory. This shall be claimed as known (3). We rather want to define the meaning of physical reality.

To show this, we start with the so called basic domain $\mathcal{G}$, containing the physical data and basic facts, which together with their mathematical description form the starting point for the formation of the theory.

The connection between $\mathcal{MT}$ and $\mathcal{G}$ is given by the mapping principles, which attach mathematical symbols to the elements of the basic domain and translate their physical relations into mathematical relations.

By adding axioms, which cannot be deduced directly from experience but only are suggested by experience, it is possible to embed the basic domain into the reality domain and to extend the mapping principles to $\mathcal{W}$.

We are now going to describe the basic domain for quantum mechanics in detail. Of course there is no unique basic domain. We therefore have to define at first, which experimental facts belong to this domain. After having set forth this domain, we have to define the mapping principles, which allow a mathematical description of this domain.

As the basic domain we choose reproducible experiments consisting of single experiments which can be divided into two macroscopically describable parts, the preparing and the effected part. The preparing part can influence the effected part via the "microscopic channel". The effected part has only two possibilities of responses (yes-no experiments). By "microscopic channel" we describe that influence of the preparing part on the effected part which is not caused by macroscopic connections. It is important to stress that there should be no possibility that the effected parts influence the behaviour of the preparing parts of the single experiment. This would be the case eq., if one triggers the preparing part by the effected part.

In our definition an experiment consists of a series of same single-experiments. Before speaking about a repetition of the same single-experiment, one has to clarify under which circumstances single-experiments are of the same series. To do this, we have to characterize the preparing parts and the effected parts.

We begin with the characterisation of the preparing parts. The preparing part can be specified by its macroscopic properties, i.e. the preparing part is describable by its technical equipment and signals, which may or may not respond on the preparing part. As an example you may think of an electron-producing-apparatus equipped with a signal, which indicates, whether the spin is up or down. Every preparing part has to be seen in a space-time reference system, which can be fixed in the preparing part itself. The total catalog, specifying the preparing part according to its macroscopic properties, is called a preparing procedure. Speaking about macroscopic properties, we include all properties which can be analyzed by well-known classical theories, such as for example electrodynamics, statistical mechanics, etc (4).

For the mathematical description of the preparing procedures we introduce a set P, whose elements are symbols for different preparing parts. All preparing procedures define a subset S_P of $\mathcal{R}(P)$, $\mathcal{R}(P)$ being the set of all subsets of P. For the elements $s_P \in S_P$ we postulate the following axiom:

$$SI \quad \alpha) \quad s_{P_1}, s_{P_2} \in S_P \Rightarrow s_{P_1} \cap s_{P_2} \in S_P$$

$$\beta) \quad s_{P_1}, s_{P_2} \in S_P \Rightarrow s_{P_2} \ominus s_{P_1} \in S_P$$
$$s_{P_1} \subset s_{P_2}$$

$s_P \in S_P$ states: s_P is a preparing procedure described above; s_P is given by $\{ p \in P,\ p$ fulfilling the criteria of the preparing procedure $s_P \}$.

$\alpha)$ guarantees that one gets a preparing procedure, if one selects according to two preparing procedures simultaneously. $s_{P_1} \subset s_{P_2}$ expresses the fact, that s_{P_1} is a finer preparing procedure than s_{P_2}. Now $\beta)$ states that one obtains a new preparing procedure if one considers from all preparing parts, selected according to a preparing procedure s_{P_2} only those which are left after having selected part of them according to a finer preparing procedure s_{P_1}. As a consequence of axiom SI, every subset in S_P of the form

$$S(s_{P_0}) = \{ s_P \mid s_P \subset s_{P_0} \}$$

is a Boolean ring with unit element s_{P_0}.
It is a typical feature of quantum mechanics that $P \notin S_P$ and therefore S_P is not boolean.

In describing the effected parts and the selecting procedures for effected parts, we proceed in an analogous way.
We introduce the set E. The elements of E are symbols for the different effected parts. Let S_E be a subset of $\mathcal{R}(E)$. $s_E \in S_E$ corresponds to s_E is a selecting procedure for effected parts. These selecting procedures are called measuring procedures. A measuring procedure specifies the technical équipment of the effected part, its possibilities of response and its space-time relation. Since the effected part is a macroscopic apparatus of a single-experiment, we consider its space-time relation within the space-time reference

system of the preparing apparatus. This includes that a measuring procedure cannot be treated independent of the preparing part.

An experiment consists of the repetition of single-experiments, whose preparing parts are chosen by a certain preparing procedure and whose effected parts by a certain measuring procedure. Since only reproducible experiments belong to the basic domain, the quotient N_+/N is constant for large N, where N is the number of single experiments N_+ is the number of responses of the effected parts. On $S_P \times S_E$ we define a function

$$\mu : S_P \times S_E \to [0,1]$$

for which the following holds:

α.) For every s_P there exists s_E with

$$\mu(s_P, s_E) = 1$$

β.) There exists $s_E = s_{E_0}$ with

$$\mu(s_P, s_{E_0}) = 0 \quad \forall s_P \in S_P \,.$$

Since $\mu \approx N_+/N$, α.) states that it is possible to count the number of single experiments in an experiment. β.) is more or less formal.

For the foundation of quantum mechanics we are only interested in the description of the interaction, which is transferred by the so-called "microscopic channel" from the preparing to the effected part. Therefore we are not going to distinguish those preparing procedures, which lead to the same interaction, id est which lead to the same frequencies for all measuring procedures. At the same we will not distinguish measuring procedures which select effected parts which react with the preparing part in the same way.
We therefore introduce the following relations for the sets S_P and S_E:

$$s_{P_1} \,\pi\, s_{P_2} \iff \mu(s_{P_1}, s_E) = \mu(s_{P_2}, s_E) \quad \forall s_E \in S_E$$

$$s_{E_1} \,\varrho\, s_{E_2} \iff \mu(s_P, s_{E_1}) = \mu(s_P, s_{E_2}) \quad \forall s_P \in S_P$$

It follows immediately that π and ρ are equivalence relations. We denote

$$S_P/\pi = \underline{K} \quad , \quad S_E/\rho = \underline{L} \quad .$$

If we restrict μ to $\underline{K} \times \underline{L}$ the following holds:

$$\mu : \underline{K} \times \underline{L} \to [0,1]$$

α.) For every $\underline{V} \in \underline{K}$ there exists $\underline{F} \in \underline{L}$ with $\mu(\underline{V},\underline{F}) = 1$.

β.) There exists $\underline{F} \in \underline{L}, \underline{F} = 0$ with $\mu(\underline{V},0) = 0 \quad \forall \underline{V} \in \underline{K}$.

γ.) For $\quad \mu(\underline{V_1},\underline{F}) = \mu(\underline{V_2},\underline{F}) \quad \forall \underline{F} \in \underline{L} \Rightarrow \underline{V_1} = \underline{V_2}$.

δ.) From $\quad \mu(\underline{V},\underline{F_1}) = \mu(\underline{V},\underline{F_2}) \quad \forall \underline{V} \in \underline{K} \Rightarrow \underline{F_1} = \underline{F_2}$.

From this it follows, that one can identify $\underline{V}$ with the function $\mu_{\underline{V}}(\underline{F})$, a function defined on $\underline{L}$ and $\underline{F}$ with the function $\mu_{\underline{F}}(V)$ defined on $\underline{K}$. The elements of $\underline{K}$ are called ensembles, the elements of $\underline{L}$ are called effects.

We now enter into a more detailed description of the preparing procedures.
As mentioned above the preparing procedures characterize the technical equipment of the preparing parts. Besides that they give the possibility of selecting preparing parts more specifically according to certain signals, appearing on the preparing part. We consider a preparing procedure $\mathfrak{a}_{p_1}$ which selects N_1 preparing parts $p_1, p_2 \cdots p_{N_1}$.
From these preparing parts we select those preparing parts, on which an additional signal has appeared. Let $\mathfrak{a}_{p_2}$ be the corresponding preparing procedure and $p_1', p_2' \cdots p_{N_2}'$ the selected preparing parts.
It turns out that for a large number of preparing parts selected according to $\mathfrak{a}_{p_1}$ the quotient N_2/N_1 tends to a constant. N_2/N_1 gives the frequency of the selection of a finer preparing procedure relative to a coarser preparing procedure.

The relative frequencies of two preparing procedures $\mathfrak{a}_{p_1}$ and $\mathfrak{a}_{p_2}$ with $\mathfrak{a}_{p_1} \supset \mathfrak{a}_{p_2}$ are introduced in the following way:

Let

$$\mathcal{T} = \{ (\Delta\rho_1, \Delta\rho_2) \in S_\rho \times S_E ,\ \Delta\rho_1 \supset \Delta\rho_2 ,\ \Delta\rho_1 \neq \emptyset \} .$$

Then we postulate the axiom:

$S\,\underline{II}$ There exists a function $\lambda : \mathcal{T} \to [0, 1]$ satisfying:

1.) For $\Delta\rho_1, \Delta\rho_2, \Delta\rho_3 \in S_\rho ,\ \Delta\rho_1 \cap \Delta\rho_2 = \emptyset ;\ \Delta\rho_3 = \Delta\rho_2 \cup \Delta\rho_3$

$$\Rightarrow \lambda(\Delta\rho_3, \Delta\rho_1) + \lambda(\Delta\rho_3, \Delta\rho_1) = 1 \quad .$$

2.) For $\Delta\rho_1, \Delta\rho_2, \Delta\rho_3 \in S_\rho ,\ \Delta\rho_1 \supset \Delta\rho_2 \supset \Delta\rho_3$

$$\Rightarrow \lambda(\Delta\rho_1, \Delta\rho_3) = \lambda(\Delta\rho_1, \Delta\rho_2)\,\lambda(\Delta\rho_2, \Delta\rho_3) .$$

3.) For $\Delta\rho_1, \Delta\rho_2 \in S_\rho ,\ \Delta\rho \neq \emptyset ,\ \Delta\rho_1 \supset \Delta\rho_2$

$$\Rightarrow \lambda(\Delta\rho_1, \Delta\rho_2) \neq 0$$

Let N_1 be the number of preparing parts selected by $\Delta\rho_1$, let N_2 resp. N_3 be the number of these preparing parts, which can also be selected by $\Delta\rho_2$ resp. $\Delta\rho_3$. Then

$$\lambda(\Delta\rho_1, \Delta\rho_i) \approx N_i / N_1 .$$

If $\Delta\rho_1 \cap \Delta\rho_2 = \emptyset ,\ \Delta\rho_3 = \Delta\rho_1 \cup \Delta\rho_2 \Rightarrow N_3 = N_1 + N_2 \ (S\,\underline{II}\,1)$.

If $\Delta\rho_1 \supset \Delta\rho_2 \supset \Delta\rho_3 \Rightarrow N_1 > N_2 > N_3 \quad (S\,\underline{II}\,2)$.

If $\Delta\rho_2 \neq \emptyset ,\ N_2 \neq 0 \Rightarrow N_2/N_1 \neq 0 \quad (S\,\underline{II}\,3)$.

In defining

$$\hat{\lambda}(\Delta\rho_1, \Delta\rho_2) = \lambda(\Delta\rho_1, \Delta\rho_1 \cap \Delta\rho_2)$$

we can extend the function λ to $S_\rho \times S_E$. The restriction of $\hat{\lambda}$ to $\mathcal{T}$ coincides with λ defined above. $\lambda(\Delta\rho_0, \cdot)$ is a positive effective measure on the boolean ring $S(\Delta\rho_0)$.

We now introduce the concept of mixture and decomposition of preparing procedures. The physical content of this concept is immediately clear from the mathematical definition.

Let $\Delta\rho, \Delta\rho_1 \in S_\rho$; $\Delta\rho$ is called "statistically independent" with respect to $\Delta\rho_1$ if for all $\Delta\rho_2 \in S_\rho$ with $\Delta\rho_2 \subset \Delta\rho_1$,

$$\lambda(\Delta\rho_1, \Delta\rho_2) = \lambda(\Delta\rho \cap \Delta\rho_1, \Delta\rho \cap \Delta\rho_2)$$

holds. This means, $\Delta\rho$ is statistically independent with respect to $\Delta\rho_1$ if $\Delta\rho \cap \Delta\rho_1$ is in principle the same preparing procedure as $\Delta\rho_1$, but only selects an arbitrary set of preparing parts selected according to $\Delta\rho_1$.

Let $\Delta\rho, \Delta\rho_1, \Delta\rho_2, \Delta\rho', \Delta\rho'' \in S_\rho$.

$\Delta\rho$ is a mixture of $\Delta\rho_1$ and $\Delta\rho_2$ if $\Delta\rho'$ resp. $\Delta\rho''$ is statistically independent with respect to $\Delta\rho_1$ resp. $\Delta\rho_2$, and in addition we have also that

$$\Delta\rho = (\Delta\rho' \cap \Delta\rho_1) \cup (\Delta\rho'' \cap \Delta\rho_2)$$

$$\text{and} \quad (\Delta\rho' \cap \Delta\rho_1) \cap (\Delta\rho'' \cap \Delta\rho_2) = \emptyset$$

From this it follows:

$$\lambda(\Delta\rho, \Delta\rho' \cap \Delta\rho_1) + \lambda(\Delta\rho, \Delta\rho'' \cap \Delta\rho_2) = 1$$

In other words $\Delta\rho$ is a mixture of $\Delta\rho_1$ and $\Delta\rho_2$ with ratio $\lambda/1-\lambda$.

If $\Delta\rho$ is of the form

$$\Delta\rho = \bigcup_i \Delta\rho_i, \quad \Delta\rho_i \cap \Delta\rho_j = \emptyset \quad i \neq j$$

we say $\Delta\rho$ is decomposed into components $\Delta\rho_i$ with ratio $\lambda_i = \lambda(\Delta\rho, \Delta\rho_i)$. Symbolically we write:

$$\Delta\rho = \sum \lambda_i \Delta\rho_i .$$

If we repeat, that $S(\Delta\rho)$ is a boolean ring, we see, that a decomposition of $\Delta\rho$ is equivalent to a disjoint decomposition of unit element of $S(\Delta\rho)$.
Concerning mixtures of preparing procedures we postulate the following axiom:

For every α with $0 < \alpha < 1$ and for $\Delta p_1, \Delta p_2 \in \mathcal{S}p$ there exists a preparing procedure Δp, which is a mixture of Δp_1 and Δp_2 with ratio $\alpha / 1-\alpha$.

If the preparing procedure is of the form

$$\Delta p = \sum \lambda_i \Delta p_i$$

we postulate the following axiom:

$$\mu(\Delta p, \Delta_E) = \sum \lambda_i \mu(\Delta p_i, \Delta_E)$$

The content of this axiom is immediately clear, if one again remembers that

$$\lambda_i = \lambda(\Delta p, \Delta p_i) \approx N_i / N$$

If one defines mixtures of measuring procedures similar to mixtures of preparing procedures, one gets the mathematical result, that $\underline{K}$ and $\underline{L}$ are convex. Further structure of the sets $\underline{K}$ and $\underline{L}$ is effected by additional axioms, which are given in the next lecture.

References

(1) G.Ludwig — Deutung des Begriffs "physikalische Theorie" und axiomatische Grundlegung der Hilbertraumstruktur der Quantenmechanik durch Hauptsätze des Messens
Lecture Notes in Physics, Vol. 4
Springer Verlag, Berlin-Heidelberg 1970

(2) A.Hartkämper — A comparison of various axiomatic quantum mechanical systems;
given at the Symposion "On Contemporary Physics", University Park Pennsylvania 1971
to appear

(3) N.Bourbaki — Elements of Mathematics, Theory of Sets
Reading, Mass. 1968

(4) G.Ludwig — Measuring and Preparing Processes
contained in this volume

THE STRUCTURE OF ORDERED BANACH SPACES IN AXIOMATIC QUANTUM MECHANICS

H. Neumann
Fachbereich Physik der Universität Marburg
Marburg, Germany

The structure of ordered Banach spaces as applied in axiomatic quantum mechanics will be discussed in this lecture. The starting point will be a set $\underline{K}$ of ensembles and a set $\underline{L}$ of effects with a probability function μ defined on the product $\underline{K} \times \underline{L}$. The function μ is assumed to fulfil the following requirements:

1. $0 \leq \mu(v,f) \leq 1$ for all $v \in \underline{K}$, $f \in \underline{L}$.
2. For all $v \in \underline{K}$ there is an $f \in \underline{L}$ with $\mu(v,f) = 1$. (It is possible to count particles.)
3. There is $0 \in \underline{L}$ with $\mu(v,0) = 0$ for all $v \in \underline{K}$.
4. μ separates points of both $\underline{K}$ and $\underline{L}$.
5. $\underline{K}$ and $\underline{L}$ are approximately convex. ($\underline{K}$ approximately convex: If $v_1, v_2 \in \underline{K}$, $0 \leq \lambda \leq 1$, $\varepsilon > 0$ there is a $v_3 \in \underline{K}$ and $0 \leq \lambda' \leq 1$ such that $|\lambda - \lambda'| < \varepsilon$ and $\mu(v_3,f) = \lambda'\mu(v_1,f) + (1-\lambda)\mu(v_2,f)$ for all $f \in \underline{L}$.)
6. $\underline{K}$ and $\underline{L}$ are countable.

A comparison between experiment and theory can be made only by a finite number of effects and ensembles. Thus as a first step towards an idealized description the sets $\underline{K}$ and $\underline{L}$ are assumed to be countable. Now, the following theorem holds.

<u>Theorem</u>. If μ is a real-valued function on the set $\underline{K} \times \underline{L}$ satisfying the assumptions 1,...,6 there is an embedding of $\underline{K} \times \underline{L}$ into a pair of real Banach spaces $B \times B'$, B' being the dual of B, such that

1. the canonical real-valued bilinear form defined on $B \times B'$ coincides with the probability function μ on $\underline{K} \times \underline{L}$,
2. B is normed by a base K of a closed cone B_+ where K is the norm closure of $\underline{K}$,
3. the linear hull of $\underline{L}$ is $\sigma(B',B)$-dense in B'.

Moreover B is separable, and this embedding is unique up to isomorphisms.

Main steps of the proof of this theorem can be found in [1] . The canonical bilinear form on B x B' will be denoted by μ , too. The norm closure K of $\underline{K}$ coincides with the $\sigma(B,\underline{L})$-closure of $\underline{K}$, where $\sigma(B,\underline{L})$ is the initial topology generated by the set $\underline{L}$ of real-valued mappings on B. Since $\sigma(B,\underline{L})$ serves to describe the inaccuracy of the correspondence between reality and the mathematical picture K is called the set of ensembles. Likewise the weak closure L of $\underline{L}$ which coincides with the $\sigma(B',\underline{K})$-closure of $\underline{L}$ is called the set of effects.

It is a consequence of statement 2 that B' is an order unit space. Since $0 \leq \mu(v,f) \leq 1$ for all $v \in K$ and $f \in L$ we have $L \subset [0,1]$, where 1 denotes the order unit in B'.

It is one of the goals of an axiomatic investigation to see which assumptions have to be imposed on the sets K and L in order to get the well known representation of B and B' by spaces of operators in a separable Hilbert space, i.e. to deduce the Hilbert space structure of quantum mechanics. One would like to impose axioms which can, in principle, be checked by experience. In the sequel I shall cover the attempt of Ludwig to solve this problem [1,2] .

It is another far more delicate problem to ask which modifications of the axioms might lead to a more general and useful physical theory than quantum mechanics.
The following axiom, the first one concerning the structure of ordered vector spaces, plays a central role in the further development of the theory. The order relation on L given by $\mu(v,f_1) \leq \mu(v,f_2)$ for all $v \in K$ will be referred to as 'f_2 is more sensitive than f_1'.
$K_0(f) = \{ v \in K / \mu(v,f) = 0 \}$ is the set of ensembles which can't produce the effect f.

Axiom V1a. (Sensitivity increase of two effects):
If $f_1,f_2 \in L$ there is an $f_3 \in L$ such that $f_3 > f_1,f_2$ and $K_0(f_1) \cap K_0(f_2) \subset K_0(f_3)$.

The physical meaning of this axiom can be discussed in several examples, among which the example of polarization filters for light is most instructive.

If $A \subset K$ the set $L_0(A) = \{ f \in L / \mu(v,f) = 0$ for all $v \in A \}$ is a weakly closed face of L. Axiom V1a is equivalent to the assumption that the sets $L_0(A)$, $A \subset K$, have greatest, i.e, most sensitive, elements $\varphi_1(L_0(A))$ which are called decision effects. The set of decision effects will be denoted by G. $\varphi_1(L_o(A))$ is an extreme point of the face $L_o(A)$ of L. Hence $\varphi_1(L_o(A))$ is also an extreme point of L and we have $G \subset \partial_e L$.

Since for a class $\{A_\alpha\}$ of subsets of K $L_o(\bigcup_\alpha A_\alpha) = \bigcap_\alpha L_o(A_\alpha)$ the class of weakly closed faces $\{L_o(A) \,/\, A \subset K\}$ is a complete lattice, where the order relation is given by the set theoretical inclusion and the infimum coincides with the set theoretical intersection. As the mapping $\varphi_1 : \{L_o(A) \,/\, A \subset K\} \longrightarrow G$ is bijective and bipositive G, too, is a complete lattice with respect to the ordering induced by B'.

Let us now discuss an axiom V1b rather similar to axiom V1a. Consider an effect f with $\|f\| = \sup_{v \in K} \mu(v,f) < 1$. Physical experience shows that in this case the sensivity of f can be increased (by multiplying filters for light or increasing the sensitivity of particle counters in various manners) such that an effect f' is constructed with the following properties

1. $K_o(f) \subset K_o(f')$ 2. $f' \geq \|f\|^{-1} \cdot f$

Requirement 2. can be interpreted as follows: Ensembles with maximal absorption by f, i.e. $\mu(v,f) \sim \|f\|$, are approximately totally absorbed by f', i.e. $\mu(v,f') \sim 1 \sim \|f\|^{-1} \cdot \mu(v,f)$.
Thus in this case the sensitivity increase is $\|f\|^{-1}$. For ensembles with not maximal absorption the sensitivity increase is better than for ensembles with maximal absorption, i.e.

$$\mu(v,f') \geq \|f\|^{-1} \cdot \mu(v,f) \qquad \text{for all } v \in K.$$

If $N = \{\lambda f \,/\, f \in L,\ 0 \leq \lambda \leq \|f\|^{-1}\}$ this assumption of sensitivity increase of one effect can be formulated as follows: For all $f \in N$ there is an $f' \in L$ with $K_o(f) \subset K_o(f')$ and $f' \geq f$.

Taking into account the inaccuracy of the correspondence between reality and mathematical picture we extend this assumption to the weak closure of N.

Axiom V1b. (Sensitivity increase of one effect):
For all $f \in \overline{N}^{\sigma}$ there is an $f' \in L$ with $K_o(f) \subset K_o(f')$ and $f' \geq f$.

We shall now turn to an axiom concerning the components of an ensemble. The closed face $C(v)$ generated by an ensemble v is called the set of components of v. This expression is chosen in analogy to the finite dimensional case in which v can be decomposed with respect to all $v' \in \mathcal{C}(v)$. Consider ensembles v_1, v_2 such that v_1 has strictly more components than v_2. We assume that in this case there is an effect f which can't be produced by v_2 but can be produced by v_1.

Axiom V2. If $v_1, v_2 \in K$ such that $\mathcal{C}(v_1) \supsetneq \mathcal{C}(v_2)$ there is an $f \in L$ with $v_2 \in K_o(f)$ but $v_1 \notin K_o(f)$.

Axiom V2 is equivalent to the assumption $K_o(L_o(\mathcal{F})) = \mathcal{F}$ for all norm closed faces $\mathcal{F}$ of K. Thus all norm closed faces $\mathcal{F}$ of K are semi-exposed and the existence of greatest elements in the sets $L_o(\mathcal{F})$ implies that they are exposed (comp. [3]).
If a mapping φ_2 is defined by $\varphi_2(\mathcal{F}) = L_o(\mathcal{F})$ for a closed face $\mathcal{F}$ of K, $\varphi_1 \circ \varphi_2$ is an antiisomorphism of the lattice of norm closed faces of K onto the lattice G of decision effects.

As a consequence of the axioms V1a,b and V2 the equation $L = [0,1]$ can be proved [2] . Thus in axiomatic quantum mechanics a separable base normed Banach space B with a closed cone B_+ is considered where the base K of B_+ is the set of ensembles and the order interval $[0,1]$ is the set L of effects. While axiom V1b is obviously fulfilled if $L = [0,1]$ axioms V1a and V2 yield an enrichment of this structure.

It is an open question which physical assumption yields that B has the minimal decomposition property.

I should like to stress that the axioms V1a,b and V2 can be equivalently formulated in terms of the sets $\underline{K}$ and $\underline{L}$, where in addition the inaccuracy of the correspondence between reality and the mathematical picture is taken into account.

As we pointed out in the beginning it is one of the goals of this axiomatic approach to deduce the Hilbert space structure of quantum mechanics. For the lattice theoretic approach to quantum mechanics a representation theory is already developed. Some further postulates with respect to the lattice G of decision effects make it possible to apply these mathematical tools also within the vector space approach. Of course this method of deriving a representation of B and B' by spaces of operators in a Hilbert space looks like a detour. A method more closely related to the structure of ordered vector spaces would be more adequate.

By axiom V1b we have $\| e \| = 1$ for all decision effects e but the set $K_1(e) = \{ v \in K \,/\, \mu(v,e) = 1 \}$ can be empty. A decision whether $K_1(e)$ is empty or not can't be made on the basis of physical experiments. As an idealization of the mathematical structure we postulate

Axiom VId. $K_1(e) \neq \emptyset$ for all $e \in G$.

Axiom VId is equivalent to $(1-e) \in G$ for all $e \in G$. The correspondence $e \mapsto 1-e$ is an orthocomplementation of the lattice G. G is readily proved to be an orthocomplemented orthomodular lattice.

Another axiom the physical meaning of which is not so easily established and which is treated in [4] reads as follows:

Axiom V3. If $v_1, v_2, v_3 \in K$ such that $C(v_1) \subset C(v_3) \subset C(\frac{1}{2}v_1 + \frac{1}{2}v_2)$ and if $C(v_2)$ and $C(v_3)$ are strictly separated then $C(v_1) = C(v_3)$.

Here $C(v_2)$ and $C(v_3)$ are called strictly separated if

$$\Delta(C(v_2), C(v_3)) = \frac{1}{2} \inf \{ \mu(v,e_3) \ / \ v \in C(v_2) \} + \frac{1}{2} \inf \{ \mu(v,e_2) \ / \ v \in C(v_3) \} \neq 0,$$

where e_i is the greatest element of $L_o(C(v_i))$ for $i = 2,3$, i.e. $e_i = \varphi_1 \circ \varphi_2 (C(v_i))$.

In order to show that the lattice G can be represented as the lattice of all closed subspaces of a Hilbert space two more assumptions are necessary.

G has to be irreducible. This well known assumption is discussed in the literature in connection with the notions: coexistence of effects and centre of a physical system.

G has to be atomic. Since in the classical case of the phase space Γ ($B = L^1(\Gamma)$, $B' = L^\infty(\Gamma)$) the lattice G is not atomic there are some doubts concerning the physical significance of atomicity. The extreme points of K being minimal non-empty faces of K are mapped by $\varphi_1 \circ \varphi_2$ onto atoms of G. Thus atomicity of G can be achieved by the assumption that K has enough extreme points.

This is ensured by

Axiom V4. For all $v \in K$ there is a sequence $v_k \in K$ such that $C(v_k) \subset C(v_{k+1})$, $C(v) = \bigvee_k C(v_k)$ and $C(v_k)$ finite dimensional faces.

If V4 is assumed axiom V3 implies the so called covering condition for the lattice G. Therefore axiom V3 is called the general covering condition.

Now, these assumptions imply that for a lattice dimension greater than four G can be represented as the lattice of all closed subspaces of a separable Hilbert space H over the field of real or complex numbers or the field of quaternions [5,6] . The question which physical assumption (concerning the union of physical systems or the representation of groups) uniquely determines the complex field can't be treated here.

The famous Gleason theorem then proves B to be the Banach space of Hermitean trace class operators in H. Hence B' is the space of Hermitean bounded operators in H.

References

[1] G. Ludwig, Deutung des Begriffs "physikalische Theorie" und axiomatische Grundlegung der Hilbertraumstruktur durch Hauptsätze des Messens.
Berlin, Heidelberg, New York 1970

[2] G. Ludwig, An improved formulation of some theorems and axioms in the axiomatic foundation of the Hilbert space structure of quantum mechanics.
Commun.Math.Phys. 26, 1972

[3] A.J. Ellis, Order Ideals in Ordered Banach Spaces, this volume

[4] G. Ludwig, A Physical Interpretation of an Axiom within an Axiomatic Approach to Quantum Mechanics and a New Formulation of this Axiom as a General Covering Condition.
Notes in Math.Phys. 1, Marburg 1971

[5] P. Stolz, Attempt of an Axiomatic Approach of Quantum Mechanics and More General Theories VI.
Commun.Math.Phys. 23 (1971)

[6] V.S. Varadarajan, Geometry of Quantum Theory I, Princeton, New Jersey 1968

MEASURING AND PREPARING PROCESSES

G. Ludwig
Fachbereich Physik der Universität Marburg, Marburg, Germany

A complex of physical theories sometimes has a similarity with a circle. For instance, quantum theory (q.t.) is to be interpreted by well known macroscopic phenomena such as signals in a counter or bubbles in a bubble chamber. On the other hand, these measuring processes should be explained or at least partially explained by q.t. Thus q.t., the theories of macroscopic systems and the theory of the measuring processes have to be compatible. Is this possible? In order to give a consistent formulation of the theories appearing in this circle of compatibility one can start at several points. In this paper I do not want to start with the axiomatic foundation of quantum mechanics as I have done in [1] (comp. the articles of Hartkämper and Neumann, this volume). Instead I shall begin with the description of the measuring process taking q.t. for granted. It turns out that we will get a more detailed description of the physical structures on which an axiomatic foundation of q.t. as given in [1] is based.

At first I will describe the so-called measuring collisions. This problem is very old and well described by many authors as J. von Neumann, Wigner a.s.o.. I will give the theory of measuring collisions in a new formulation using base norm spaces and order unit spaces. Therefore I will analyze some well known structures of q.t. in these terms.

In q.t. the basic expression tr(EW) is interpreted as the probability or frequency of the yes-no measurement E in the ensemble W. E is a projection operator in the Hilbert space $\mathcal{H}$ of the system, and W is a Hermitean operator $W \geq 0$ with $tr(W) = 1$.

All Hermitean operators V of the trace class form a Banach space B with the norm $\|V\|_{tr} = tr(\sqrt{V^2})$. B is a base norm space with the positive cone B_+ consisting of all $V \geq 0$. The base of B is the set K of all $W \geq 0$ with $tr(W) = 1$. The dual Banach space B' of B is the space of all (bounded) Hermitean operators A normed by the usual operator norm. The bilinear form of this duality is $tr(VA)$. B' is an order unit space, the order unit being the "unit operator" $\mathbb{1}$.

1. Measuring Collisions *

A very illustrating example for a measuring collision is the following (Fig.1): We want to observe an electron by a microscope. For this purpose a photon is scattered by the electron in the object plane of the microscope. The scattered photon can blacken silver grain on a photoplate put in the image plane or in the focal plane. If we have the "signal" of a black grain in the image plane we can determine the position of the electron (Fig. 1). If we have the "signal" of a black grain in the focal plane we can calculate the change of momentum of the electron by the collision with the photon. However, these conclusions are wrong if one photoplate is put in the image plane and another in the focal plane because the photoplate in the focal plane disturbs the photon on its way to the image plane. We will say: the two measuring processes

a) only a photoplate in the image plane

b) only a photoplate in the focal plane

are not "coexistent".

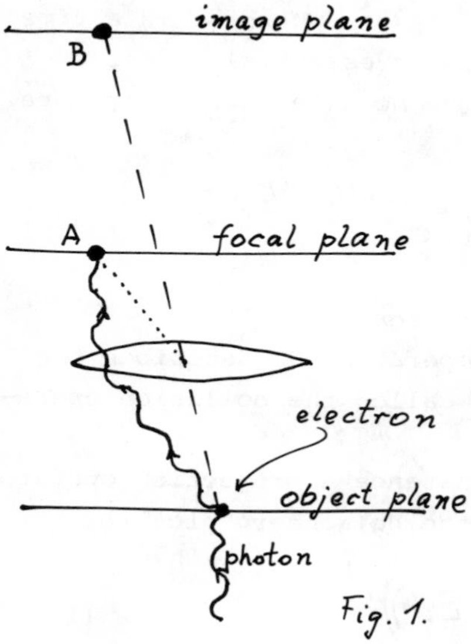

Fig. 1.

There is no principal difficulty in describing the collision of the photon (2) and the electron (1) by q.t.. After the collision the photon (2) is measured by the photoplate (case a or b). Therefore we will give a general framework of such measuring collisons:

System (1) shall be measured by a collision with system (2) and a measurement of system (2) after this collision. The Hilbert spaces of systems (1) and (2) will be denoted by $\mathcal{H}_1$ and $\mathcal{H}_2$ respectively. The Hilbert space of the composed system is $\mathcal{H}_1 \times \mathcal{H}_2$.

---

* For details see [5]

The initial ensemble, the ensemble before the collision takes place, is

$$W^i(t) = e^{-i(H_1+H_2)t}(W_1 \times W_2)e^{i(H_1+H_2)t} \tag{1.1}$$

The ensemble is considered in the Schrödinger picture with H_1 and H_2 as the free Hamiltonians of the systems 1 and 2 respectively, i.e. without interaction V_{12} of the two systems.

$W^i(t)$ is valid for times t with $-\infty < t < \tau^i$ where τ^i is a time before the interaction of the systems becomes essential. After a time τ^f, i.e. for $\tau^f < t < \infty$ there will be no interaction any more, so that we can write:

$$W^f(t) = e^{-i(H_1+H_2)t} S(W_1 \times W_2)S^+ e^{i(H_1+H_2)t} \tag{1.2}$$

There is no need of an expression for W(t) for times $\tau^i < t < \tau^f$ where the interaction is essential. The operator S describes the total interaction, the collision, and is called the collision operator. S is a unitary operator.
After the collision we can measure for instance a projection operator E_2 of system (2). The probability is in the Heisenberg picture:

$$\mathrm{tr}\left(S(W_1 \times W_2)S^+(1 \times E_2)\right). \tag{1.3}$$

This is the basic expression concerning measuring collisions. Now I will give some useful definitions:
An ensemble W of the composite system (1)(2) is an element of the base K_{12} of the Banach space B_{12}. If $1 \times A$ denotes the product operator in B'_{12} with $A \in B'_2$ there is a reduction mapping R_2: $B_{12} \longrightarrow B_2$ defined by

$$\mathrm{tr}\left((R_2 W)A\right) = \mathrm{tr}\left(W(1 \times A)\right) \tag{1.4}$$

such that $R_2 \; K_{12} \subset K_2$.

A linear mapping of one base of a base norm space into the base of another base norm space will be called a mimorphism. Mimorphisms are norm continuous. R_2 is a mimorphism. Likewise $W_1 \times W_2 \xrightarrow{\Sigma_{12}} S(W_1 \times W_2)S^+$ defines a mimorphism $\Sigma_{12}: K_{12} \longrightarrow K_{12}$. The following composed mapping is a mimorphism $\tilde{T}(1, 2; W_2): K_1 \longrightarrow K_2$:

$$\tilde{T}(1, 2; W_2): W_1 \longrightarrow W_1 \times W_2 \longrightarrow S(W_1 \times W_2)S^+ \longrightarrow R_2[S(W_1 \times W_2)S^+].$$

The adjoint operator T (1, 2;W_2) maps B_2' into B_1' and the order interval $L_2 = [0_2, 1_2]$ into the order interval $L_1 = [0_1, 1_1]$.

Therfore we have

$$tr\left(S(W_1 \times W_2)S^+(1 \times E_2)\right) = tr(W_1 F) \tag{1.5}$$

with $F = T(1, 2; W_2)\, E_2 \in L_1$.

The physical interpretation of (1.5) is as follows: The measurement of E_2 on system (2) is equivalent to an "effect" F of system (1). In general F will <u>not</u> be a projection operator. It is very difficult to realize collisions experimentally such that F will be a projection operator. How shall an experimental physicist know whether he measured a projection operator F or not? Since this is not a trivial question I have proposed to consider all "effects" $F \in L_1$ as possible yes-no measurements. As the word yes-no measurement can be misunderstood as reserved for projection operators I will only use the expression "effect". The effects F which are projection operators will be called "decision effects".

It is clear that E_2 in (1.3) and (1.5) can also be replaced by a general effect, and T(1, 2;W_2) will determine an effect F in the same way.

The mapping T(1,2;W_2) will be called the measuring collision morphism.

If one has several commuting projection operators $E_2^{(\lambda)}$ (commensurable projection operators) one can easily construct the Boolean ring generated by these $E_2^{(\lambda)}$. Suppose the set $\{E_2^{(\lambda)}\}$ itself is a Boolean ring.

Then the mapping $T(1,2;W_2)$: $E_2^{(2)} \longrightarrow F^{(2)}$ is an additive mapping, a so-called additive measure on the Boolean ring. This situation is generalized by means of the following definition (see [1]):
A set M of effects is called coexistent if there is a Boolean ring Q and an additive mapping $Q \longrightarrow L$ such that M is contained in the range of this mapping. We easily see that $T(1, 2; W_2)$ maps a set of coexistent effects into a set of coexistent effects. Coexistence is preserved by measuring collision morphisms.
Also it can be proved that, given a set of coexistent effects $F^{(2)} \in L_1$, there exists W_2, a set of commensurable projections $E_2^{(2)}$ and a unitary operator S such that there is a mapping $E_2^{(2)} \longrightarrow F^{(2)}$ of the form $T(1, 2;W_2)$. The physical problem, however, is much more difficult since we are not free in choosing the collision operator S. The collision operators are completely determined by the structure of nature.
A second property of expression (1.3) is very essential for measuring processes. Putting

$$W^f = S(W_1 \times W_2)S^+ \tag{1.6}$$

W^f is the ensemble of the composed systems (1)(2) after the collision. The probability for measuring an effect F_1 on system (1) <u>after</u> the collision is

$$tr(W^f(F_1 \times 1)) = tr((R_1 W^f) F_1) \tag{1.7}$$

where R_1 is the reduction operator concerning system (1) and $W_1{}^f = R_1 W^f$ is the ensemble of system (1) after the collision.
It is possible to decompose the ensemble $W_1{}^f$ by measurements on system (2) alone as, for instance, the example of the microscope shows. To see this, we consider a measurement of an effect $F_1 \times F_2$ after the collision. The effect $F_1 \times F_2$ corresponds to a measurement of the occurence of effect F_1 <u>and</u> effect F_2. The probability for this combined effect is $tr(W^f(F_1 \times F_2))$.
We are interested in the following experiment: We select all those cases, i.e. all those systems (1), where F_2 responds with "yes" and look for the frequency, that in <u>these</u> cases the effect F_1 gives "yes", too. This frequeny is

$$\frac{tr\,(W^f\,(F_1 \times F_2))}{tr\,(W^f(1 \times F_2))}. \tag{1.8}$$

If $tr\,(W^f(1 \times F_2)) \neq 0$ it will be shown that there is an ensemble $\overline{W}_1^f \in K_1$ such that

$$\frac{tr\,(W^f(F_1 \times F_2))}{tr(W^f(1 \times F_2))} = tr\,(\overline{W}_1^f F_1). \tag{1.9}$$

Moreover

$$W_1^f = \lambda \overline{W}_1^f + (1-\lambda)\, \overline{\overline{W}}_1^f \tag{1.10}$$

with $\overline{\overline{W}}_1^f$ determined by

$$\frac{tr\,(W^f\,(F_1 \times F_2))}{tr(W^f(F_1 \times (1-F_2)))} = tr(\overline{\overline{W}}_1^f F_1) \tag{1.11}$$

and

$$\lambda = tr\,(W^f(1 \times F_2)) = tr\,((R_2 W^f) F_2).$$

Hence we have decomposed the ensemble W_1^f into two parts $\overline{W}_1^f$ and $\overline{\overline{W}}_1^f$ corresponding to the occurrence or not-occurrence of F_2. Note that this decomposition only depends on a measurement on system (2) after the collision!

In order to prove this we define the following mapping

$$\phi(F_2):\; W \longrightarrow (1 \times F_2^{1/2})\, W (1 \times F_2^{1/2}) \quad \text{for} \quad W \in K_{12}.$$

$\phi(F_2)$ is a linear mapping $B_{12} \longrightarrow B_{12}$ which in general does not map K_{12} into K_{12}. Defining the cap $\check{K}$ of the cone B_+ generated by K to be

$$\check{K} = \bigcup_{0 \leq \lambda \leq 1} \lambda K$$

we have $K_{12} \xrightarrow{\phi} \check{K}_{12}$ and therefore also $K_{12} \xrightarrow{R_1 \phi} \check{K}_1$. The statements given above are proved by defining

$$\lambda \overline{W}_1^f = R_1 \Phi(F_2) W^f,$$
$$(1-\lambda) \overline{\overline{W}}_1^f = R_1 \Phi(1-F_2) W^f. \tag{1.12}$$

In the context of this decomposition of W_1^f another mapping is of great physical importance:

$$\mathcal{P}(1,F_2,W_2): \; W_1 \longrightarrow W_1 \times W_2 \longrightarrow S(W_1 \times W_2)S^+ = W^f \longrightarrow$$
$$\longrightarrow \Phi(F_2)W^f \longrightarrow R_1 \Phi(F_2) W^f = \lambda \overline{W}_1^f. \tag{1.13}$$

$\mathcal{P}(1,F_2,W_2)$ is a linear mapping of B_1 into B_1 with $K_1 \longrightarrow \check{K}_1$. In general a linear map $B_1 \longrightarrow B_2$ with $K_1 \longrightarrow \check{K}_2$ will be called an "operation". We note that there is no sudden jump from W_1 to $\lambda \overline{W}_1^f$. The considered physical process can completely be described in the framework of q.t. provided the measurement on system (2) is assumed to be known.
The factor λ is given by

$$\lambda = tr\left(\mathcal{P}(1,F_2,W_2)W_1\right) = tr(W_1 \, T(1,2;W_2)F_2)$$
$$= tr\left((R_2 W^f)F_2\right) = tr\left(W^f(1 \times F_2)\right). \tag{1.14}$$

Sometimes the mapping

$$W_1 \longrightarrow \frac{\mathcal{P}(1,F_2,W_2)W_1}{tr(\mathcal{P}(1,F_2,W_2)W_1)} \tag{1.15}$$

is of great interest, too. This mapping, defined only on the subset of K_1 where $\lambda \neq 0$ will be called a "normed operation". These normed operations are the starting point of Pool's approach to an axiomatic foundation of q.t..

The most troubling thing concerning the operations and the decompositions of ensembles is that there are different possibilities of decomposing the same W_1^f by a selection process according to the occurrence of effects of system (2). The example of the microscope

discussed at the beginning shows this feature:
If one selects according to the appearance of black silver grains on the photoplate in the image plane on can decompose W_1^f into subensembles with practically exact positions of the electrons. If W_1 practically has an exact value of momentum and one selects according to the appearance of black silver grains on a photo plate in the focal plane all decomposed subensembles of the <u>same</u> W_1^f practically have exact values of momentum, too. It is, however, not possible to realize these two decompositions together.
For together or not together realized decompositions we can give a mathematical description:
Two decompositions

$$W = \lambda W' + (1-\lambda) W'' \quad \text{and} \quad W = \mu \overline{W}' + (1-\mu) \overline{W}''$$

are called <u>compatible</u> if there is a Boolean ring Q and an additive mapping: $Q \rightarrow \check{K}$ with $e \rightarrow W$, e denoting the unit of Q, so that $\lambda W'$, $(1-\lambda) W''$, $\mu \overline{W}'$, $(1-\mu) \overline{W}''$ are contained in the range of this mapping.

Likewise a set M of operations $\mathcal{P}$ is called compatible if there is a Boolean ring Q and an additive mapping of Q into the set of positive mappings of B into B with norm ≤ 1, so that M is contained in the range of this mapping.

To consider a set of coexistent effects of system (2) let $F_2(q)$ be an additive measure on a Boolean ring Q with $F_2(e) = 1$. For $W_2 \in K_2$ fixed $\mathcal{P}(q) = \mathcal{P}(1, F_2(q), W_2)$ is an additive measure on Q. Therefore the $\mathcal{P}(q)$, $q \in Q$, are compatible operations and all decompositions

$$W_1^f = \mathcal{P}(e) W_1 = \mathcal{P}(q) W_1 + \mathcal{P}(e \dotplus q) W_1$$

are compatible.
In general, however, there are non compatible operations $\mathcal{P}(1, F_2, W_2)$ with the same W_2 and S and corresponding non compatible decompositions of W_1^f. After the collision of systems (1) and (2), i.e. when the collision has been already finished, one can choose between non coexistent effects, non compatible operations and non compatible decompositions!

Finally we may say that the measuring collisions cannnot solve all problems of the measuring process:

1. It is impossible to see whether a measuring process is completely finished because it is still necessary to measure with respect to the scattered system (2). In order to measure the scattered system (2) one has to carry out a new collision with a system (3) and to measure with respect to the system (3), and so on without coming to an end of this chain.

2. After the measuring collision there are various possibilities of non coexistent effects, various possibilities of non compatible decompositions and operations. However, if in nature a measuring process is finished there are no such possibilities at all. For instance, if a silver grain is "blackened" by a photon this effect is fixed, is registered, and there are no other possibilities than those concerning the registered events. By the measuring collision we get no solution of the problem of "registration".

There are various solutions offered. One solution, for instance, is that there is no completion of the measuring process and no registration other than in our consciousness. I do not think this a correct explanation of the problem. I can't see why, for instance, a registration of measuring results in a computer cannot be taken as objectively given without any knowledge of the results in my consciousness. Therefore I will try to give another solution, namely the solution that the measuring process is completed in the macroscopic region. We have, however, to pay much for this explanation, or rather, we gain much by this explanation (whatever you prefer). We have to pay by abandoning q.t. as the most comprehensive theory describing also macroscopic systems. We gain the possibility of finding new physical theories and of getting a better insight into the structure of the world.

2. Macroscopic Systems *

Without looking at q.t. we will give a general frame for describing macroscopic systems corresponding to the experiments carried out with these systems. The experiments with the macroscopic systems are described in an objective manner, in a so-called classical way. The question arises whether it is possible to give a general framework for such classical descriptions.

---

* For details see [6]

We start by postulating a state space $\mathcal{Z}$. The points of this state space are assumed to represent objective qualities of the systems. For instance, in classical mechanics of n mass points $\mathcal{Z}$ is the 3n-dimensional space of the 3n position coordinates of the mass points. In the mechanics of an imcompressible fluid the points of $\mathcal{Z}$ are the functions of the velocity field $\vec{u}(\vec{r})$ and $\mathcal{Z}$ is a function space. In usual q.t. there is no possibility to describe a single microscopic system in such a state space (see [1] and [7] , volume 3). In macroscopic physics, however, we describe a single macroscopic system by a trajectory $z(t)$ in $\mathcal{Z}$, t being the time parameter. For a system of n mass points we have trajectories: $x^i_\nu(t)$, i = 1,...n, ν = 1, 2,3. For the incompressible fluid we have trajectories $\vec{u}(\vec{r},t)$.

It is a general feature of physics that a set, the elements of which have a physical interpretation, has a physically relevant uniform structure p (see [1] and [6]). The physical relevance of this uniform structure rests upon the fact that the mathematical set is an inexact picture of the situation in nature: A vicinity of the uniform structure is a good measure for this inexact relation between mathematics and nature.
$\mathcal{Z}$ equipped with the uniform structure describing this inexactness will be denoted by $\mathcal{Z}_p$. Let $\hat{\mathcal{Z}}_p$ be the completion of $\mathcal{Z}_p$. We assume $\hat{\mathcal{Z}}_p$ to be compact (thus $\mathcal{Z}_p$ is precompact). Also in general this is a physically reasonable assumption. For almost all cases of physical theories $\mathcal{Z}$ carries an additional uniform structure g , finer than p , yielding the same topology in $\mathcal{Z}$. We assume $\mathcal{Z}_g$ to be complete. The uniform structure g does not have a direct physical meaning. g is an idealization, resulting from other physically meaningful but mathematically idealized structures as, for instance, group structures.

In order to see what is meant by p and g we consider the state space $\mathcal{Z}$ of n mass points. Then $\mathcal{Z} = X^n$, X being the three-dimensional Euklidean space. g shall be the usual uniform structure of this space and $\mathcal{Z}_g = X^n_g$. p can be constructed as follows: Consider a four-dimensional sphere and the projection of X onto the sphere (see fig. 2). The corresponding uniform structure on the sphere gives a uniform structure p on X. We put $\mathcal{Z}_p = X^n_p$. X_p and $\mathcal{Z}_p$ are not complete but precompact.

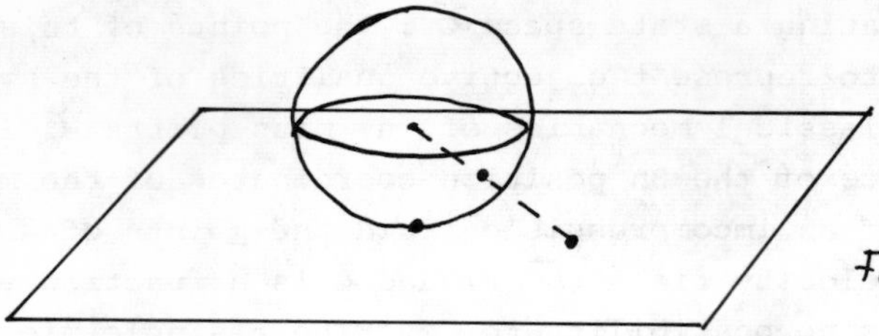

Fig. 2.

The uniform structure p is physically meaningful because it reflects the difficulty in measuring large distances.

It is not possible to give all the essential physical reasons here to justify the following postulates. (These reasons are very general and have nothing to do with the special situation of the state space $\mathcal{Z}$.) We postulate: $\mathcal{Z}_p$ is separable (then the same holds for $\mathcal{Z}_g$) and the uniform structures p and g have a denumerable base, so that $\mathcal{Z}_g$ and $\mathcal{Z}_p$ are metrizable. It is easy to show: for every (separable and metrizable) $\mathcal{Z}_g$ there is a uniform structure p which fulfills all the mentioned conditions relative to g.

A physical system is described by a trajectory z (t) in $\mathcal{Z}$. Since we know the feature of irreversibility, we only consider the time $t > 0$. For times $t \leq 0$ we don't make assumptions concerning the behaviour of the macroscopic system. It can happen that the system did not exist as a system to be described in $\mathcal{Z}$ for times $t \leq 0$ because, for instance, it has been built up by an experimental physicist. Therefore we take for the timescale only the set $\Theta = \{t \mid 0 < t < \infty\}$. We want to construct a "physically meaningful" trajectory space. There are many physical reasons not to take $\mathcal{Z}^{\Theta}$; but we don't have the time to explain these reasons.

As trajectory space we take the space $\mathcal{Y} = C(\Theta, \mathcal{Z})$ of all continuous z (t). What could be the physically useful uniform structures of $\mathcal{Y}$? We begin with the uniform structure p of $\mathcal{Z}$, the uniform structure of the physical inaccuracy.

A fundamental system of vicinities of a uniform structure on $\mathcal{Y}$ is defined by

$$\tilde{\mathcal{U}}_{\vartheta} = \{(y, y') \mid \mathcal{G}_{\vartheta}(y) \subset \mathcal{U}\mathcal{G}(y') \text{ and } \mathcal{G}_{\vartheta}(y') \subset \mathcal{U}\mathcal{G}(y)$$
with $\mathcal{U} \in \mathcal{N}$, $\vartheta \in$ set of all compact subsets of $\Theta\}$;
$\mathcal{G}$ (y) is the graph of y = z (t) in $\Theta \times \mathcal{Z}$, $\mathcal{G}_{\vartheta}(y)$ the graph of y = z (t) in $\vartheta \times \mathcal{Z}$, ϑ being a subset of Θ. $\mathcal{N}$ is the vicinity filter of the uniform structure of $\Theta \times \mathcal{Z}_p$.

This uniform structure will also be called p and $\mathcal{Y}$ equipped with this structure will be denoted by $\mathcal{Y}_p$. $\mathcal{Y}_p$ can be proved to be pre-compact and metrizable [6].

Starting with g on Z we can introduce another uniform structure g on $\mathcal{Y}$: $\mathcal{Y}_g = C_{gc}(\Theta, \mathcal{Z})$ with gc as the uniform structure of uniform convergence (relative to $\mathcal{Z}_g$) on all compact subsets of Θ.

Also in $\mathcal{Y}$ the uniform structure g is finer than p. $\mathcal{Y}_g$ is complete; $\mathcal{Y}_g$ and $\mathcal{Y}_p$ have the same topology; $\mathcal{Y}_g$ and $\mathcal{Y}_p$ are metrizable and separable. Hence we have the same situation in $\mathcal{Y}$ and in $\mathcal{Z}$. That is very satisfactory.

The aim of all these mathematical descriptions is a physically very significant problem namely to describe quite generally the dynamical laws of macroscopic systems. We know examples of these laws: the canonical equations for point mechanics, the Navier-Stokes equations for fluid mechanics, the Master equations for the statistical development of some systems and so on. The method of description we want to propose is mathematically similar to that in q.t. We will define "ensembles" and "effects" and the "probability" of effects in ensembles. In this way, the mathematical structure of base norm spaces and the dual order unit spaces will be very essential also for the description of macroscopic systems.

The first step into this direction is to define "trajectory effects". For this purpose we introduce a set M_m with the physical interpretation: M_m is the set of all individual macroscopic (therefore the index m) systems describable in the same state space $\mathcal{Z}$. Then there will be a mapping h of M_m into Y with the physical meaning: h (x) is the trajectory of the system x. Let H be the graph of the mapping h.

Now we introduce the concept of "registration procedures" by a set R. An element r of R is a pair (ρ, η) of functions. ρ is a mapping $\rho : M_m \longrightarrow \{0, 1\}$ and $\rho(x) = 1$ means the registration procedure r is employed on the system x. η is defined only for those y with $\varrho(x) = 1$ and $(x,y) \in H$. Also $\eta(y)$ has only the values 0,1 with the physical interpretation: $\eta(y) = 1$ means the registration procedure has given a "positive" registration of the system.

As we will see later these registration procedures are very essential for the application of macroscopic systems to measure microscopic systems.

Let $C(\hat{\mathcal{Y}})$ be the Banach space of all continuous real-valued functions on $\hat{\mathcal{Y}} = \hat{\mathcal{Y}}_p$. $C_+^1(\hat{\mathcal{Y}})$ denotes the order interval $[0, 1]$ of $C(\hat{\mathcal{Y}})$.

We don't have the time to give all the axioms (introduced for physical reasons) for the set $\mathcal{R}$ of registration procedures. One of these axioms, however, is necessary to understand the measuring process. This axiom is the following: There is a surjective mapping $\mathcal{R} \longrightarrow C_+^1(\hat{\mathcal{Y}})$ with the physical interpretation: $r = (\varrho, \eta) \longrightarrow f(y) \in C_+^1(\hat{\mathcal{Y}})$ means $f(y)$ is the frequency of the "registration-values" $\eta(y) = +1$ for the <u>same</u> trajectory y. In practice, most of all registration procedures will be characterized by functions $f(y)$ which are very similar to characteristic functions of subsets of $\hat{\mathcal{Y}}$: the set $\{y \mid f(y) = 1\}$ is the set of all y, which will be registered with certainty; $\{y \mid f(y) = 0\}$ is the set of all y, which will not be registered. There will be only a small region $\{y \mid f(y) \neq 1, 0\}$ where the registration is uncertain.

We define $C_+^1(\hat{\mathcal{Y}})$ is the set of "trajectory effects".
The second step is to define "macroscopic ensembles". In general the mapping $h: M_m \rightarrow Y$ will not be surjective because not all mathematically possible trajectories will be "real trajectories in nature". The so-called dynamical laws are axioms for the "real trajectories". How to describe these laws in general?

Similarly to the set R of "registration procedures" we introduce a set Q of "preparing procedures". This concept of "preparing procedures" is a very basic one for all statistical theories. Q is a subset of $\mathcal{P}(M_m)$, satisfying some axioms not given here (see [1], [2] and the article of Hartkämper, this volume). A subset $q \subset M_m$ with $q \in Q$ has the physical meaning that q is an experimental method to construct and select a "large" number of individual systems in order to observe the frequencies of the various trajectories of these systems. In classical mechanics, for instance, a gun is such a "preparing procedure" for projectiles; and one can be interested in the frequencies of the various trajectories of these projectiles.

A subset $w \subset M_m$ defined by:

$$w = \{ x \mid x \in q \in Q, \; \varrho(x) = 1, \; r = (\varrho, \eta) \in R \}$$

will be called the experiment determined by the preparing procedure q and the registration procedure r.

A physicist can perform such an experiment with N systems $x_1, x_2, \ldots, x_N$. He will get a number $N_+ = \sum_{i=1}^{N} \eta(y_i)$ of positively registered systems where y_i is the trajectory of the system x_i, that is $(x_i, y_i) \in H$. So he can measure the frequency N_+/N. These frequencies are the experimental results which describe the statistical and dynamical behaviour of the observed systems. In the example of the gun as preparing procedure , for instance, we can take several registration procedures r for the trajectories and so we can experimentally find the frequencies of the various trajectories y of the projectiles. In the sequel we shall give a mathematical description of these frequencies N_+/N.

We postulate that there is a real-valued function $0 \leq \mu_m(q, r) \leq 1$ on $Q \times \mathbf{R}$ such that the experimental result will be $N_+/N \approx \mu_m(q, r)$ for large numbers N. The physical meaning of $r = (\varrho, \eta)$ is that only the function $\eta(y)$ depends on the form of registration. Thus $\mu_m(q, r)$ is assumed to be a function only of that $f(y) \in C^1_+(\hat{\mathcal{Y}})$ which is associated with the registration procedure. Therefore we will write $\mu_m(q, f)$ instead of $\mu_m(q, r)$. We assume that $\mu_m(q, f)$ is a function on $Q \times C^1_+(\hat{\mathcal{Y}})$. It is possible to weaken this assumption but, at the moment, it is more convenient to introduce this strong formulation.
By $\mu_m(q_1, f) = \mu_m(q_2, f)$ for all $f \in C^1_+(\hat{\mathcal{Y}})$ an equivalence relation $q_1 \sim q_2$ is defined. Let K_m be the set of equivalence classes. μ_m can be considered as a function on $K_m \times C^1_+(\hat{\mathcal{Y}})$. An element $u \in K_m$ will be called "ensemble" and K_m will be called the set of "macroscopic ensembles". Because of the physical meaning of $f(y) \in C^1_+(\hat{\mathcal{Y}})$ we postulate

$$\mu_m(u, f_1 + f_2) = \mu_m(u, f_1) + \mu_m(u, f_2) \text{ if } f_1 + f_2 \in C^1_+(\hat{\mathcal{Y}}).$$

Thus $\mu_m(u, f)$ defined on the entire Banach space $C(\hat{\mathcal{Y}})$ is a positive linear functional, and therefore μ can be identified with an element of the dual Banach space $C'(\hat{Y})$. Denote by K' the base of the base norm space $C'(\hat{Y})$, that is K' is the set of all normed positive measures. Thus K_m can be identified with a subset of K', and μ_m can be considered as the canonical bilinear form of the duality $C'(\hat{Y}), C(\hat{Y})$.

Axioms for the set Q of preparing procedures yield that K_m is convex and norm separable. Without loss of physical generality K_m can be assumed to be norm closed.

K_m will be a proper subset of K'. It is this subset K_m which determines the statistical <u>and</u> the dynamical laws of the systems.

The method of introducing the dynamical laws by a subset K_m of K' is so unusual that it seems necessary to explain in what form the dynamical laws are incorporated in the subset K_m.

In general K_m is not a separating subset of C'(Y). Therefore we introduce the Banach space:

$$D_m = C(\hat{Y})/K_m^{\perp}$$

where $K_m^{\perp}$ is the subspace of $C(\hat{Y})$ orthogonal to K_m.

By means of the canonical mapping $\chi : C(\hat{Y}) \longrightarrow C(\hat{Y})/K_m^{\perp}$ we define $\underline{L}_m = \chi(C_+^1(\hat{Y}))$. $\underline{L}_m$ will be called the set of "macroscopic effects".

The dual Banach space D'_m can be identified with the subspace $K_m^{\perp\perp}$ of $C'(\hat{Y})$. $K_m^{\perp\perp}$ is the $\sigma(C', C)$-closed subspace spanned by K_m. The <u>norm closed</u> subspace spanned by K_m is denoted by B_m : B_m is norm separable since K_m is norm separable. D'_m is the $\sigma(C', C)$-closure of B'_m. We can identify D_m with a subspace of B'_m. The $\sigma(B'_m, B_m)$-closure L_m of $\underline{L}_m$ in B'_m will be called the set of macroscopic effects, too.

We postulate: $K_m = B_m \cap K'$. The particular physical meaning of this assumption will not be explained here. As a consequence K_m is the base of the base norm space B_m and L_m is the order interval $[0, 1]$ in B'_m.

We now have a "very similar" structure B_m, B'_m for macroscopic systems as in q.t. for microscopic systems. Though the theory of macroscopic systems cannot be deduced from an extrapolated q.t. for many particles there should be compatibility of the extrapolated q.t.

and the theory of macroscopic systems. It is a profound and difficult problem to show that the extrapolated q.t. and the theory of macroscopic systems are compatible. To solve this problem the similarity of the structure of dual spaces B_m, B'_m for macroscopic systems and B,B' for microscopic systems is of great significance. Prosperi, Lanz and coworkers developed the main approaches to show this compatibility [3] . We do not have the time to sketch this problem (see also [6]). We will only try to clarify in what form K_m determines the dynamical laws of the systems.

For this purpose it is necessary to introduce the following sets: Let $\overline{K_m}^{\sigma}$ be the $\sigma(C',C)$-closure of K_m in $C'(\hat{y})$. We have $K_m \subset \overline{K}_m^{\sigma} \subset K'$. Every $f(y) \in C_+^1(\hat{y})$ defines an open set $O_f = \{y \mid f(y) > 0\}$ and a complementary set $A_f = \{y \mid f(y) = 0\}$. The closure of O_f in $\hat{y}$ is called the support S_f of f.

Every $u \in K'$ defines sets $l_u = \{f \mid f \in C_+^1(\hat{y}), \mu_m(u,f) = 0\}$ and $\mathcal{U}_u$ as the greatest open set with $\{f \in C_+^1(\hat{y}), S_f \subset \mathcal{U}_u \Rightarrow f \in l_u\}$. The complementary set S_u of $\mathcal{U}_u$ is called the support of u.

Since K_m is norm separable, there is a denumerable set u_ν norm dense in K_m. With $\bar{u} = \sum_\nu \lambda_\nu u_\nu$, $\lambda_\nu > 0$, $\sum_\nu \lambda_\nu = 1$ we have $\bar{u} \in K_m$ and $\mu_m(\bar{u},f) = 0 \Longrightarrow \mu_m(u,f) = 0$ for all $u \in K_m$. $\hat{S}_m \overset{def}{=} S_{\bar{u}}$ is called the support of K_m. The physical meaning is obvious: Only trajectories $y \in \hat{S}_m$ can be realized; other trajectories $y \notin \hat{S}_m$ have probability "zero".

Now it is not difficult to go over from $\hat{y}$ to the compact subset $\hat{S}_m$ and from $C(\hat{y})$ to $C(\hat{S}_m)$. We find K_m as a subset of the base K'_s of $C'(\hat{S}_m)$ and $\mu_m(\bar{u}, f) = 0$ for $f \in C_+^1(\hat{S}_m)$ implies $f = 0$. Hence K_m is separating with respect to $C(\hat{S}_m)$. $\overline{K}_m^{\sigma}$ is a subset of K'_s.

All measures $u \in K'_s$ can be considered as σ-additive measures on the Borelring $\mathcal{L}(\hat{S}_m)$ of $\hat{S}_m$.

The extremal points of K'_s can be identified with the points of $\hat{S}_m$ (the extremal points of K'_s are "point measures"). The topology of $\hat{S}_m$ then equals to $\sigma(C',C)$ restricted to the set of the extremal points of K'_s.

In general $\overline{K}_m^{\sigma} \neq K'_s$, but if $\overline{K}_m^{\sigma} = K'_s$ every extremal point of K'_s can be physically approximated by a measure $u \in K_m$. Hence it is possible to select the systems by a "preparing procedure" in such a manner that all systems have practically the same trajectory. Therefore we give the definition: If $\overline{K}^{\sigma}_{m} = K'_{s}$ the dynamical laws are called "deterministic".

In general the dynamics of the systems are "statistical" in the following form: The dynamics of the systems are determined by the extremal points $\mathcal{U}_e$ of $\overline{K}^{\sigma}_{m}$. If the support $S_{\mathcal{U}_e}$ does not consist of one point only, $\mathcal{U}_e$ gives a probability distribution of trajectories: As discussed above in connection with $\hat{S}_m$ $\mathcal{U}_e$ can be considered as a probability measure on the Borelring $\mathcal{L}(S_{\mathcal{U}_e})$.

It would be an "exercise", to formulate classical mechanics corresponding to this general framework with $\mathcal{Z}$ as X. $\mathcal{Y}$ is the space $C(\theta, \mathcal{Z})$, that is the space of trajectories in $\mathcal{Z}$ (not in the phase-space Γ). K_m can be identified with the Borel measures in Γ, that is with the set of all measurable $\varrho(p_i, q_i) \geq 0$ with $\int \varrho(p_i, q_i)\, dp_1 \ldots \ldots dq_{3N} = 1$. $\hat{S}_m$ can be identified with a special compactification of Γ, where the points of Γ are the initial values of the trajectories in $\mathcal{Z}$ for time t = 0.

3. Macroscopic effects, caused by microobjects *

The first task is to find a description of how microobjects can influence the dynamical behaviour of macroscopic systems. Therefore we start with a general problem: The description of a system composed of a microsystem 1 and a macrosystem 2.

Let $\mathcal{Z}_2$ be the state space and $\mathcal{Y}_2$ the trajectory space of the macrosystem 2. $\hat{\mathcal{Y}}_2$ denotes the completion of $\mathcal{Y}_2$.

* For details see [5]

Let $\mathcal{H}_1$ be the Hilbert space of the microsystem 1. As described in the beginning of this article B_1 shall be the base norm space of all Hermitean trace class operators, B_1' the dual order unit space of all bounded Hermitean operators. The base K_1 in B_1 is the set of ensembles of system 1 and the order interval $L_1 = [0,1]$ in B_1' is the set of effects of system 1.

We need an additional physically relevant structure for the microsystem 1: As shown in [1] (see also the article of Neumann in this volume) there is a norm separable Banach subspace $D_1 \subset B_1'$ such that $D_1 \cap L_1$ is $\sigma(B_1', B_1)$-dense in L_1.

The next steps are very similar to the description of a macrosystem alone as sketched in section 2. The set $C(\hat{Y}_2, D_1)$ of all continuous mappings $\hat{Y}_2 \xrightarrow{F(y)} D_1$ (D_1 endowed with the norm-topology) is a Banach space with the norm

$$\|F(y)\|_{12} = \sup\{\|F(y)\| \mid y \in \hat{Y}_2\}$$

where $\|F(y)\|$ is the operator norm. If $C_+^1(\hat{Y}_2, D_1)$ denotes the order interval $[0,1]$ in $C(\hat{Y}_2, D_1)$ we have $C_+^1(\hat{Y}_2, D_1) = C(\hat{Y}_2, D_1 \cap L_1)$. We will write the bilinear form of the duality $C'(\hat{Y}_2, D_1), C(\hat{Y}_2, D_1)$ as $\mu_{12}(u, F(y))$ with $u \in C'(\hat{Y}_2, D_1)$. K_{12}' denotes the base of $C'(\hat{Y}_2, D_1)$.

The elements of $C_+^1(\hat{Y}_2, D_1)$ will be called the effects of the composed system. $\mu_{12}(u, F(y))$ is the probability of the effect $F(y) \in C_+^1(\hat{Y}_2, D_1)$ for $u \in K_{12m} \subset K_{12}'$. The subset K_{12m} determines the dynamics and the statistics of the composed system.

In connection with the measuring process one is interested in the following special effects: $F(y) = f(y)\,1$ with $f(y) \in C_+^1(\hat{Y}_2)$.

These are effects concerning only the macrosystem. By $\mu_{12}(u, f(y)1) = \mu_m(u_2, f(y))$ there is defined a mimorphism $R_2 : u \to u_2$ of $K'_{12} \to K'_2$ and $C'(\hat{Y}_2, D_1) \to \to C'(\hat{Y}_2)$.

In general the set $R_2 K_{12m}$ will not be the set K_{2m} of ensembles of the macrosystem without interaction with a microsystem. The microobject 1 may influence the trajectories of the system 2 and we are interested in such influences.

If we measure together a trajectory effect $f(y)$ of the macrosystem and an effect F of the microsystem (after the interaction between the microsystem and the macrosystem!) then $F(y) = f(y)F$.

By $\mu_{12}(u, f(y)F(y)) = \mu_{12}(u', F(y))$

for all $F(y) \in C(\hat{Y}_2, D_1)$ a norm continuous mapping $C'(\hat{Y}_2, D_1) \to C'(\hat{Y}_2, D_1)$ is defined which will be denoted by f, too. If $f \in C^1_+(\hat{Y}_2)$ then $fK'_{12} \subset \check{K}'_{12}$ where $\check{K}'_{12}$ is the cap of the cone $C'_+(\hat{Y}_2, D_1)$ determined by K'_{12}. With this definition we get the following equations:

$$\mu_{12}(u, f(y)F) = \mu_{12}(fu, F) = \mu_1(R_1 fu, F) \tag{3.1}$$

with R_1 as the reduction mimorphism of K'_{12} to the base of D'_1.

The expression

$$\frac{\mu_1(R_1 fu, F)}{\mu_{12}(u, f(y)1)} = \frac{\mu_1(R_1 fu, F)}{\mu_2(R_2 u, f)} \tag{3.2}$$

is the probability to measure the effect F on those microobjects selected by positive signals for the trajectory effect $f(y)$. Since this should be a meaningful ensemble of microobjects we postulate:

If $u \in K_{12m}$ and $f(y) \in C^1_+(\hat{Y}_2)$ then $R_1 fu \in \check{K}_1$ where $\check{K}_1$ is the cap defined by B_{1+} and the base K_1.

We see that the ensemble $R_1 u$ of the microobjects can be decomposed after the interaction with the macrosystem by means of a trajectory effect f of the macrosystem:

$$R_1 u = R_1 f u + R_1 (1-f) u.$$

$R_1 f$ is an operation: $K_{12m} \longrightarrow \check{K}_1$.

If $u \in K_{12m}$ is fixed $f \longrightarrow R_1 f u$ defines a mapping $C_+^1(\hat{y}_2) \longrightarrow \check{K}_1$. This mapping can be called a positive measure with values in $\check{K}_1$. It is possible to extend this measure to the Borel algebra $\mathcal{L}(\hat{y}_2)$. By $\sigma \rightarrow R_1 \eta_\sigma u$ with the characteristic function η_σ of $\sigma \in \mathcal{L}(\hat{y}_2)$ we get a σ-additive (in the norm topology of B_1) measure on $\mathcal{L}(\hat{y}_2)$ with values in $\check{K}_1$. A decomposition σ_i of $\hat{y}_2$ $(\sigma_i \in \mathcal{L}(\hat{y}_2), \bigcup_i \sigma_i = \hat{y}_2, \sigma_i \cap \sigma_j = \phi$ if $i \neq j)$ provides a decomposition

$$R_1 u = \sum_i R_1 \eta_{\sigma_i} u.$$

We can write

$$R_1 f u = \int f(y)\, d u_1(y) \qquad \text{with} \quad u_1(\sigma) = R_1 \eta_\sigma u.$$

All decompositions of the form

$$R_1 u = \sum_i R_1 \eta_{\sigma_i} u \tag{3.3}$$

are compatible. Only compatible decompositions of $R_1 u$ are possible by selections according to the trajectories of the macrosystem. Also all operations $R_1 \eta_\sigma$ are compatible.

The next step is to describe collisions between a microsystem 1 and a macrosystem 2. Before the collision the ensemble of the microsystem may be W_1^i and that of the macrosystem $u_2^i \in K_{2m}$. u_2^i would describe macrosystem 2 <u>without interaction</u> with the microsystem 1. Then $W_1^i \times u_2^i$ is an element of K'_{12} defined by

$$\mu_{12}(W_1^i \times u_2^i, F(y)) = \mu_2(u_2^i, tr(W_1^i \cdot F(y)))$$

So we can identify $K_1 \times K_{2m}$ with a subset of K'_{12} .

Similar to the case of the collision of two microsystems we describe the collision of the microsystem and the macrosystem by a "collision operator" Σ , i.e. a bilinear mapping Σ_{12} of $K_1 \times K_{2m}$ into $K_{12\,m}$. However, there is an essential difference:

$u = \Sigma_{12}(W_1^i, u_2^i) \in K_{12m}$ shall describe the macrosystem also <u>during</u> the interaction, i.e. during the total time Θ . The microsystem can be measured only "after" the collision, i.e. after the interaction. It is also possible to describe the case, where the microsystem is absorbed in the macrosystem, but we will not do this here.

After the collision, the ensemble of the microobjects is

$W^f = R_1 \Sigma_{12}(W_1^i, u_2^i)$. As described above it is possible to decompose the ensemble W^f by trajectory effects of the macrosystem $(\sigma_i \in \mathcal{L}(\hat{y}_2))$:

$$W^f = \sum_i R_1 \eta_{\sigma_i} \Sigma_{12}(W_1^i, u_2^i). \tag{3.4}$$

If f and u_2^i are fixed

$$W_1^i \longrightarrow R_1 f \Sigma_{12}(W_1^i, u_2^i) \in \check{K}_1 \tag{3.5}$$

is an operation $K_1 \rightarrow \check{K}_1$. <u>All these operations are compatible</u>. Caused by the interaction with a macrosystem <u>only compatible opera-</u>

<u>tions are possible</u> (if the same $\mathcal{U}_2^i$ is considered). Other $\mathcal{U}_2^i$ (i.e. other macrosystems) can give different decompositions not compatible with the foregoing decomposition.

Now, we will show that every macrosystem defines an observable of the microsystem. The special effects $f(y)1$ have the probability:

$$\mu_{12}(\Sigma_{12}(W_1^i, u_2^i), f(y)1) = \mu_2(R_2\Sigma_{12}(W_1^i, u_2^i), f(y)). \tag{3.6}$$

If $\mathcal{U}_2^i$ and f are fixed (3.6) defines a positive linear functional on K_1; therefore we get a linear mapping: $T(1,2;\ u_2^i)$ of $C_+^1(\hat{\mathcal{Y}}_2)$ into L_1:

$$T(1,2;u_2^i)f(y) = F$$

with
$$\mu_2(R_2\Sigma_{12}(W_1^i, u_2^i), f(y)) = \mu_1(W_1^i, F) = tr(W_1^i F). \tag{3.7}$$

This mapping $T(1,2;u_2^i)$ can be extended to the characteristic functions η_σ of the elements $\sigma \in \mathcal{L}(\hat{\mathcal{Y}}_2)$:

$$T(1,2;\mathcal{U}_2^i)\eta_\sigma(y) = F(\sigma). \tag{3.8}$$

$F(\sigma)$ is a σ-additive (in the $\sigma(\mathcal{B}_1', \mathcal{B}_1)$ -topology) measure on $\mathcal{L}(\hat{\mathcal{Y}}_2)$ with values in L_1. If $\mathcal{J}$ is the ideal of all sets of measure 0, then $Q = \mathcal{L}(\hat{\mathcal{Y}}_2)/\mathcal{J}$ is a Boolean algebra and by

$$q \in Q:\quad F(q) = F(\sigma) \text{ if } \sigma \in q$$

we get an effective additive measure on Q. Q is complete in the measure topology and as a Boolean algebra.

The <u>general definition</u> of an observable (see [1]) is given by a pair $(Q, F(q))$ of a complete Boolean algebra Q and an additive measure $F(q): Q \longrightarrow L_1$.

We see that — u_2^i being fixed - a macrosystem as a measuring system defines one and only one observable. There is no possibility to measure any other observable by system 2 than this one determined above. So we can say that a measuring process is finished, if a collision with a macrosystem has taken place.

The problem of calculating the collision operator Σ_{12} can not be solved in general. On one hand the experimental physicists try to do this by combining quantum theoretical considerations for the first part of the interaction between the microsystem and some atoms of the macrosystems with phenomenological and statistical considerations and known macroscopic theories as for instance the theory of electronics. On the other hand the theoretical physicists try to show that this description by a collision operator Σ_{12} is compatible with the extrapolated q.t. (see [4] and the article of Prosperi in this volume).
Let us consider now another problem: What are the physical structures, by which we are allowed to introduce the concept of atoms, electrons and so on? To answer this question it is necessary to describe also the preparing processes.

A preparing system is a system composed of a macrosystem and a microsystem where no microsystem is absorbed but a microsystem is emitted. If we registrate such macro-preparing systems it can happen that we do not detect the microsystems. If one considers, for instance, an X-ray-tube, it can happen that nothing is observed except the macrosystem. To describe that thesemacrosystems emit microsystems, we can employ the general description of systems composed of a macrosystem and a microsystem given in the beginning. We will see that one can introduce a preparing operator, by which the preparing process will be described mathematically.

Let $u \in K_{12\,m}$ be an ensemble of the composed system (1: microsystem, 2: macrosystem). $u_2 = R_2 u \in K_2'$ describes completely the probabilities for trajectory effects of the macrosystem. u_2 and these trajectory effects are the only possibilities to select preparing processes for statistical experiments. Therefore one can hope that the following postulate describes the physical situation:

The mapping R_2: $K_{12m} \longrightarrow K_2'$ is injective. This means that all the statistics of possible measurements are already determined by $R_2 u = u_2$.

We define in the sense of the description of macrosystems:

$$R_{2m} = R_2 K_{12m}.$$

Since R_2 is injective, there exists a mimorphism $\Gamma : K_{2m} \longrightarrow K_{12m}$ which is the reciprocal of R_2. We call Γ the "preparing operator".

We see immediately that the preparing system prepares the ensemble $W_1 = R_1 \Gamma u_2$, if one does not select according to possible trajectory effects of the system 2. By selections according to trajectory effects of the system 2 this W_1 can be decomposed, as for instance $(f \in C_+^1(\hat{y}_2))$:

$$W_1 = R_1 \Gamma u_2 = R_1 f \Gamma u_2 + R_1 (1-f) \Gamma u_2 .$$

In this way, it is also possible to prepare the normed ensemble (if $\mu_2(u_2, f) \neq 0$):

$$\bar{W}_1 = \frac{R_1 f \Gamma u_2}{\mu_2(u_2, f)}$$

by selecting all those cases, where the trajectory effect f has happened.

The calculation of the "preparing operator" Γ is a problem as complicated as that of calculating the "collision operator" Σ_{12}. All what has been said above about the problem of calculating Σ_{12} could be repeated here.

We want to summarize the essential results of this section, because these will be the starting point for the investigations of section 4, where we hope to show the physical structures, which allow to speak of microsystems: microsystems as the "mediators of interaction" between the preparing systems and the measuring systems. These essential results are:

The preparing system is described macroscopically by $u_2 \in K_{2m}$; the "preparing operator" Γ gives a $u = \Gamma_2 u_2 \in K_{12m}$, which determines the total statistical description. If one selects the prepared microsystem according to a trajectory effect $f(y) \in C^1_+(\hat{y}_2)$ one gets the prepared ensemble $(\text{if } \mu(u_2, f) \neq 0)$:

$$W_1 = \frac{R_1 f \Gamma u_2}{\mu(u_2, f)}. \tag{3.9}$$

As a measurement we define a collision between macrosystems (3) of an ensemble u_3^i and microsystems (1) of an ensemble W_1^i (for the measuring macrosystems we use now the index 3):
This collision can be described by a "collision operator" Σ_{13} .

$u = \Sigma_{13}(W_1^i, u_3^i) \in K_{13m}$ is the ensemble of the composed system. The signals of a trajectory effect $f(y)$ of the macrosystem are equivalent to an "effect" $F \in L_1$ of the microsystem "induced" by the macrosystem: The mapping $T(1,2;\ u_3^i)$ of $C^1_+(\hat{y}_3)$ in L_1 is determined by:

$$T(1,2;\ u_3^i) f(y) = F$$

$$\text{with } \mu_3(\Sigma_3(W_1^i, u_3^i), f(y)) = \mu_1(W_1^i, F) \tag{3.10}$$

and $\Sigma_3 \overset{def}{=} R_3 \Sigma_{13}$. In section 4 we will no longer be interested in $R_1 \Sigma_{13}(W_1^i, u_3^i)$. We will use only the mapping

$$\Sigma_3 = R_3 \Sigma_{13} : \quad K_1 \times K_{3m} \xrightarrow{\Sigma_3} K_3'$$

where K_{3m} is the set of ensembles of system 3 without interaction with other systems. It is also possible to describe the more general case, but this would be much more complicated, because it would be necessary to take into consideration more than two interacting macrosystems.

4. Experiments composed of a preparing system and a measuring system *

We will now describe the following experiment: A preparing system prepares an ensemble of microobjects, and these microobjects collide with a measuring system. It is not difficult to describe this experiment by the preparing operator and collision operator. Let 2 be the preparing system, 1 the microsystem and 3 the measuring system. 2 and 3 are macrosystems.

The preparing process can be described by (3.9):

$$u_{12} = \Gamma_2 u_2 , \quad W_1 = (\mu_2(u_2, f))^{-1} R_1 f \Gamma_2 u_2 ; \tag{4.1}$$

the measuring process can be described by:

$$u_{13} = \Sigma_{13} (W_1^i, u_3^i). \tag{4.2}$$

In order to extend the definition of Σ_{13} we define on $K_{2m} \times K_1 \times K_{3m}$:

$$\Sigma_{13} (u_2, W_1, u_3) = (u_2, \Sigma_{13} (W_1, u_3)) \in K'_{123} .$$

This definition can be extended to the whole set $K_{12m} \times K_{3m}$ since K_{12m} is contained in the norm closed convex hull of $K_{2m} \times K_1$ in K'_{12}. In this way Σ_{13} is defined for $(\Gamma_2 u_2, u_3^i)$.

$$u = \Sigma_{13} (\Gamma_2 u_2 , u_3^i)$$

is an ensemble $u \in K'_{123}$ which describes the preparing system 2 and the measuring system 3 for the full time of the experiment and describes the microsystem 1 after the collision of 1 and 3.

---

* For details see [5]

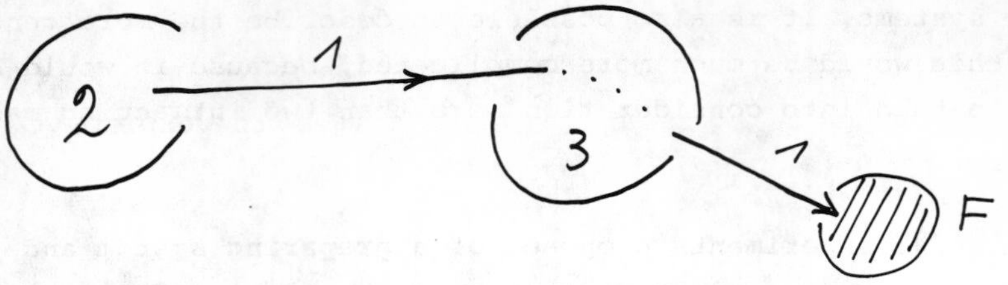

If we don't measure an effect F on the microsystems 1 after the collision with 3 the total system can be looked at as a composed interacting macrosystem, composed of 2 and 3. This composed system can be described in the statespace $\mathcal{Z}_{23} = \mathcal{Z}_2 \times \mathcal{Z}_3$ and in the trajectory space $\hat{\mathcal{Y}}_{23} = \hat{\mathcal{Y}}_2 \times \hat{\mathcal{Y}}_3$. The statistics of the trajectories is given by

$$u_{23} = R_{23}\, u = R_{23} \sum_{13} (\Gamma_2 u_2, u_3^{i}) \in K'_{23}$$

with R_{23} as the reduction operator $K'_{123} \longrightarrow K'_{23}$.

By this equation a bilinear mapping $\mathcal{W}_{23}$ is defined:

$$\mathcal{W}_{23}: K_{2m} \times K_{3m} \longrightarrow K'_{23} ,$$

$$u_{23} = R_{23} \sum_{13} (\Gamma_2 u_2, u_3^{i}) = \mathcal{W}_{23} (u_2, u_3^{i}). \tag{4.4}$$

Because of $R_2 \dot{\Gamma}_2 u_2 = u_2$ we have

$$R_2 \mathcal{W}_{23} (u_2, u_3^{i}) = u_2 . \tag{4.5}$$

(4.5) reflects the fact that 2 acts on 3 but that 3 does not act on 2. 2 acts on 3 via the "microchannel" 1, i.e. by emitting microobjects 1 which collide with 3. The microobjects have disappeared in the latter description of the interaction by means of the interaction operator $\mathcal{W}_{23}$. <u>This is the description of the real, experimental</u>

<u>physical situation:</u> Only trajectories of the macrosystems 2 and 3 are registered. No microsystem appears in the results of the experiment. The microsystem only is a "theoretical invention" to "explain" the experiment. The question arises whether this invention is a discovery of physically real systems, the microsystems.
To answer this question more rigorously I quite generally discussed the problem: What is a physical theory and especially, what is the meaning of "physically real" [1, chap II]. (An improved version of this chapter II which eliminates mistakes and clears up some misunderstandings will be published in the second edition of [1]). At the moment, however, it is sufficient to take some of the physical expressions intuitively.

Concerning the composed macrosystem 2, 3 it is sufficient to register only trajectory effects of the form $f(y_2)g(y_3)$. This restriction simplifies our physical discussions considerably and the statistics of all trajectory effects $h(y_2, y_3)$ is completely determined by the statistics of the special effects $f(y_2)g(y_3)$. The product effects $f(y_2)g(y_3)$ describe the physical situation where the preparing system 2 and the measuring system 3 are registered without correlation between the two registration procedures concerning 2 and 3.

The probability for measuring such effects is

$$\mu_{23}(u_{23}, f(y_2)g(y_3)) = \mu_3(\Sigma_3(R_1 f(y_2)\Gamma_2 u_2, u_3^i), g(y_3)) \quad (4.6)$$

with $\Sigma_3 = R_3\Sigma_{13}$. (4.6) can be interpreted as follows:
2 prepares the ensemble W_1 with $\lambda W_1 = R_1 f(y_2)\Gamma_2 u_2$ of the microsystem 1 and the microsystem collides with system 3 which then has to be described by $\Sigma_3(W_1, u_3^i)$.

We define a mapping $\Xi : C_+^1(\hat{y}_2)\times K_{2m} \longrightarrow \check{K}_1$

$$\text{by} \quad (f(y_2), u_2) \longrightarrow R_1 f(y_2)\Gamma_2 u_2 . \quad (4.7)$$

$\Xi(f(y_2), u_2)$ is interpreted to be an ensemble of "interaction-mediators" called microobjects which mediate an interaction between system 2 and system 3. What kind of conditions have to be fulfilled by the mapping

$$\mathcal{W}_{23} : K_{2m} \times K_{3m} \longrightarrow K'_{23}$$

defined by $u_{23} = \mathcal{W}_{23}(u_2^i, u_3^i)$

so that we can regain the mapping Ξ and in this sense regain the microsystems as "interaction-mediators"? The answer would be an essential step to the justification for speaking of "physically real" microobjects.

A single preparing system, a macrosystem, is characterized by a pair $(x,y) \in H$ with $x \in \mathcal{M}_m$, $y \in \hat{\mathcal{Y}}$ ($(x, y) \in H$ is to be interpreted as "y is the trajectory of x"). All the systems $x \in \mathcal{M}_m$ are of the "same sort", i.e. can be described in the same state space $\mathcal{Z}$. In nature, however, there are several sorts of systems. Since we have no theory describing what sorts of systems are possible in nature we take an index set Λ such that the indices $\lambda \in \Lambda$ describe the various sorts of systems. The corresponding spaces will be denoted by $\mathcal{M}_m^{(\lambda)}$, $\hat{\mathcal{Y}}^{(\lambda)}$, $H^{(\lambda)}$. Let $P = \bigcup_{\lambda \in \Lambda} \mathcal{M}_m^{(\lambda)}$.

$x \in P$ is the symbol for a single preparing system and P is called the set of preparing systems for microobjects.

In section 2 we defined the set $Q^{(\lambda)}$ of preparing procedures of macroobjects and the set $\mathcal{R}^{(\lambda)}$ of registration procedures. For a sort λ, a preparing procedure $q \in Q^{(\lambda)}$ and a registration procedure $r = (\varrho, \eta) \in R^{(\lambda)}$ we define the following subsets of P:

$$\begin{aligned} s_P(\lambda,q,r) &= \{x \mid x \in q,\ (x,y) \in H^{(\lambda)},\ \varrho(x) = 1\}, \\ s_P(\lambda,q,r,+) &= \{x \mid x \in q,\ (x,y) \in H^{(\lambda)},\ \varrho(x) = 1,\ \eta(y) = 1\}. \end{aligned} \tag{4.9}$$

The subset s_P of P is called a preparing procedure for microsystems and the set $\mathcal{S}_P$ of all $s_P \neq \emptyset$ is called the set of such preparing procedures for microsystems.

By $q \rightarrow u^{(\lambda)}$ and $r = (\varrho, \eta) \rightarrow f^{(\lambda)}(y)$ a surjective mapping α is defined

$$\alpha : \mathcal{S}_P \longrightarrow \bigcup_{\lambda \in \Lambda} C^1_+(\hat{y}^{(\lambda)}) \times K^{(\lambda)}_m \tag{4.10}$$

with $\quad \alpha\, s_P(\lambda, q, r) = (1, u^{(\lambda)})$

and $\quad \alpha\, s_P(\lambda, q, r, +) = (f^{(\lambda)}(y), u^{(\lambda)})$.

Recall the definition of the mapping $\Xi^{(\lambda)}$:

$$(f^{(\lambda)}(y), u^{(\lambda)}) \xrightarrow{\ \Xi^{(\lambda)}\ } R_1 f^{(\lambda)}(y) \Gamma^{(\lambda)} u^{(\lambda)}.$$

Since $s_P \neq \emptyset$ we get the relation

$$tr\left(\Xi^{(\lambda)}(f^{(\lambda)}(y) \Gamma^{(\lambda)} u^{(\lambda)})\right) = \mu_2(u^{(\lambda)}, f^{(\lambda)}(y)) \neq 0$$
$$\text{for} \quad \alpha(s_P) = (f^{(\lambda)}, u^{(\lambda)}).$$

We now define the mapping $\beta : \alpha(s_P) \rightarrow K_1$ by:

$$\beta(f^{(\lambda)}(y), u^{(\lambda)}) = tr\left(\Xi^{(\lambda)}(f^{(\lambda)}(y), u^{(\lambda)})\right)^{-1} \Xi^{(\lambda)}(f^{(\lambda)}(y), u^{(\lambda)}). \tag{4.11}$$

Let $k : \mathcal{S}_P \rightarrow K_1$ be the composed mapping $k = \beta \circ \alpha$. (4.12)

Denoting the subset of $\bigcup_{\lambda \in \Lambda} C^1_+(\hat{y}^{(\lambda)}) \times K^{(\lambda)}_m$ with $\mu_2(u^{(\lambda)}, f^{(\lambda)}) \neq 0$ by $\left[\bigcup_{\lambda \in \Lambda} C^1_+(\hat{y}^{(\lambda)}) \times K^{(\lambda)}_m\right]'$ we can give a survey of the mappings defined in (4.10), (4.11), (4.12) in the following diagram:

$$P\ ,\ \mathcal{S}_P \subset \mathcal{R}(P);$$

$$\begin{array}{ccccc} & & \left[\bigcup_{\lambda \in \Lambda} C^1_+(\hat{y}^{(\lambda)}) \times K^{(\lambda)}_m\right]' & & \\ & \overset{\alpha}{\nearrow} & & \overset{\beta}{\searrow} & \\ \mathcal{S}_P & & \xrightarrow{\qquad\quad k \qquad\quad} & & K_1 \end{array} \tag{4.13}$$

Similarly the effected parts can be treated: We consider an index set J for the various sorts of effected parts: $x \in M^{(\iota)}_m$, $(x, y) \in H^{(\iota)}$.

Let $Q^{(\iota)}$ be the set of preparing procedures and $R^{(\iota)}$ the set of registration procedures of the effected parts of sort (ι).

We define $E = \bigcup_{\iota \in J} M_m^{(\iota)}$ and for a sort (ι), a preparing procedure $q \in Q^{(\iota)}$ and a registration procedure $r \in R^{(\iota)}$ the following subsets of E:

$$s_E(\iota, q, r) = \{x \mid x \in q,\ (x, y) \in H^{(\iota)},\ \varrho(x) = 1\}$$

and $s_E(\iota, q, r, +) = \{x \mid x \in q,\ (x, y) \in H^{(\iota)},\ \varrho(x) = 1,\ \eta(y) = 1\}$.

The set of all these s_E will be called $\mathcal{S}_E$. By $q \longrightarrow u^{(\iota)} \in K_m^{(\iota)}$, $r = (\varrho, \eta) \longrightarrow g^{(\iota)} \in C_+^1(\hat{y}^{(\iota)})$ and $s_E \longrightarrow (g^{(\iota)}, u^{(\iota)})$ a surjective mapping

$$\gamma : \mathcal{S}_E \longrightarrow \bigcup_{\iota \in J} C_+^1(\hat{y}^{(\iota)}) \times K_m^{(\iota)} \tag{4.15}$$

is defined with $g^{(\iota)} = 1$ for $s_E(\iota, q, r)$.

The composite system , composed of a preparing and an effected system, can be described by a subset $M \subset P \times E$ of all those pairs which constitute the composed system (M is the graph of a bijective mapping $P \longrightarrow E$).
An experiment is described by two subsets of M determined by a preparing procedure $s_P \in \mathcal{S}_P$ and a pair

$$s_E(\iota, q, r),\quad s_E(\iota, q, r, +)$$

of elements of $\mathcal{S}_E$.
In order to distinguish the elements of P and E we will write v for the elements of P and e for the elements of E. Then an experiment is characterized by the two subsets of M:

$$\tilde{M} = \{(v, e) \mid (v, e) \in M,\ v \in s_P,\ e \in s_E(\iota, q, r)\},$$
$$\tilde{M}_+ = \{(v, e) \mid (v, e) \in M,\ v \in s_P,\ e \in s_E(\iota, q, r, +)\}. \tag{4.16}$$

If an experimentalist performs such an experiment he has N pairs $(v, e) \in \widetilde{M}$ and N_+ of these pairs are elements of $\widetilde{M}_+$, namely all those $(v, e) \in \widetilde{M}$ for which $\eta(y) = 1$ for the effected part e. The experimental frequency N_+/N is to be compared with the theory as follows ($q \longrightarrow u^{i(\iota)} \in K_m^{(\iota)}$, $r \longrightarrow g^{(\iota)} \in C_+^1(\hat{y}^{(\iota)})$, $\Sigma_\iota = R_\iota \Sigma_{1\iota}$):

$$N_+/N \approx \mu_\iota(\Sigma_\iota(W_1^i, u^{i(\iota)}), g^{(\iota)}) = tr(W_1^i F)$$

with $F = T(1, \iota; u^{i(\iota)})\, g^{(\iota)}(y)$

and $W_1^i = k\, \Delta_P$.

A mapping δ is defined by

$$(g^{(\iota)}(y), u^{i(\iota)}) \longrightarrow T(1, \iota; u^{i(\iota)})\, g^{(\iota)}(y). \qquad (4.17)$$

Let $\ell : \mathcal{S}_E \longrightarrow L_1$ be the composed mapping $\ell = \delta \circ \gamma$. (4.18)

Now we can state the very familiar result:

$$N_+/N \approx tr(WF) \quad \text{with} \quad W = \beta(f^{(2)}(y), u^{(2)})$$
$$\text{and} \quad F = \delta(g^{(\iota)}(y), u^{i(\iota)}). \qquad (4.19)$$

Hence

$$tr(WF) = \mu_2(u^{(2)}, f^{(2)})^{-1} \mu_{2\iota}(u_{2\iota}, f^{(2)} g^{(\iota)}) \qquad (4.20)$$

with $u_{2\iota} = \vartheta_{2\iota}(u^{(2)}, u^{i(\iota)})$.

This formula is the basis of all following considerations.

On the right side we only find symbols concerning the two macrosystems, and on the left side only the symbols W, F of the microsystems. The question arises whether it is possible to deduce the left side from the structure of the right side alone. Here we deduced this formula <u>accepting</u> the existence of the microobjects, the preparing operator T and the collision operator Σ.

A survey of the mappings defined in (4.15), (4.17), (4.18) is given in the following diagram:

$$E\,,\quad \mathcal{S}_E \subset \mathcal{P}(E)\,;$$

$$\begin{array}{ccccc} & & \bigcup\limits_{\iota\in J} C^1_+(\hat{y}^{(\iota)})\times K^{(\iota)}_m & & \\ & \overset{\gamma}{\nearrow} & & \overset{\delta}{\searrow} & \\ \mathcal{S}_E & & \xrightarrow{\qquad\quad \ell \quad\qquad} & & L_1 \end{array} \tag{4.21}$$

The relation between the two diagrams (4.13) and (4.21) is given by (4.19), (4.20), i.e.

$$\begin{aligned} & tr\left(\beta(f^{(2)}, u^{(2)})\cdot\delta(g^{(\iota)}, u^{i(\iota)})\right) \\ & = \mu_2(u^{(2)}, f^{(2)})^{-1}\,\mu_{2\iota}\left(\mathcal{W}_{2\iota}(u^{(2)}, u^{i(\iota)}), f^{(2)} g^{(\iota)}\right) \end{aligned} \tag{4.22}$$

where the interaction operator $\mathcal{W}_{2\iota}$ is defined by (4.4), i.e.

$$\mathcal{W}_{2\iota}(u^{(2)}, u^{i(\iota)}) = R_{2\iota}\,\Sigma_{1\iota}(\Gamma_2 u^{(2)}, u^{i(\iota)}). \tag{4.23}$$

In the sequel we will try to find the structures which allow to recover the microobjects starting from the description of the two macrosystems, namely the preparing and the effected part.
It is easy to see that the following sets and mappings discussed in the preceding development are defined by the macrosystems alone:
P, the set of the preparing parts,
E, the set of the effected parts,
$\mathcal{S}_P$, the set of the preparing procedures,
$\mathcal{S}_E$, a set of selected subsets of E,
M, the set of the composite systems, composed of a preparing and an effected part, the mappings α and γ in (4.13) and (4.21).

But how to regain the sets K_1, L_1 and the mappings β, δ, k, ℓ in (4.13) and (4.21)?

We define the following subset of $\mathcal{P}(M)$:

Q_M the set of all $\{(v,e) \mid (v,e) \in M, v \in s_P\}$ (4.24)

and the following set of function pairs:

R_M the set of all pairs of functions (ϱ_M, η_M) on M:

$$\varrho_M(v,e) = \begin{cases} 1 & \text{if } e \in s_E(\iota, q', r') \\ 0 & \text{otherwise} \end{cases}, \qquad (4.25)$$

$$\eta_M(v,e) = \begin{cases} 1 & \text{if } e \in s_E(\iota, q', r', +) \\ 0 & \text{otherwise} \end{cases}.$$

The following central theorem can be proved: Q_M, R_M are structures of the species "preparing procedures" (Q_M) and "registration procedures" (R_M) on the base M (or shorter: M is endowed with the structure Q_M, R_M of preparing and registrating procedures) if the following two assumptions hold:

Assumption 1): A macroscopic interaction operator $\mathcal{W}_{2\iota}$ is given such that $R_2 \mathcal{W}_{2\iota}(u^{(2)}, u^{i(\iota)}) = u^{(2)}$ with $s_P(2, q, \dots)$ and $q \rightarrow u^{(2)}$, q' as in (4.25) and $q' \rightarrow u^{i(\iota)}$.

Assumption 2): The sets $\{(v,e) \mid (v,e) \in M, v \in s_P, \varrho_M(v,e) = 1\}$ are not empty.

Assumption 2) can be weakened to a more physical form.

The concepts of "structure", "species of structure" and "base" in the formulation of the theorem are used in the mathematical sense (see Bourbaki, I, 4). As mentioned in section 2 we have not given the axiomatic relations characterizing the species of structures "preparing and registrating procedures", but we discussed the intuitive meaning of these structures in the example of the set M_m of macroscopic systems.

We want to stress that the elements of Q_M, the preparing procedures on M, are the subsets of all experiments selected by the subsets s_P of P. However, according to (4.9) and (4.24) s_P itself is given by a preparing procedure q and a registrating procedure r on P. Similarly: the elements of R_M, the registrating procedures on M are given by a preparing procedure q' and a registrating procedure r' on E according to (4.14) and (4.25). We cannot prove the central theorem here but we will give a survey of the essential features of this proof:

By assumption 1) it is expressed that the system of the sort $\mathfrak{z}$ (the preparing part) can influence the system of sort ι (the effected part), but the inverse is impossible. A consequence of assumption 1) is (with $u_{\mathfrak{z}\iota} = \mathfrak{w}_{\mathfrak{z}\iota}(u^{(\mathfrak{z})}, u^{i(\iota)})$):

$$0 \le \mu_{\mathfrak{z}}(u^{(\mathfrak{z})}, f^{(\mathfrak{z})})^{-1} \mu_{\mathfrak{z}\iota}(u_{\mathfrak{z}\iota}, f^{(\mathfrak{z})} g^{(\iota)}) \le 1. \tag{4.26}$$

Assumption 2) permits to define a function μ on $S_P \times S_E$:

$$\mu(s_P, s_E) = \mu_{\mathfrak{z}}(u^{(\mathfrak{z})}, f^{(\mathfrak{z})})^{-1} \mu_{\mathfrak{z}\iota}(u_{\mathfrak{z}\iota}, f^{(\mathfrak{z})} g^{(\iota)}) \tag{4.27}$$

with $(f^{(\mathfrak{z})}, u^{(\mathfrak{z})}) = \alpha s_P$, $(g^{(\iota)}, u^{i(\iota)}) = \gamma s_E$.

According to (4.26) we have

$$0 \le \mu(s_P, s_E) \le 1. \tag{4.28}$$

Considering the two diagrams (4.13), (4.21) and equation (4.22) the function $\mu(s_P, s_E)$ should be compatible with the mappings k and ℓ:

$$\mu(s_P, s_E) = tr(k(s_P) \cdot \ell(s_E)).$$

The mapping k, ℓ as well as the sets K_1, L_1, however, are not yet defined if we start with the macroscopic systems alone. We will introduce the sets K_1, L_1 and the mappings k, ℓ in a new way.

By $\mu(s_{P1}, s_E) = \mu(s_{P2}, s_E)$ for all $s_E \in S_E$ an equivalence relation is defined on S_P, and K will denote the set of equivalence classes of S_P. We define the mapping k by the canonical mapping $k: \mathcal{S}_P \rightarrow \underline{K}$. Since $\mu(s_P, s_E)$ only depends on $\alpha s_P = (f^{(\lambda)}, u^{(\lambda)})$ there is a mapping β so that $k = \beta \circ \alpha$. Thus we get the following diagram similar to (4.13):

$$
\begin{array}{ccccc}
 & & \left[\bigcup\limits_{\lambda \in \Lambda} C^1_+(\hat{y}^{(\lambda)}) \times K^{(\lambda)}_m\right]' & & \\
 & \overset{\alpha}{\nearrow} & & \overset{\beta}{\searrow} & \\
\mathcal{S}_P & & \xrightarrow{\quad\quad k \quad\quad} & & \underline{K}
\end{array}
\tag{4.29}
$$

The function $\mu(k s_P, s_E) = \mu(s_P, s_E)$ is well defined on $\underline{K} \times \mathcal{S}_E$. In the same way an equivalence relation is defined on $\mathcal{S}_E$ by $\mu(w, s_{E1}) = \mu(w, s_{E2})$ for all $w \in \underline{K}$. The set of equivalence classes of $\mathcal{S}_E$ will be called $\underline{L}$ and l is defined to be the canonical mapping $l: \mathcal{S}_E \rightarrow \underline{L}$. Since $\mu(w, s_E)$ only depends on $\gamma s_E = (g^{(\iota)}, u^{i(\iota)})$ there is a mapping δ with $l = \delta \circ \gamma$. Thus we get the following diagram similar to (4.21):

$$
\begin{array}{ccccc}
 & & \bigcup\limits_{\iota \in J} C^1_+(\hat{y}^{(\iota)}) \times K^{(\iota)}_m & & \\
 & \overset{\gamma}{\nearrow} & & \overset{\delta}{\searrow} & \\
\mathcal{S}_E & & \xrightarrow{\quad\quad l \quad\quad} & & \underline{L}
\end{array}
\tag{4.30}
$$

Instead of equation (4.22) we have:

$$
\begin{aligned}
\mu(k s_P, l s_E) &= \mu\left(\beta(f^{(\lambda)}, u^{(\lambda)}), \delta(g^{(\iota)}, u^{i(\iota)})\right) \\
&= \mu_\lambda(u^{(\lambda)}, f^{(\lambda)})^{-1} \mu_{\lambda\iota}\left(W_{\lambda\iota}(u^{(\lambda)}, u^{i(\iota)}), f^{(\lambda)} g^{(\iota)}\right).
\end{aligned}
\tag{4.31}
$$

Because of the central theorem we can introduce the following concepts: By $Q_M \longleftrightarrow S_P \xrightarrow{k} \underline{K}$ every <u>preparing procedure</u> determines an "ensemble" $w \in \underline{K}$. Several preparing procedures, however, can determine the <u>same</u> ensemble w.

By $R_M \longleftrightarrow S_E \xrightarrow{\ell} \underline{L}$ every registrating procedure determines an "effect" $f \in \underline{L}$. Several registrating procedures, however, can determine the same effect f. (Remark: Many wrong conclusions in q.m. are caused by not distinguishing between the elements of Q_M and the "ensembles". See [1] and [7], volume 3. The elements of Q_M are subsets of M while the ensembles are not).

The central theorem provides the basis for calling M the set of "physically real microsystems". To make this sentence meaningful it is necessary to define the words "physically real" and "physical system". As mentioned above it is not possible to explain the meaning of "physically real" here (see [1]). The meaning of "physical system", however, can be defined by:

A set X endowed with a structure of species "preparing and registrating procedures" is called a set of "physical systems" if the physical interpretation of the preparing and registrating procedures is well defined.

In our special case we call the set M a set of "microsystems". The basic assumptions 1) and 2) permit to consider the set M as a "discovery" of microobjects. Only keeping in mind this way of rediscovery it is possible to avoid intuitive mistakes in thinking of "microobjects". E.g. the usual opinion that momentum and position cannot be measured "simultaneously" can be proved to be wrong if simultaneous measurement is interpreted as measuring $P(t)$ and $Q(t)$ at the same time t. It is true, however, that $P(t)$ and $Q(t)$ cannot be measured together, i.e. with the same "effected part"; but neither can $Q(t_1)$ and $Q(t_2)$ (see [1] and [7] volume 3) for $t_1 \neq t_2$ be measured together.

Finally, we want to show that the introduction of the sets $\underline{K}$ and $\underline{L}$ and the mappings β and δ in the diagrams (4.29), (4.30) permit to reconstruct the "preparing operator" Γ and the "collision operator" Σ if the macroscopic interaction operator $\mathcal{W}_{\lambda\iota}$ is given. In this way it is possible to get a "compatibility circle" of the description of measuring and preparing processes.

The set $\underline{K}$ can be embedded (see [1]) in a vector space $\underline{B}$. We define a mapping by

$$(f^{(\lambda)}, u^{(\lambda)}) \xrightarrow{\xi} \mu_\lambda(u^{(\lambda)}, f^{(\lambda)}) \cdot \beta(f^{(\lambda)}, u^{(\lambda)}).$$

It can be shown that ξ is a bilinear mapping

$$\xi : C^1_+(\hat{y}^{(\lambda)}) \times K^{(\lambda)}_m \longrightarrow \check{\underline{K}}$$

with $\check{\underline{K}} = \bigcup_{0 \leq \lambda \leq 1} \lambda \underline{K} \subset \underline{B}$.

Hence

$$\mu_{\lambda\iota}(\mathcal{W}_{\lambda\iota}(u^{(\lambda)}, u^{i(\iota)}), f^{(\lambda)} g^{(\iota)})$$
$$= \mu(\xi(f^{(\lambda)}, u^{(\lambda)}), \delta(g^{(\iota)}, u^{i(\iota)}))$$
$$= \mu(\mu_\lambda(u^{(\lambda)}, f^{(\lambda)}) \cdot w, f)$$

with $w = \beta(f^{(\lambda)}, u^{(\lambda)})$ and $f = \delta(g^{(\iota)}, u^{i(\iota)})$.

By means of this formula it is easy to show that δ is a bilinear mapping $\delta : C^1_+(\hat{y}^{(\iota)}) \times K^{(\iota)}_m \longrightarrow \underline{L} \subset \underline{D}$ where $\underline{B}$, $\underline{D}$ are two spaces in duality by the bilinear form $\mu(w, f)$ (see [1]). The spaces $\underline{B}$, $\underline{D}$ can be embedded (see [1]) in Banach spaces B respectively B' (B' being the dual of B). Let K denote the norm closed convex hull of $\underline{K}$, let L denote the σ(B',B)-closed convex hull of $\underline{L}$ and let D denote the norm closure of $\underline{D}$. Since $\mu(\xi(f^{(\lambda)}, u^{(\lambda)}), f)$ is a bilinear form ($u^{(\lambda)}$ fixed) of $f^{(\lambda)}$ and f we can write:

$$\mu(\xi(f^{(\lambda)}, u^{(\lambda)}), f) = \mu_{1\lambda}(u_{1\lambda}, f^{(\lambda)} f)$$

and define in this way the ensemble $u_{1\lambda} \in C'_+(\hat{y}^{(\lambda)}, D)$.

By $$u_{1\lambda} = \Gamma^{(\lambda)} u^{(\lambda)} \qquad (4.32)$$

we <u>define</u> the <u>preparing operator</u> $\Gamma^{(\lambda)}$.

$\mu(w, \delta(g^{(\iota)}, u^{i(\iota)}))$ is linear in $g^{(\iota)}$ so that there exists a $u^{(\iota)} \in K'^{(\iota)} \subset C'(\hat{y}^{(\iota)})$ with $\mu(w, \delta(g^{(\iota)}, u^{i(\iota)})) = \mu_\iota(u^{(\iota)}, g^{(\iota)})$.

By $$u^{(\iota)} = \Sigma_\iota(w, u^{i(\iota)}) \qquad (4.33)$$

we define a collision operator Σ_c. Having introduced the operators $\Gamma^{(2)}$ and Σ_c one can deduce the identities (!):

$$\xi(f^{(2)}, u^{(2)}) = R_1 f^{(2)} \Gamma^{(2)} u^{(2)}$$

and $$\delta(g^{(c)}, u^{i(c)}) = T(1, c, u^{i(c)}) g^{(c)}$$

with $T(1, c; u^{i(c)})$ defined by this equation.

Therefore it is justified to say that the interaction of the preparing systems v $\in$ P and the effected systems e $\in$ E, described by $\mathcal{W}_{2c}$ can be interpreted as follows:

The preparing systems v of a subset $s_P \in S_P$ prepares a subset $q_M \subset M$ ($q_M \in Q_M$) of microsystems; q_M determines an ensemble ($q_M \xrightarrow{k} v$)

$$v = \mu_2(u^{(2)}, f^{(2)})^{-1} R_1 f^{(2)} \Gamma^{(2)} u^{(2)} \in K \tag{4.34}$$

($\Gamma^{(2)}$ defined by (4.32)).

The microsystems of q_M collide with the effected systems and generate together with the effected systems the probability for trajectory effects $g^{(c)}$ on the effected parts:

$$\mu_c(\Sigma_c(v, u^{i(c)}), g^{(c)})$$

which can also be given in the form

$$\mu(v, f) \quad \text{with} \quad f = \delta(g^{(c)}, u^{i(c)}) = T(1, c, u^{i(c)}) g^{(c)} \tag{4.35}$$

i.e. as the probability of the effect f in the ensemble v.
It was the goal of the present paper to show this consistency.
Of course other axioms have to be added (see [1]) to show for microobjects, as for instance atoms, that the sets K and L can be represented by the well known sets of operators in a Hilbert space, namely K by the base of the base norm space B of Hermitean trace class operators and L by the order interval [0,1] of the space

B' of all Hermitean bounded operators such that the probability function $\mu(v, f)$ coincides with $tr(v \cdot f)$.

I would like to thank A.Hartkämper and H.Neumann for critical reading of the manuscript and for useful suggestions.

References

[1] Ludwig,G.,'Deutung des Begriffs "physikalische Theorie" und axiomatische Grundlegung der Hilbertraumstruktur der Quantenmechanik durch Hauptsätze des Messens', Lecture Notes in Physics, volume 4, Springer-Verlag, Berlin and Heidelberg (1970). A second improved edition is in preparation.

Stolz,P.,'Attempt of an Axiomatic Foundation of Quantum Mechanics and More General Theories,VI' Commun.Math.Phys. 23, 117-126(1971).

Ludwig,G.,'A physical interpretation of an axiom within an axiomatic approach to quantum mechanics and a new formulation of this axiom as a general covering condition', Notes in Math.Phys. 1 (Marburg 1971).

Ludwig,G.,'An Improved Formulation of Some Theorems and Axioms in the Axiomatic Foundation of the Hilbert Space Structure of Quantum Mechanics',Commun.Mathem.Phys. 26, 78-86 (1972)

Neumann,H. 'Classical Systems and Observables in Quantum Mechanics', Commun.Math.Phys. 23, 100 - 116 (1971)

[2] Ludwig,G., 'Transformationen von Gesamtheiten und Effekten', Notes in Math.Phys. 4 (Marburg 1971).

[3] Lanz,L., Ramella,G. 'On the Deduction of a Markoffian Master Equation', Physica 44, 499-531 (1969).

Lanz,L., Lugiato,L.A. 'On the Asymptotic-Behaviour of the Solutions of the Liouville-von Neumann Equation', Physica 44, 532-554 (1969).

Lanz,L., Lugiato,L.A. 'A Rigorous Approach to a Markoffian Master Equation', Physica 47, 345-372 (1970).

Lanz,L., Lugiato,L.A., Ramella,G. 'On the Existence of Independent Subdynamics in Quantum Statistics', Physica 54, 1, 94-136 (1971)

For other literature see the articles of Prosperi and Lanz in this volume.

[4] Prosperi,G.M. 'Macroscopic Physics, Quantum Mechanics and Quantum Theory of Measurement' Proceedings of the Conference Quantum Theory and Beyond , (Cambridge 1968).

Lanz,L., Prosperi,G.M., Sabbadini,A. 'Time Scales and the Problem of Measurement in Quantum Mechanics'. Nuovo Cimento B2, 184 (1971).

Hepp,K. 'Quantum Theory of Measurement and Macroscopic Observables', Helv.Phys.Acta 45, 237-248 (1972)

For other Literature see the article of Prosperi in this volume.

[5] G.Ludwig, 'Meß- und Präparierprozesse', Notes in Math.Phys. 6 (Marburg 1972).

[6] G.Ludwig, 'Makroskopische Systeme und Quantenmechanik', Notes in Math.Phys. 5 (Marburg 1972).

[7] G.Ludwig, 'Einführung in die Grundlagen der Theoretischen Physik (four volumes, volume 1 and 2 in print), Bertelsmann Universitätsverlag, Düsseldorf.

MODELS OF THE MEASURING PROCESS AND OF MACRO-THEORIES

Giovanni Maria Prosperi
Istituto di Fisica, Università degli Studi di Milano
Istituto Nazionale di Fisica Nucleare, Sezione di Milano

INTRODUCTION

In the first part of these lectures I want to discuss the process of measurement in quantum mechanics and the related problem of reconciling the quantum and the classical description of the macroscopic behaviour of a large body according to the lines I already followed in preceding occasions (1).

Precisely, first I shall assume

a) that the initial state of the large body is chosen in an appropriate way;

b) that certain two characteristic times t_m and t_M, refering to the duration of elementary microscopic processes occuring in the body and to the evolution of the macroscopic quantities respectively, are well separated.

Then I show that under these conditions, even in the framework of quantum mechanics it is possible to derive a closed set of deterministic equations involving the macroscopic quantities alone and consequently, as long as we restrict our consideration to such quantities, to neglect "interference terms" between macroscopically distinguishable states. Secondly I shall idealize the measurement apparatus as a large body which possesses one macroscopic quantity J with a relaxation time t_J much larger then the characteristic time t_M of the other ones and assume that it is such quantity J which is coupled to the state of the object. Then the separation between t_J and t_M produces the disappearance of the interference terms between states of the compound system object + apparatus corresponding to different values of the quantity for the object we are measuring, and the separation between t_m and t_M produces as before the disappearance of the interference terms between the macroscopically distinguishable states of the apparatus.

This treatment suffers from some important limitations; the deterministic macroscopic behaviour and the disappearance of the interference terms is obtained as a consequence of

i) certain definite approximations,

ii) the "a priori" restriction of our consideration to an appropriate set of observables alone,

iii) the large body or the compound system object + apparatus is treated as an isolated system.

It should be stressed in this connection that all usual derivations of irreversible thermodynamic equations from quantum mechanics have this same characteristic and use essentially the same ingredients in spite of the different techniques involved.

The fact that the results involve approximations is not a serious limitation in my opinion; any unified description of small and large bodies must necessarily contain the purely macroscopic description of the large ones as a limit case.

The situation is different for what concerns the "a priori" restriction to the set of the so called "macroscopic observables". If we admit that an actual observable R for the large system or for an apparatus incompatible with the macroscopic quantities exists, then interference terms between macroscopically distinguishable states can be produced and brought into evidence.

For what I understand, this difficulty is the main motivation for Ludwig's attempts to formulate a new theory in which microscopic and macroscopic systems are treated on a different basis and which does not contain this inconsistency from the beginning (2).

In the same connection various models and discussions of the problem are also interesting in which the apparatus or the large body is treated as a system with infinitely many degrees of freedom (cf. e.g.ref. (3) (4) (5) (6) and the other ones ibiquoted). Also even if the research in this field must be considered yet in a somewhat preliminary stage, such models and discussions indicate interesting new possibilities. It seems that in the limit of infinitely many degrees of freedom the undesired interference terms can be made to vanish without any a priori restriction to a set of privileged observables, that the macroscopic quantities can be made compatible with any other observable and that consequently a superposition of macroscopically distinguishable states can be made always equivalent to the corresponding mixture in a strict mathematical sense.

I shall devote the last part of these lectures to a brief account and a discussion of two papers in this last direction. A paper of C. George et al. (3), bringing the idea of independent subdynamics, and a paper of K. Hepp (6) which contains the to my knowledge most comprehensive discussion of the measurement problem in the so called C^*-algebra formalism.

We note that, having in mind a criticism by Zeh (7), the infinite system could perhaps be intended as an idealization of the system apparatus + rest of the world.

2 - STATEMENT OF THE PROBLEM

Quantum mechanics consists essentially of a set of mathematical rules which should give, at least in principle, the solution of the following problem: given specified certain conditions on the system of interest (elementary particle, atom, molecule etc.), or more schematically, given the result of a first observation made on it at a time t = 0, calculate the probability that a second observation gives some other result after a time t. The conceptually most significant part of the mentioned set of mathematical rules can be briefly summarized in the following three postulates:

i) To every physical system S is associated an appropriate Hilbert space $\mathcal{H}_S$. The "state" of the system is represented by a unit vector $|\psi(t)\rangle$ in $\mathcal{H}_S$ which evolves in time according to the Schroedinger equation

(2.1) $$i\frac{\partial}{\partial t}|\psi(t)\rangle = H|\psi(t)\rangle \quad \text{or} \quad |\psi(t)\rangle = e^{-iHt}|\psi_0\rangle .$$

ii) An "observable quantity" A is represented by a self-adjoint operator. The eigenvalues $\{\alpha_r\}$ of such operator are the possible outcomes of the observations of A and the probability that an observation of A at the time t gives a specific result α_r, is given by

(2.2) $$P(A=\alpha_r/t) = \sum_j |\langle\varphi_{rj}|\psi(t)\rangle|^2 ,$$

where $\{|\varphi_{rj}\rangle\}$ is a complete orthonormal set of eigenvectors of A (the subscript r referring to the eigenvalue α_r, the subscript j being a degeneration label).

iii) Let A stand for a complete set of observables (i.e. we assume that the spectrum of A is nondegenerate), let us restrict our consideration to first kind observations alone (x) and assume that we have performed a measurement of A and found the result A = α_r. Then the state vector immediately after the observation is given by

(2.3) $$|\psi'\rangle = |\varphi_r\rangle$$

If we use the formalism of the statistical operators, rather than the formalism of the vectors, the basic equations (2.1), (2.2) and

(x) By a first-kind measurement we mean an observation which, when immediately repeated, reproduces always the same result. For convenience throughout this paper we shall deal only with this kind of measurement.

(2.3) can be restated

(2.1') $$i \frac{\partial W(t)}{\partial t} = [H, W(t)] \quad \text{or} \quad W(t) = e^{-iHt} W_0 \, e^{iHt}$$

with

(2.4) $$W_0 = |\psi_0\rangle\langle\psi_0| \,,$$

(2.2') $$P(A = \alpha_r | t) = Tr(P_r W(t)) \,, \qquad P_r = \sum_j |\varphi_{rj}\rangle\langle\varphi_{rj}|$$

and

(2.3') $$W' = |\varphi_r\rangle\langle\varphi_r|$$

respectively. As it is well known the latter formalism has the advantage that the "pure state" case, for which W_0 is a projector (eq. (2.4)), and the mixture, which corresponds to an incomplete specification of the state of the system,

(2.5) $$W_0 = \sum_s p_s |\psi_s\rangle\langle\psi_s| \qquad (p_s \geq 0 \,,\ \sum_s p_s = 1)$$

can be treated in the same way . This is particularly convenient for our discussion.

The postulates i) - iii) have certain important consequences.

Let A and B be two observables for a certain physical system to which we shall refer hereafter as the "object". We assume for simplicity that the spectra of A and B are purely discrete and nondegenerate,

(2.6) $$A\,|\varphi_r\rangle = \alpha_r\,|\varphi_r\rangle \,, \qquad B\,|\chi_s\rangle = \beta_s\,|\chi_s\rangle$$

and we write as

(2.7) $$|\psi_0\rangle = \sum_r c_r\, |\varphi_r\rangle$$

the "state vector" of the system at the time $t = 0$.

According to postulate ii) we have

(2.8) $$P(A = \alpha_r \,|\, t) = |c_r|^2 = T_r\,(|\varphi_r\rangle\langle\varphi_r|\, W_0)\,.$$

Then, if a measurement of the quantity B is performed at the subsequent time t, in connection with postulate iii) we have the following three different characteristic situations:

a) No measurement is performed on the object at the time $t = 0$; we have

(2.9) $$P(B = \beta_s \,|\, t) = |\langle\chi_s|\, e^{-iHt}\, |\psi_0\rangle|^2 = \\ = T_r\,(|\chi_s\rangle\langle\chi_s|\, e^{-iHt}\, W_0\, e^{iHt})\,.$$

b) At the time $t = 0$ a measurement of A has been performed and the value $A = \alpha_r$, has been found; then

(2.10) $$P\,(B = \beta_s \,|\, t) = |\langle\chi_s|\, e^{-iHt}\, |\varphi_r\rangle|^2 = \\ = T_r\,(|\chi_s\rangle\langle\chi_s|\, e^{-iHt}\, |\varphi_r\rangle\langle\varphi_r|\, e^{iHt}).$$

c) At the time $t = 0$ a measurement of A has been performed, in the sense that the object has interacted with the appropriate measuring apparatus, but the result of the experiment has not been read by the operator; in this case, combining eqs. (2.8) and (2.10), we have

(2.11) $$P(B = \beta_s \,|\, t) = \sum_r |c_r|^2\, |\langle\chi_s|\, e^{-iHt}\, |\varphi_r\rangle|^2 = \\ = T_r\,(|\chi_s\rangle\langle\chi_s|\, e^{-iHt}\, \overline{W}\, e^{iHt})\,,$$

where

(2.12) $$\overline{W} = \sum_{r} |c_r|^2 |\varphi_r\rangle\langle\varphi_r| \;.$$

We now note that

(2.13) $$W_o = \overline{W} + \sum_{r \neq r'} c_r c^*_{r'} |\varphi_r\rangle\langle\varphi_{r'}| \;.$$

The last term on the right-hand side is the so-called "interference term". We recall that the occurence of this interference term is what makes the case a) and the case c) essentially different and it is responsible for the most paradoxical aspect of quantum mechanics. In particular it prevents from ascribing a definite value to the quantity A independently of an actual observation of it and it is consequentely responsible for the so called irreducibly statistical character of the theory.

In conclusion, according to eqs. (2.10) and (2.11), as a result of the measurement of A the statistical operator is modified as

(2.14) $$W_o \longrightarrow |\varphi_r\rangle\langle\varphi_r|$$

in case b), and as

(2.15) $$W_o \longrightarrow \overline{W}$$

in case c).

The process (2.14) and sometimes the process (2.15) are called "the collapse of the wave function".

By the problem of measurement in quantum mechanics one usually means the problem of a correct understanding of such collapse.

Process (2.14) can be thought of as resulting from two separate steps, the first one expressed by eq. (2.15), the second one by

$$\bar{W} \longrightarrow |\varphi_r\rangle\langle\varphi_r| . \tag{2.16}$$

This second step can be simply interpreted as a choice among a set of alternatives resulting from the information we have gained by the observation. Consequently the real point is to understand the meaning of eq. (2.15).

In connection with eq. (2.15) at least three main problems can be distinguished.

Problem 1). It must be shown that the basic postulates i), ii) and iii) are internally consistent; precisely that an appropriate idealization of the apparatus is possible such that, if also this is explicitely included in the treatment, eq. (2.15) follows from the postulates i) and ii) alone as applied to the compound system object + apparatus.

Problem 2). The macroscopic behaviour of a large body falls under the domain of classical physics and can be described in terms of a limited number of variables M_1, M_2, ... to which at every time well defined values are ascribed and which evolve deterministically usually according to certain differential equations [(x)]

$$\frac{dM_i}{dt} = \alpha_i(M_1, M_2, \ldots) . \tag{2.17}$$

(To a given set of values for M_1, M_2, ... I shall refer in the following as to a macroscopic state of the system). If one believes that quantum mechanics can be applied to large bodies the possibility of such description must follow from quantum mechanics itself, when we

---

(x) The characterization of the macroscopic description of a large body which I use here is obviously less general than that one given by Ludwig.

restrict our consideration to the appropriate set of observable quantities (the macroscopic quantities) and to a certain scale of observation.

In particular it must be shown that interference terms between macroscopically distinguishable states are negligible and consequently that an observation does not affect appreciably the macroscopic evolution of the body.

Problem 3). If the large body character of the apparatus is explicitly taken into account, it mustbe possible to improve the idealization of it in such a way that, at the end of the interaction with the object, the apparatus behaves again as a classical system and its macroscopic evolution can follow a set of alternatives in one-one correspondence with the values of the quantity one is measuring on the object.

Problem 1 has been already considered and completely solved by von Neumann (8) in this famous book on Foundations of Quantum Mechanics. I shall give an account of von Neumann's treatment in section 3. In section 4 and section 5 I shall treat problem 2 and 3, with the limitation which I have already discussed in the introduction. In section 6 and 7 I finally shall try to give a brief account and discussion of the recent work on systems with infinite components.

3 - VON NEUMANN'S IDEALIZATION OF THE MEASUREMENT PROCESS

If we denote by I the object and by II the apparatus, in order that II acts as a measurement apparatus for the quantity A relative to I the following circumstances must be verified: 1) the interaction beween I and II must be effective for a sufficiently short time τ ; 2) a quantity G^{II} relative to II must exist such that if immediately before the interaction I is in a state for which the quantity A^{I} has a definite value α_r and II is in a state for which G^{II} has a certain definite value γ_0 at the end of the interaction G^{II}

must have a definite value γ_r uniquely related to α_r, while the value of A^I must be left unchanged.

Let then be

(3.1) $$G^{II}|\Phi_r^{II}\rangle = \gamma_r|\Phi_r^{II}\rangle$$

the eigenvalue equation for G^{II}. If $|\varphi_r^I\rangle$ is the state of I at the time $t = 0$, according to von Neumann we must have

(3.2) $$e^{-iH\tau}|\varphi_r^I\rangle|\Phi_o^{II}\rangle = |\varphi_r^I\rangle|\Phi_r^{II}\rangle$$

as result of the interaction (cf. on this eq. ref. 9).

If the object I is in an arbitrary state at time $t = 0$ from (3.2) we have

(3.3) $$e^{-iH\tau}\sum_r c_r|\varphi_r^I\rangle|\Phi_o^{II}\rangle = \sum_r c_r|\varphi_r^I\rangle|\Phi_r^{II}\rangle$$

or, using the statistical operator,

(3.3') $$e^{-iH\tau}W_0^{I+II}e^{iH\tau} \equiv e^{-iH\tau}|\psi_0^I\rangle\langle\psi_0^I| \otimes |\Phi_o^{II}\rangle\langle\Phi_o^{II}|e^{iH\tau} = $$
$$= W_\tau^{I+II} \equiv \sum_{rr'} c_r c_{r'}^* |\varphi_r^I\rangle\langle\varphi_{r'}^I| \otimes |\Phi_r^{II}\rangle\langle\Phi_{r'}^{II}|.$$

If we now eliminate the system II, using postulate ii) and performing the appropriate contraction on the expression of W_τ^{I+II} we obtain

(3.4) $$W_{\tau\,red}^I = Tr^{II}\,W_\tau^{I+II} = \sum_r |c_r|^2|\varphi_r^I\rangle\langle\varphi_r^I| = \overline{W}^I.$$

This equation provides the required "explanation" of eq. (1.15). We note also

(3.5) $$P(G^{II} = \gamma_r|\tau) = |c_r|^2 = P(A^I = \alpha_r|0).$$

Naturally we observe that, if at the time $t = \tau$ we read the apparatus and find $G^{II} = \gamma_r$, according to postulate iii) W_τ^{I+II} must be replaced by

$$W'^{I+II} = |\varphi_r^I\rangle\langle\varphi_r^I| \otimes |\Phi_r^{II}\rangle\langle\Phi_r^{II}| \tag{3.6}$$

and, if we read the apparatus but do not take into account the result, by

$$\overline{W}^{I+II} = \sum_r |c_r|^2 \, |\varphi_r^I\rangle\langle\varphi_r^I| \otimes |\Phi_r^{II}\rangle\langle\Phi_r^{II}| \, . \tag{3.7}$$

Now again we have $\overline{W}^{I+II} \neq W_\tau^{I+II}$. Consequently the simple "reading" of the apparatus results in the following "modification" of the statistical operator of the compound system

$$W_\tau^{I+II} \longrightarrow \overline{W}^{I+II} \tag{3.8}$$

and the problem of the meaning of the collapse (2.15) is reproduced at the level of the compound system. Again process (3.8) could be "explained", if we want, in terms of a new system III acting as measurement apparatus for I + II.

4 - DERIVATION OF THE DETERMINISTIC EQUATION OF CLASSICAL PHYSICS FROM QUANTUM MECHANICS

We consider a macroscopic isolated system with a definite macroscopic energy. We denote by the collective symbol M the set of the macroscopic quantities and write the eigenvalue equation for them as

$$M\,|\Omega_{\nu j}\rangle = \mu_\nu\,|\Omega_{\nu j}\rangle \quad , \quad j = 1, 2, \ldots s_\nu . \tag{4.1}$$

We also set

$$P_\nu = \sum_j |\Omega_{\nu j}\rangle\langle\Omega_{\nu j}| \, . \tag{4.2}$$

We find convenient to use the formalism of the so-called Hilbert space of operators (or Liouville space). As is well known, this space is defined as the set $\mathcal{L}$ of the Hilbert-Schmidt operators structured with the inner product

$$(W_1, W_2) = Tr\,(W_1^* W_2). \tag{4.3}$$

In this formalism the Liouville-von Neumann equation for the statistical operator W(t) can be written as (cf. eq. (1.1'))

$$i\,\frac{\partial W(t)}{\partial t} = \mathcal{H}\, W(t), \tag{4.4}$$

where the operator $\mathcal{H}$ is defined by

$$\mathcal{H} W = [H, W] \tag{4.5}$$

and it is self-adjoint in $\mathcal{L}$.

Let us then introduce the projection operator

$$\mathcal{P} W = \sum_\nu Tr\,(P_\nu W)\,\frac{1}{s_\nu}\,P_\nu \tag{4.6}$$

and set

$$\tilde{W}(t) = \mathcal{P}\, W(t). \tag{4.7}$$

We have

$$Tr\,\big(f(M)\,W(t)\big) = Tr\,\big(f(M)\,\tilde{W}(t)\big). \tag{4.8}$$

Setting

$$W(t) = \tilde{W}(t) + (1 - \mathcal{P})\,W(t) \tag{4.9}$$

in eq. (4.4) we obtain two coupled equations. If then we assume an initial condition of the form

(4.10) $$W(0) = \sum_{\nu} \mu_{\nu}^{o} \frac{1}{s_{\nu}} P_{\nu} ,$$

i.e. an initial condition such that

(4.10') $$(1-\mathcal{P})\, W(0) = 0 ,$$

we can eliminate $(1 - \mathcal{P})\, W(t)$ between the two equations and obtain the <u>generalized master equation</u> (10)

(4.11) $$\frac{\partial \widetilde{W}(t)}{\partial t} = - \int_0^t dt' \, \mathcal{P} \mathcal{M} \, e^{-i \hat{\mathcal{M}} t'} \mathcal{M} \mathcal{P} \, \widetilde{W}(t-t') ,$$

where

(4.12) $$\hat{\mathcal{M}} = (1-\mathcal{P})\, \mathcal{M}\, (1-\mathcal{P})$$

The distribution of probability μ_{ν}^{o} occuring in equation (4.10) represents the information that we have obtained by an observation of M at the time $t = 0$. We note that to represent the initial information by a distribution of probability rather then by a single set of values μ_{ν_o} fits better with the fact that macroscopically we must regard M as a set of continuous rather then discrete variables.

For a system bounded in a finite region of the space the Hamiltonian H of the system and consequently the operator $\mathcal{M}$ have a purely discrete spectrum. In the thermodynamic limit, however, as the volume of the system becomes infinite, this spectrum becomes continuous. The same characteristic can be expected also for the spectrum of $\hat{\mathcal{M}}$. For times short as compaired with the Poincaré recurrence time the spectrum of $\hat{\mathcal{M}}$ can be treated as continuous also for an actually finite system. If we do so, the kernel of equation (4.11)

(4.13) $$\mathcal{K}(t) = -\mathcal{P}\mathcal{H}\, e^{-i\hat{\mathcal{H}}t}\, \mathcal{H}\mathcal{P}$$

has the important characteristic of decaying fastly as t increases. In fact if $\{\Psi_\omega\}$ are the eigenstates of $\hat{\mathcal{H}}$, we can write

(4.13') $$\mathcal{K}(t)\, P_\nu = \sum_{\nu'} \frac{1}{s_{\nu'}} P_{\nu'} \int d\omega\, Tr(P_{\nu'} \mathcal{H} \Psi_\omega)\, e^{-i\omega t}\, Tr(\Psi_\omega^* \mathcal{H} P_\nu).$$

Then, denoting by $\delta\omega$ the order of magnitude of the values of ω for which expressions like $Tr(\Psi_\omega^* \mathcal{H} P_\nu)$ are appreciably different from 0, we find that $\mathcal{K}(t)$ vanishes for $t \gg t_m \equiv 1/\delta\omega$. The time t_m usually equals the duration of some elementary process, e.g. the duration of a collision in the case of dilute gas.

We now introduce the macroscopic characteristic time t_M by

(4.14) $$t_M \frac{d\langle M\rangle}{dt} \sim \langle M\rangle ,$$

where

(4.15) $$\langle M\rangle = Tr(M\, W(t)) = Tr(M\, \widetilde{W}(t)) ,$$

and assume

(4.16) $$t_m \ll t_M .$$

Then one expects that for $t \gg t_m$, in eq. (4.15) $\widetilde{W}(t)$ can be replaced by an appropriate solution of the equation

(4.17) $$\frac{\partial \widetilde{W}^{(0)}(t)}{\partial t} = \mathcal{Q}\, \widetilde{W}^{(0)}(t) ,$$

where

(4.18) $$\mathcal{Q} = -\int_0^{+\infty} dt\, \mathcal{P}\mathcal{H}\, e^{-i\hat{\mathcal{H}}t}\, \mathcal{H}\mathcal{P}.$$

As is apparent eq. (4.17) is obtained from eq. (4.11) by replacing $\tilde{W}(t-t')$ by $\tilde{W}(t)$ and letting the range of the integration go to infinity. Proceeding in a more systematic way one could write

(4.19) $$\tilde{W}(t) = \tilde{W}^{(0)}(t) + \tilde{W}^{(1)}(t) + \ldots$$

with $\tilde{W}^{(0)}(t)$ as the solution of eq. (4.17), $W^{(n)}(t)$ defined iteratively by the equation

(4.20) $$\frac{\partial \tilde{W}^{(n+1)}(t)}{\partial t} = \int_0^t dt' \, \mathcal{K}(t') \tilde{W}^{(n)}(t-t') - \frac{\partial \tilde{W}^{(n)}(t)}{\partial t}$$

For $\tilde{W}^{(1)}(t)$ one can obtain the following estimate

(4.21) $$Tr \, (M \, \tilde{W}^{(1)}(t)) \sim \frac{t_m}{t_M} \, Tr\,(M \, \tilde{W}^{(0)}(t)) \, ;$$

so one can write

(4.22) $$\langle M \rangle = Tr\,(M \tilde{W}^{(0)}(t)) \, (1 + o(\frac{t_m}{t_M}))$$

We note that eq. (4.22) remains certainly valid if M is replaced by a sufficiently smooth function $f(M)$. This is not necessarily true if $f(M)$ is a sharp function. In this case the characteristic time for $f(M)$ may be much smaller than t_M and even than t_m. In particular the "distribution of probability" $u_\nu^{(0)}(t) = Tr\,(P_\nu \tilde{W}^{(0)}(t))$ derived by $\tilde{W}^{(0)}(t)$ may be not a good approximation for $u_\nu(t) = Tr\,(P_\nu \tilde{W}(t))$. This fact is, however, of no physical significance and depends strictly on the technical definition of M.

Note that we can write for any t and t_0

(4.23) $$\tilde{W}^{(0)}(t) = e^{\mathcal{L}(t-t_0)} \, \tilde{W}^{(0)}(t_0) \, ;$$

for this reason eq. (4.17) is called a markoffian master equation. The operator $\mathcal{L}$ can be shown to be dissipative (i.e. the real part of its expectation value is non positive), it has a null eigenvalue and the corresponding eigenvectors correspond to the equilibrium states of the system.

In terms of the distribution function $\mu_\nu^{(0)}(t)$ eq. (4.17) takes the form

(4.24) $$\frac{d\mu_\nu^{(0)}(t)}{dt} = \sum_{\nu'} Q_{\nu\nu'}\, \mu_{\nu'}^{(0)}(t) ,$$

where
$$Q_{\nu\nu'} = \frac{1}{s_{\nu'}} Tr\,(P_\nu\, 2\, P_{\nu'}) ,$$

and in the approximation in which M is treated as a continuous variable

(4.24') $$\frac{\partial \mu^{(0)}(\mu,t)}{\partial t} = \int d\mu' \left(Q(\mu,\mu')\, \mu^{(0)}(\mu',t) - Q(\mu',\mu)\, \mu^{(0)}(\mu,t)\right)$$

(where the relation $\sum_\nu Q_{\nu\nu'} = 0$ has been taken into account).

From

(4.22') $$\langle M\rangle_t \cong Tr\,(M\, \mathcal{W}^{(0)}(t)) = \int d\mu\; \mu\; \mu^{(0)}(\mu,t)$$

we then obtain

(4.25) $$\frac{d\langle M\rangle_t}{dt} = \langle \alpha(M)\rangle_t ,$$

where

(4.26) $$\alpha(\mu) = \int d\mu'\; Q(\mu',\mu)\,(\mu'-\mu) ,$$

and, under the assumption

(4.27) $$\langle (M-\langle M\rangle)^2\rangle \frac{\partial^2\alpha}{\partial\mu^2} \ll \alpha ,$$

finally

(4.28) $$\frac{d\langle M\rangle_t}{dt} = \alpha(\langle M\rangle_t)$$

which is of the form (2.17). On the other hand assumption (4.27) can be justified deriving in an analogous way an equation of motion for the second moment $\langle (M - \langle M\rangle)^2\rangle_t$ (cf. ref. 1).

5 - IDEALIZATION OF THE MEASUREMENT APPARATUS AND INTERACTION WITH THE OBJECT

We can idealize an apparatus in the following way.

An apparatus is a large body the macroscopic quantities of which can be divided in three categories which we shall denote by E, J and M'.

The symbol E stands for the set of the macroscopic quantities which are constants of the motion as long as the system remains isolated; E includes the energy, and its values specify the different possible equilibrium states; J stands for a quantity or a set of quantities with a relaxation time t_J ; M' for a set of quantities with a relaxation time t_M.

We assume that

(5.1) $$t_M \ll t_J$$

Then, at a certain initial time $t=0$ the apparatus is assumed to be in an equilibrium state corresponding to a definite value of E. As a consequence of the interaction with the object the quantities J and M' are changed and the apparatus is removed from its equilibrium state; in general the modification of E can be considered negligible; J is assumed to be modified in a way which is in a one-one correspondence with the value of the quantity A of the object we are measuring. Due to the above assumptions the apparatus evolves towards the equilibrium following different paths corresponding to the different values of A . As a limit case J can be a constant of the motion itself and then it is the final equilibrium state of the apparatus to be in a one-one correspondence to the value of A ; in this case the apparatus itself includes the permanent record of the result, usually, however, this record belongs to the macroscopic level already and only the <u>transient</u> behaviour at the apparatus is relevant.

To show that our idealization is reasonable, let us consider some examples of actual apparatuses.

Let us start with the proportional counter (Fig.1). We assume that the counter is originally in equilibrium, charged at a potential U_0 equal to the e.m.f. of the generator and with no charged particle in it. If a charged particle goes through the counter it produces N ions; the average value $N(\omega)$ of N is related to the energy ω of the particle. Then we have a discharge and if λ is the ions multiplication coefficient of the instrument and C the capacity of the condenser, the difference of potential decreases by the value $\frac{e\lambda}{C}N(\omega)$ and the pointer of the electrometer assumes a new position which is related to the energy of the particle. Finally, the generator restores the original difference of potential in the counter. Here J may be identified with the expression $U - \frac{e\lambda}{C}N$, where U and N are the difference of potential and the number of ions at a given time. If a charged particle goes through the counter, J changes from its original value $J = U_0$ to $J = U_0 - \frac{e\lambda}{C}N(\omega)$; it remains practically constant during the discharge and so, when this is finished $(N=0)$, we have $U = J = U_0 - \frac{e\lambda}{C}N(\omega)$; afterwards J starts to increase again to the equilibrium value U_0 . Here t_M is essentially the

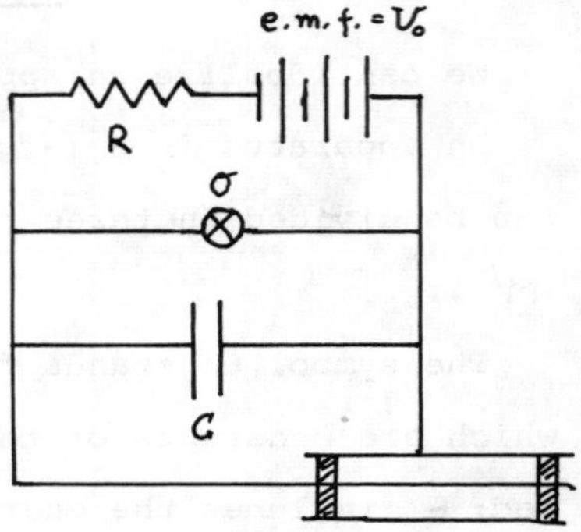

Fig. 1

time of discharge, whilst t_J is the time of recharge, that is maintained appropriately long by the resistance R . A strictly similar analysis can be made for the cases of a scintillation counter, of a system of Geiger counters or of a photographic plate.

For the system of Geiger counters J stands for the set of dichotomic variables specifying whether a counter is charged or not at a certain time, for the photographic plate J specifies similarly the states of the various grains.

In the case of the Wilson chamber or the bubble chamber J may be identified with the variables specifying the position in which the ions are created, t_M specifies the time necessary in order a drop or a bubble is built around an ion, t_J indicates the diffusion time and the recombination time of an ion. In the spark chamber, finally, J specifies the position in which the ions are produced near the plates, t_M denotes the migration time of the electrons in the direction of the electrostatic field and t_J is the diffusion time of the electrons transverse to the field or again the recombination time of the ions.

Let us now go back to the formal discussion.
The symbol M of section 4 represents the set (E, J, M') in the case of the apparatus and the index ν represents a set of indices (a, k, ς) specifying the eigenvalues of E, J and M' respectively; the two times t_M have obviously to be identified.

As in sections 2 and 3 we shall denote by I the object and by II the apparatus and write

$$
(5.2) \quad \begin{cases} P^{II}_{ak\varsigma} \equiv P^{II}_{\nu} = \sum_j |\Omega^{II}_{ak\varsigma j}\rangle\langle\Omega^{II}_{ak\varsigma j}| \\ P^{II}_{ak} = \sum_\varsigma P^{II}_{ak\varsigma} \\ P^{II}_{a} = \sum_{k\varsigma} P^{II}_{ak\varsigma} \end{cases}
$$

(5.3) $$\widetilde{W}^{II} = \mathcal{P}^{II} W = \sum_{\nu} \frac{1}{s_{\nu}} P_{\nu}^{II} \, T_{r}^{II} (P_{\nu}^{II} W^{II}).$$

We may assume that at the initial time $t = 0$ the state of the system is represented by the microcanonical statistical operator corresponding to a given value of E,

(5.4) $$W_0^{II} = \frac{1}{S_{a_0}} P_{a_0}^{II} \quad , \quad S_{a_0} = \sum_{\rho k} s_{a_0 \rho k} \; ,$$

or more realistically

(5.4') $$W_0^{II} = \sum_a u_a^0 \frac{1}{S_a} P_a^{II}$$

with u_a^0 sharply peaked for $E \sim E_{a_0}$.

For the initial state of the compound system $I + II$ we then write

(5.5) $$W^{I+II}(0) = W_0^{I} \otimes W_0^{II} = \sum_{r r'} c_r c_{r'}^{*} \, |\varphi_r^{I}\rangle\langle\varphi_{r'}^{I}| \otimes W_0^{II} .$$

Let be

(5.6) $$H = H^{I} + H^{II} + H_{int} = H_0 + H_{int}$$

the total hamiltonian. We introduce the hypothesis that the interaction between I and II is effective only for a short time assuming that the operator $e^{iH_0 t} e^{-iHt}$ has a limit for $t \to +\infty$:

(5.7) $$\lim_{t \to +\infty} e^{iH_0 t} e^{-iHt} = U$$

and that this limit is actually achieved at a time τ_{int} sufficientely small

(5.8) $$\tau_{int} \ll t_M .$$

The operator U has to be essentially indentified with the operator $e^{-iH\tau}$ of eq. (3.2) and this equation may be now replaced by

$$(5.9) \qquad U\left(|\varphi_r^I\rangle \otimes |\Omega^{II}_{a_0 k\rho j}\rangle\right) = |\varphi_r^I\rangle \otimes U_r^{II}\, |\Omega^{II}_{a_0 k\rho j}\rangle .$$

For $t \gg \tau_{int}$ the statistical operator of the compound system can then be written as

$$(5.10) \qquad W^{I+II}(t) = e^{-iHt}\, W^{I+II}(0)\, e^{iHt} =$$

$$= e^{-iH_0 t}\left(e^{iH_0 t}\, e^{-iHt}\right) W^{I+II}(0) \left(e^{iHt}\, e^{-iH_0 t}\right) e^{iH_0 t} \longrightarrow$$

$$\longrightarrow e^{-iH_0 t}\, U\, W^{I+II}(0)\, U^*\, e^{iH_0 t} =$$

$$= \sum_{rr'} c_r c_{r'}^*\, e^{-iH^I t} |\varphi_r^I\rangle\langle\varphi_{r'}^I| e^{iH^I t} \otimes$$

$$\otimes\, e^{-iH^{II} t}\, U_r^{II}\, W_0^{II}\, U_{r'}^{II*}\, e^{iH^{II} t} .$$

If, for what concerns the apparatus II, we restrict our considerations to the macroscopic quantities alone, $W^{I+II}(t)$ may be replaced by

$$(5.11) \qquad \tilde{W}^{I+II}(t) \equiv \left(1^I \otimes \mathcal{P}^{II}\right) W^{I+II}(t) =$$

$$= \sum_{rr'} c_r c_{r'}^*\, e^{-iH^I t} |\varphi_r^I\rangle\langle\varphi_{r'}^I| e^{iH^I t} \otimes \tilde{W}^{II}_{rr'}(t) ,$$

where

$$(5.12) \qquad \tilde{W}^{II}_{rr'}(t) = \mathcal{P}^{II}\, e^{-i\mathcal{H}^{II} t} \left(U_r^{II}\, W_0^{II}\, U_{r'}^{II*}\right) =$$

$$= \sum_{k\rho} \frac{1}{s_{a_0 k\rho}}\, P^{II}_{a_0 k\rho}\, Tr^{II}\left(P^{II}_{a_0 k\rho}\, e^{-iH^{II} t}\, U_r^{II}\, W_0^{II}\, U_{r'}^{II*}\, e^{iH^{II} t}\right).$$

The fact that the interaction between I and II establishes a correlation between the values of A^{I} and the values of J^{II} can be expressed saying that the quantity

$$(5.13)\qquad T_r^{II}\left(P_{a_0k\rho}^{II}\, U_r^{II} W_0^{II} U_r^{II*}\right) = \frac{1}{S_{a_0}}\, T_r^{II}\left(P_{a_0k\rho}^{II}\, U_r^{II} P_{a_0}^{II} U_r^{II*}\right)$$

is different from zero only for k in the neighbourhood of a certain value k_r . From this assumption, by Schwarz inequality

$$\left\{T_r^{II}\left(P_{a_0k\rho}\, U_r^{II} P_{a_0}^{II} U_{r'}^{II*}\right)\right\}^2 = \left\{T_r^{II}\left(P_{a_0k\rho} U_r^{II} P_{a_0}^{II} P_{a_0}^{II} U_{r'}^{II*} P_{a_0k\rho}^{II}\right)\right\}^2 \le$$

$$\le T_r^{II}\left(P_{a_0k\rho}^{II}\, U_r^{II} P_{a_0}^{II} U_r^{II*}\right) \cdot T_r^{II}\left(P_{a_0k\rho}^{II}\, U_{r'}^{II} P_{a_0} U_{r'}^{II*}\right),$$

it follows that $\widetilde{W}_{rr'}^{II}(0)$ is negligible for $r \neq r'$.
Schematically we can write

$$\widetilde{W}_{rr'}^{II}(0) = \delta_{rr'}\, \widetilde{W}_r^{II}(0) \equiv$$

$$(5.14)\qquad \equiv \delta_{rr'} \sum_\rho \frac{1}{s_{a_0k_r\rho}}\, P_{a_0k_r\rho}^{II}\, T_r^{II}\left(P_{a_0k_0\rho}^{II}\, U_r^{II} W_0^{II} U_r^{II*}\right).$$

Due to assumption (5.1) we can also write

$$(5.15)\qquad \widetilde{W}_{rr'}^{II}(t) \cong \delta_{rr'}\, \widetilde{W}_r^{II}(t) \qquad \text{for} \quad t \ll t_J .$$

Proceeding as for eq. (4.11) the following generalized master equation can be derived for $\widetilde{W}_{rr'}^{II}(t)$:

$$(5.16)\qquad \frac{\partial \widetilde{W}_{rr'}^{II}(t)}{\partial t} = -\int_0^t dt'\, \mathcal{P}^{II} \mathcal{H}^{II} e^{-i\hat{\mathcal{H}}^{II}t'} \mathcal{H}^{II} \mathcal{P}^{II}\, \widetilde{W}_{rr'}^{II}(t-t') -$$

$$- i\, \mathcal{P}^{II} \mathcal{H}^{II} e^{-i\hat{\mathcal{H}}^{II}t} \left(U_r^{II} W_0^{II} U_{r'}^{II*}\right).$$

The last term in eq. (5.16) comes from the fact that $(1-\mathcal{P}^{II})\cdot(U_r^{II} W_0^{II} U_{r'}^{II*}) \neq 0$. Since we can write

(5.17) $$\mathcal{P}^{II} \mathcal{H}^{II} e^{-i\mathcal{H}^{II}t} (U_r^{II} W_0^{II} U_{r'}^{II*}) =$$

$$= \sum_\nu \frac{1}{s_\nu} P_\nu^{II} \int d\omega\, T_r^{II}(P_\nu^{II} \mathcal{H}^{II} \Psi_\omega^{II}) e^{-i\omega t} T_r^{II}(\Psi_\omega^{II*} U_r^{II} W_0^{II} U_{r'}^{II*}),$$

such term is expected to decay in a time of the order of $t_m = \frac{1}{\delta\omega}$.

Finally for $t \gg t_m$ we can replace $\tilde{W}_{rr'}^{II}(t)$ by $\tilde{W}_{rr'}^{II(0)}(t)$, the solution of the markoffian master equation

(5.18) $$\frac{\partial \tilde{W}_{rr'}^{II(0)}(t)}{\partial t} = \mathcal{Q}^{II} W_{rr'}^{II(0)}(t),$$

with an error in the evaluation of the expectation values of operators for the compound system of the form $B^I \otimes M^{II}$ (being B^I any observable for I) of the order t_m / t_M.

For $t_m \ll t_0 \ll t_J$ we have

(5.19) $$\tilde{W}_{rr'}^{II(0)}(t) = e^{\mathcal{Q}^{II}(t-t_0)} \tilde{W}_{rr'}^{II}(t_0) \cong \delta_{rr'}\, e^{\mathcal{Q}^{II}(t-t_0)} \tilde{W}_r^{II}(t_0).$$

In conclusion, under the above specifications, we can replace $W^{I+II}(t)$ by

(5.20) $$\tilde{W}^{I+II(0)}(t) = \sum_r |c_r|^2 e^{-iH^I t} |\varphi_r^I\rangle\langle\varphi_r^I| e^{iH^I t} \otimes \tilde{W}_r^{II(0)}(t) =$$

$$= \sum_r |c_r|^2 e^{-iH^I t} |\varphi_r^I\rangle\langle\varphi_r^I| e^{iH^I t} \otimes e^{\mathcal{Q}^{II}(t-t_0)} \tilde{W}_r^{II}(t_0),$$

in which all desired kinds of interference terms have disappeared.

6 - INDEPENDENT SUBDYNAMICS FORMULATION

In this section I want to quote and briefly discuss the main result of the paper of C. George et al. (3); I refer to the original paper for the details.

Starting from certain assumptions on the thermodynamic limit of many components systems, C. George et al. show that any solution of the Liouville-von Neumann equation (4.4) has for large t an asymptotic behaviour of the form:

$$W(t) \underset{t\to+\infty}{\longrightarrow} \tilde{W}(t) = \tilde{\Pi}\, W(t), \tag{6.1}$$

where $\tilde{\Pi}$ is a non orthogonal projection operator which has the property

$$\tilde{\Pi}\, e^{-i\mathcal{H}t} = e^{-i\mathcal{H}t}\, \tilde{\Pi}. \tag{6.2}$$

It follows from eq. (6.2) that $\tilde{W}(t)$ is an **exact** solution of eq. (4.4), but also satisfies the Markoffian master equation

$$\frac{d\tilde{W}(t)}{dt} = \mathcal{Q}\, \tilde{W}(t), \tag{6.3}$$

with

$$\mathcal{Q} = -i\mathcal{H}\tilde{\Pi} = -i\tilde{\Pi}\mathcal{H}. \tag{6.4}$$

The operator $\tilde{\Pi}$ plays a role similar to the operator $\mathcal{P}$ of section 4, the characteristic time of $\tilde{W}(t)$ is much larger than that of $W(t)$ and the interpretation of the formalism is the same . There are however certain important differences.

a) The operator $\tilde{\Pi}$ is not introduced a priori as $\mathcal{P}$, on the basis of physical arguments, but it depends exclusively on the mathematical structure of the hamiltonian. b) Eq. (6.1) is a much stronger result then eq. (4.22); it states that the interference terms between macroscopically distinguishable states vanish for $t \longrightarrow +\infty$ independently of any a priori restriction to a special class of observables.

The assumptions which are at the basis of the derivation of the mentioned result are essentially assumptions of analyticity in the z-plane for the operator,

$$(1-\mathcal{P}_0)\frac{1}{(1-\mathcal{P}_0)\mathcal{H}(1-\mathcal{P}_0)-iz}(1-\mathcal{P}_0) \tag{6.5}$$

Here the hamiltonian H is assumed to be written as $H = H_0 + V$ (H_0 being the free hamiltonian of the elementary components and V their interaction) and $\mathcal{P}_0$ is the projection operator which selects the diagonal part in the H_0-representation of an element $W \in \mathcal{L}$. The mentioned analyticity properties can be shown to be satisfied in significant examples (e.g. a gas with short range interactions) in the thermodynamic limit and at any order of the pertubation theory.

In conclusion the results of C. George et al. are extremely interesting; they give a very clear solution for the problem of the macroscopic description of large bodies and refine the analysis of section 5. It should, however, be mentioned that the paper is essentially heuristic, many developments are purely formal and mathematical ambiguities are present (cf. Ref. (11)).

7 - MODELS IN C*-ALGEBRAS FORMALISM

I now want to try to summarize the paper of K. Hepp with some minor changes and simplifications.

The central idea of the C*-algebra approach is that for a system with infinitely many degrees of freedom the relevant representations of the algebra of observables are not irreducible and the center is non trivial. To the center of a representation can then be associated a set of privileged quantities which are compatible among themselves and with any other observable; such quantities are called classical observables. The macroscopic quantities are assumed to be special cases of classical observables.

The basic postulates are the following:

1) The set of observables of the system generates a C^*-algebra $\mathcal{A}$ with unit.
2) This algebra is assumed to have a quasi local structure; i.e. to every bounded open region $\Lambda \subset \mathbb{R}^3$ is associated a C^*-subalgebra $\mathcal{A}(\Lambda)$ with unit generated by the set of observables referring to the region Λ.
3) $\mathcal{A}(\Lambda)$ satisfies the following conditions

$$(7.1) \quad \begin{cases} \mathcal{A}(\Lambda') \subset \mathcal{A}(\Lambda'') & \text{for } \Lambda' \subset \Lambda'' \\ [\mathcal{A}(\Lambda'), \mathcal{A}(\Lambda'')] = 0 & \text{for } \Lambda' \cap \Lambda'' = \emptyset \\ \bigvee_{\Lambda} \mathcal{A}(\Lambda) \text{ norm dense in } \mathcal{A} & \end{cases}$$

4) A "state" of the system is represented by a continuous positive linear functional ω on $\mathcal{A}$ such that $\omega(1) = 1$.
5) The time evolution of the system is expressed by an automorphism Φ_t of the algebra $\mathcal{A}$. This description of the time evolution corresponds most naturally to the Heisenberg picture of the ordinary formulation of quantum mechanics. However, the identity

$$(7.2) \quad \omega(\Phi_t(A)) = (\omega \circ \Phi_t)(A)$$

provides immediately the analogue of the Schroedinger picture.

6) For a given representation π of $\mathcal{A}$ in a Hilbert space $\mathcal{H}_\pi$ the self-adjoint elements of the center

(7.3) $$Z_\pi = \pi'(\mathcal{A}) \cap \pi''(\mathcal{A})$$

are called, as already mentioned, "classical observables". The macroscopic quantities are assumed to belong to Z_π. Among the macroscopic quantities of special interest are the <u>intensive quantities</u> which are represented by expressions of the form

(7.4) $$\varrho = w\text{-}\lim_{N\to\infty} \frac{1}{N} \sum_{n=1}^{N} \pi(A_n) ,$$

where $A_n \in \mathcal{A}(\Lambda_n)$ and $\bigcup_{n=1}^{\infty} \Lambda_n = \mathbb{R}^3$.

Note that in contrast to the ordinary formalism of quantum mechanics, which we have summarized in section 2, in the present formalism there is no Hilbert space associated a priori to the system; the states are defined as elements of the dual space of $\mathcal{A}$ and only the abstract structure of $\mathcal{A}$ matters. Hilbert spaces are introduced only with reference to specific representations of $\mathcal{A}$.

Let us now consider a given representation π of $\mathcal{A}$ in a Hilbert space $\mathcal{H}_\pi$. To every unit vector $|\psi\rangle \in \mathcal{H}_\pi$ can be associated, as in the usual formulation, a "vector state" $\omega_\psi(A) = \langle\psi\,|\,\pi(A)\psi\rangle$. The new phenomenon that occurs in a system with infinitely many degrees of freedom is that, considering the three unit vectors $|\psi_1\rangle$, $|\psi_2\rangle$ and $|\psi\rangle = c_1|\psi_1\rangle + c_2|\psi_2\rangle$ ($|c_1|^2+|c_2|^2=1$) it can happen that the vector state ω_ψ is identical to the mixture $|c_1|^2\omega_{\psi_1} + |c_2|^2\omega_{\psi_2}$, i.e. it can happen that

(7.5) $$\omega_\psi(A) = |c_1|^2\,\omega_{\psi_1}(A) + |c_2|^2\,\omega_{\psi_2}(A) \qquad \forall A \in \mathcal{A}$$

or equivalently

(7.6) $$\langle\psi_1\,|\,\pi(A)\psi_2\rangle = 0 \qquad \forall A \in \mathcal{A}$$

If eq. (7.5) or (7.6) are satisfied the vectors $|\psi_1\rangle$ and $|\psi_2\rangle$ are said to be incoherent, in the other case they are said to be coherent.

In connection with the described phenomenon the concept of the Gel'fand-Naimark-Segal- representation associated with a given state and the concept of disjointness of two states are of particular interest.

For a given state ω the corresponding G.N.S.-representation π_ω is defined in the following way; introducing the subset of $\mathcal{A}$ (left ideal)

$$\mathcal{J} = \{ Y \,/\, Y \in \mathcal{A} \,,\; \omega(Y^* Y) = 0 \}$$

one denotes by $|\xi(A)\rangle$ the class of those elements of $\mathcal{A}$ which differ from a given $A \in \mathcal{A}$ by an element of $\mathcal{J}$. The set of all classes $|\xi(A)\rangle$ realizes a linear space which can be equipped with the scalar product

$$\langle \xi(A) | \xi(B) \rangle = \omega(A^* B) \;;$$

the Hilbert space $\mathfrak{h}_\omega$ obtained by completion of this space is assumed as the representation space of the representation of $\mathcal{A}$ defined by

$$\pi_\omega(B) | \xi(A) \rangle = | \xi(BA) \rangle \,.$$

Obviously $\omega(A) = \langle \xi(1) | \pi_\omega(A) | \xi(1) \rangle$ and $|\xi(1)\rangle$ is a cyclic vector for the representation.

In general π_ω is not irreducible and possesses various subrepresentations.

Two states ω_1 and ω_2 are said to be disjoint if π_{ω_1} , and π_{ω_2} have no unitarily equivalent subrepresentations.

After such premises we are in condition to quote the following lemmas reported by Hepp.

<u>Lemma 1:</u> ω_1 and ω_2 are disjoint if and only if for every representation π of $\mathcal{A}$ for which $\omega_i(A) = \langle \psi_i | \pi(A) \psi_i \rangle$ with $|\psi_1\rangle, |\psi_2\rangle \in \mathcal{H}_\pi$, eq. (7.6) is satisfied.

<u>Lemma 2:</u> If ω_1 and ω_2 are disjoint and ϕ is an automorphism of $\mathcal{A}$, $\omega_1 \circ \phi$ and $\omega_2 \circ \phi$ are disjoint.

<u>Lemma 3:</u> Consider the sequences $\omega_{in} \xrightarrow[n\to\infty]{w} \omega_i$, i = 1,2, with ω_1 and ω_2 disjoint. Let π_n be representations of $\mathcal{A}$ such that $\omega_{in}(A) = \langle \psi_{in} | \pi_n(A) \psi_{in} \rangle$ with $|\psi_{in}\rangle \in \mathcal{H}_{\pi_n}$. Then, for all $A \in \mathcal{A}$

$$\lim_{n\to\infty} \langle \psi_{1n} | \pi_n(A) | \psi_{2n} \rangle = 0 \tag{7.7}$$

<u>Lemma 4:</u> Let $\omega_i = \sum_{n=1}^{\infty} p_{in}\, \omega_{in}$, $p_{in} \geq 0$, $\sum_{n=1}^{\infty} p_{in} = 1$ for i = 1,2 and let ω_{1m} and ω_{1n} be disjoint for all m, n. Then ω_1 and ω_2 are disjoint.

Lemma 1 establishes the equivalence between incoherentness and disjointness for vector states.

According to lemma 2 it is impossible that, as a result of the evolution during a finite interval of time, two coherent vector states become disjoint.[x]

It is, however, possible that this happens in an infinite time and then lemma 3 states that the corresponding interference terms vanish in the limit.

As matter of fact one can give very simple examples of sequences of coherent states ω_{1n} and ω_{2n} which converge weakly towards disjoint states ω_1 and ω_2. Let us identify $\mathcal{A}$ with the algebra

---

[x] However, this is possible if the time evolution does not corresponds to an automorphism of $\mathcal{A}$; Hepp discusses briefly also this case.

of observables of a system of infinitely many spins at the lattice sites

(7.8) $$[\sigma_j^h, \sigma_{j'}^k] = 2i\, \delta_{jj'}\, \varepsilon^{hkl}\, \sigma_j^l .$$

In the Hilbert space $\mathfrak{H} = \bigotimes_{j=1}^{\infty} \mathbb{C}_j^2$ let us put

(7.9) $$\begin{cases} |\psi^+\rangle = \bigotimes_{j=1}^{\infty} |\varphi_j^+\rangle \\ |\psi^-\rangle = \bigotimes_{j=1}^{\infty} |\varphi_j^-\rangle \end{cases}$$

and

(7.10) $$\begin{cases} |\psi_n^+\rangle = |\psi^+\rangle \\ |\psi_n^-\rangle = \bigotimes_{j=1}^{n} |\varphi_j^-\rangle \bigotimes_{j=n+1}^{\infty} |\varphi_j^+\rangle \end{cases} ,$$

where $|\varphi_j^\pm\rangle$ denotes the eigenstates of σ_j^3 in $\mathbb{C}_j^2$

(7.11) $$\sigma_j^3\, |\varphi_j^\pm\rangle = \pm\, |\varphi_j^\pm\rangle .$$

One has

(7.12) $$\begin{cases} \langle \psi_n^+ | \sigma_{j_1}^{k_1} \sigma_{j_2}^{k_2} \dots \sigma_{j_p}^{k_p} | \psi_n^+ \rangle \xrightarrow[n\to\infty]{} \langle \psi^+ | \sigma_{j_1}^{k_1} \sigma_{j_2}^{k_2} \dots \sigma_{j_p}^{k_p} | \psi^+ \rangle \\ \langle \psi_n^- | \sigma_{j_1}^{k_1} \sigma_{j_2}^{k_2} \dots \sigma_{j_p}^{k_p} | \psi_n^- \rangle \xrightarrow[n\to\infty]{} \langle \psi^- | \sigma_{j_1}^{k_1} \sigma_{j_2}^{k_2} \dots \sigma_{j_p}^{k_p} | \psi^- \rangle \end{cases}$$

for arbitrary $j_1, j_2, \dots j_p$, $k_1, \dots k_p$; i.e.

(7.12') $$\omega_{\psi_n^+} \xrightarrow[n\to\infty]{w} \omega_{\psi^+} , \quad \omega_{\psi_n^-} \xrightarrow[n\to\infty]{w} \omega_{\psi^-} .$$

Now $|\psi_n^+\rangle$ and $|\psi_n^-\rangle$ are coherent, since they can be obtained each from the other by application of an appropriate element of the algebra. ω_{ψ^+} and ω_{ψ^-} are disjoint, since obviously

$$\langle \psi^- | \sigma_{j_1}^{k_1} \sigma_{j_2}^{k_2} \dots \sigma_{j_p}^{k_p} | \psi^+ \rangle = 0$$

for any $j_1, j_2, \dots j_p, k_1, \dots k_p$.

There is another class of states of particular interest in connection with the macroscopic quantities, namely the class of primary states. A state ω is said to be primary if the center $Z_\omega = \pi_\omega'(\mathfrak{A}) \cap \pi_\omega''(\mathfrak{A})$ of the representation $\pi_\omega(\mathfrak{A})$ is trivial, i.e. if $Z_\omega = \{\lambda 1\}$. For a primary state all classical quantities have obviously a definite value. In particular for the intensive quantities the following lemmas are important.

<u>Lemma 5.</u> Let ω be a primary state and let A_n be as in (7.3). If

$$\lim_{N\to\infty} \frac{1}{N} \sum_{n=1}^{N} \omega(A_n) = a \tag{7.13}$$

exists, then one has in $\mathfrak{H}_\omega$

$$w\text{-}\lim_{N\to\infty} \frac{1}{N} \sum_{n=1}^{N} \pi_\omega(A_n) = a\,1 \tag{7.14}$$

<u>Lemma 6.</u> Let ω_1, ω_2 be primary and A_n as in (7.3). If

$$\lim_{N\to\infty} \frac{1}{N} \sum_{n=1}^{N} \omega_i(A_n) = a_i \quad , \quad i=1,2 \quad , \tag{7.15}$$

and $a_1 \neq a_2$, then ω_1 and ω_2 are disjoint.

In the spin lattice example considered above the vector states ω_{ψ^+}, $\omega_{\psi_n^-}$ and ω_{ψ^-} are obviously primary, since the corresponding G.N.S.-representations are irreducible and can be identified with the restriction of the algebra $\mathfrak{A}$ to the invariant subspaces cyclically generated by the vectors $|\psi^+\rangle$ and $|\psi^-\rangle$ respectively. If we introduce the spin density

$$\vec{\Sigma} = w\text{-}\lim_{N\to\infty} \frac{1}{N} \sum_{j=1}^{N} \vec{\sigma}_j \tag{7.16}$$

we find that Σ^3 has the value 1 for ω_{ψ^+}, ω_{ψ^-} and the value -1 for ω_{ψ^-} (Σ^2 and Σ^1 have the value 0 for all these three states and do not discriminate between them).

To show the actual relevance of the above results for the problem of measurement in quantum mechanics, Hepp discusses a number of simple models of measurement processes in which the undesired interference terms in the state of the compound system I + II vanish in the limit $t \longrightarrow \infty$ as a consequence of lemma 3 and of the phenomenon exemplified in eq. (7.12'). I shall report on one of these models, the Coleman model.

As object I we consider a particle of zero mass and spin $1/2$ in one dimension; then we can write $\mathfrak{h}^{I} = L^2(\mathbb{R}^1) \otimes \mathbb{C}^2$ for the Hilbert space corresponding to I, $H^{I} = P$ for the free hamiltonian and assume σ^3 to be the quantity we want to observe. As apparatus II we consider a system of infinitely many spins, just as in the example considered above, but placed now at the points of integer coordinates of the positive real half-axis. We assume that II has no free evolution, i.e. $H^{II} = 0$, $H_0 = H^{I} + H^{II} = P$ and that the interaction between I and II is expressed by

$$(7.17) \qquad H_{int} = \sum_{n=1}^{\infty} V(x-n)\, \sigma_n^1\, P_-$$

where $V(x)$ is a function with compact support with $\int_{-\infty}^{+\infty} dx\, V(x) = \frac{\pi}{2}$ and P_+ and P_- are the projectors corresponding to the two eigenvalues of σ^3 .

We write

$$(7.18) \qquad \begin{cases} \phi_t(A) = e^{i(H^I + H_{int})t}\, A\, e^{-i(H^I + H_{int})t} \\ \phi_t^{\circ}(A) = e^{iH^I t}\, A\, e^{-iH^I t} \end{cases},$$

where A is any element of the algebra $\mathfrak{a}^{I+II}$ of the observables of the compound system I + II. Then we have

$$(7.19) \qquad (\phi_t \circ \phi_{-t}^{\circ})(A) = W^+(t)\, A\, W(t) \quad ,$$

where

$$(7.20) \qquad W(t) = P_+ + U(t)\, P_-$$

and $U(t)$ is defined by the Dyson equation

(7.21) $$U(t) = 1 - i\int_0^t ds\, e^{ips} H_{int}\, e^{-ips}\, U(s)$$

$$= 1 - i\int_0^t ds \sum_{n=1}^{\infty} V(x+s-n)\, \sigma_n^1\, U(s).$$

The solution of (7.21) is

(7.22) $$U(t) = e^{-i\int_0^t ds \sum_{n=1}^{\infty} V(x+s-n)\,\sigma_n^1}$$

Then, let $|\chi\pm\rangle = |\chi\rangle \otimes |\varphi\pm\rangle$, where $|\chi\rangle \in L^2(\mathbb{R}^1)$, $|\varphi\pm\rangle \in \mathbb{C}^2$, $\sigma^3|\varphi\pm\rangle = \pm|\varphi\pm\rangle$ and assume that the support of χ is contained in the negative real halfaxis and that supp $\chi(x) \cap$ supp $V(x) = \emptyset$. Let us assume II to be initially in the vector state $|\psi\pm\rangle \in h^{II}$ defined as in eq. (7.9). We have

(7.23) $$\begin{cases} W(t)\,|x+\rangle \otimes |\psi^+\rangle = |x+\rangle \otimes |\psi^+\rangle \\ W(t)\,|x-\rangle \otimes |\psi^+\rangle = |x-\rangle \otimes U(t)\,|\psi^+\rangle = \\ = |x-\rangle \bigotimes_{n=1}^{\infty}\left(|\varphi_n^+\rangle \cos\int_0^t ds\, V(x+s-n) - i\,|\varphi_n^-\rangle \sin\int_0^t ds\, V(x+s-n)\right) \end{cases}$$

and

(7.24) $$|\varphi_n^+\rangle \cos\int_0^t ds\, V(x+s-n) - i\,|\varphi_n^-\rangle \sin\int_0^t ds\, V(x+s-n) \xrightarrow[t\to+\infty]{}$$

$$\xrightarrow[t\to+\infty]{} -i\,|\varphi_n^-\rangle.$$

From (7.19), (7.23), (7.24), arguing as in (7.12) and taking into account that

(7.25) $$\omega_\psi \circ \phi_t = (\omega_\psi \circ \phi_t \circ \phi_{-t}^0) \circ \phi_t^0 = \omega_{W(t)\psi} \circ \phi_t^0$$

we have finally

(7.26)
$$\omega_{|\chi+\rangle \otimes |\psi+\rangle} \circ \Phi_t = \omega_{|\chi+\rangle \otimes |\psi+\rangle} \circ \Phi_t^{\circ}$$
$$\omega_{|\chi-\rangle \otimes |\psi+\rangle} \circ \Phi_t \xrightarrow[t \to +\infty]{w} \omega_{|\chi-\rangle \otimes |\psi-\rangle} \circ \Phi_t^{\circ}$$

($|\psi-\rangle$ is again defined as in eq. (7.9)).

Eq. (7.26) is the analogue of eqs. (3.2) or (5.9) in the present formalism and qualifies II as a measurement apparatus of the quantity σ^3 for I. Since $|\psi+\rangle$ and $|\psi-\rangle$ are disjoint, we can also write

(7.27)
$$\omega_{(c_+|\chi+\rangle + c_-|\chi-\rangle) \otimes |\psi+\rangle} \circ \Phi_t \xrightarrow[t \to +\infty]{w} \left(|c_+|^2 \omega_{|\chi+\rangle \otimes |\psi+\rangle} + |c_-|^2 \omega_{|\chi-\rangle \otimes |\psi-\rangle} \right) \circ \Phi_t^{\circ}$$

where an interference term between $|\chi+\rangle$ and $|\chi-\rangle$ is no longer present.

8 - CRITICAL REMARKS ON THE MODELS OF APPARATUSES WITH INFINITELY MANY DEGREES OF FREEDOM

The model of the apparatus we have discussed at the end of the last section contains many oversimplifications as compared to realistic cases. Such simplifications are responsable for the lack of many of the characteristics of a measurement device we have described in section 5. Particularly the absence of an interaction between the components of II and the absence of a free evolution of this system are non realistic (for this reason all problems discussed in section 4 are lacking).

For what concerns the treatment of an apparatus or more generally of a macroscopic body as a system with infinitely many degrees of freedom two different attitudes may be taken:

1) The fact may be considered as an idealization of an actual system with a finite but extremely large number of degrees of freedom (as the greatest number of authors seems inclined to do).

In this case the exact vanishing of the interference terms in eq. (7.27) or (6.1) has no special meaning. In the real situation such terms are extremely small but finite. The choice of the formalism of sections 6 and 7 rather than that used in sections 4 and 5 may be a matter of mathematical convenience and rigour, but in principle the various treatments suffer exactly from the same limitations.

2) One may start from the observation that a macroscopic body can never be an isolated system and accept the suggestion made in various forms by many authors that a consistent theory must in some way consider the entire universe. Having this in mind the apparatus should be thought in interaction with the "rest of the world" and the system with infinitely many degrees of freedom could idealize the apparatus + the rest of the world. If this attitude is taken, eqs. (6.1) and (7.27) become relevant in "principle". For what concerns the C^*-algebra formalism it must be stressed, however, that the specific model we have discussed and the other similar ones which have been proposed are too simple to prove these ideas to be fruitful; in particular it is certainly necessary to consider non homogeneous systems and consequently macroscopic quantities which are functions of the space; it is not at all clear that these quantities can be considered as classical variables in the sense of section 7.

As a final remark I want to stress what seems to be a general character of all models with infinitely many degrees of freedom; the complete disappearence of the interference terms occurs only in the limit $t \rightarrow +\infty$.

REFERENCES

(1) G.M. Prosperi - Proc.Intern.School of Phys. E. Fermi, Course 49, 97 (1971).

cf. also

A. Daneri, A. Loinger and G.M. Prosperi - Nucl.Phys. 44, 297 (1962).

L. Lanz, G.M. Prosperi and A. Sabadini - Nuovo Cim. 2 B, 184 (1971).

L. Rosenfeld - Supp. Progr. Theor. Phys. extra numb. 222 (1965)

(2) G. Ludwig - Notes in Math. Phys. Marburg University NMP 5, NMP 6 (1972) and this issue.

(3) C. George, I. Prigogine and L. Rosenfeld - Danske Mat. fys. Medd. 38 (1972).

(4) F. Haake and W. Weidlich - Zeits. Phys., 213, 451 (1968).

(5) H. Primas - Preprint Swiss Institute of Technology, Zurich.

(6) K. Hepp - Helv. Phys. Acta, 45, 234 (1972).

(7) H.D. Zeh - Proc. Intern. School of Phys. E. Fermi, Course 49, 263 (1972).

(8) J. von Neumann - Mathematical Foundation of Quantum Mechanics, chap. V and VI, Princeton 1955.

(9) M.M. Yanase - Proc. Intern. School of Phys. E. Fermi, Course 49, 77 (1972).

(10) R. Zwanzig - Lectures in Theor. Phys., Vol. 3, 106, Boulder 1960.

E. W. Montrol - ibid. 221.

G. Ludwig - Zeits. Phys. 173, 232 (1963).

I. Prigiogine - Nonequilibrium Statistical Mechanics, New York 1962.

(11) L. Lanz, L.A. Lugiato and G. Ramella - Physica, 54, 94 (1971).

(12) J.M. Jauch - Helv. Phys. Acta, 37, 193 (1964).

THE CENTRE OF A PHYSICAL SYSTEM

C.M. Edwards

The Queen's College,

Oxford, England

1. Introduction. Axiomatic approaches to the theory of statistical physical systems fall into two broad categories, the 'ordered vector space' or 'operational' approach and the 'lattice' approach. For classical probability theory the connection between the two is fully understood and is described briefly in §3. For non-classical systems the theory is less fully developed. In both approaches it is possible to associate with the mathematical object representing a system a notion of 'centre'. The centre is an object of the same category and is supposed to represent the classical part of the system. In particular, when the mathematical object and its centre coincide, the system is deemed to be classical. It is therefore essential in any axiomatic approach to ensure that the centre is well-defined and in fact represents a classical system. The main purpose of this paper is to describe such an axiomatic approach from the operational point of view.

The rest of the introduction is taken up with a brief description of the standard operational approach. To give a complete description a list of axioms should be given and the relevant mathematical structure should be deduced. However, since the axioms used in deducing the structure described below are more or less universally agreed, they will be omitted. In §2 a statement of the main mathematical results is given and in §3 the standard classical situation is described. In §4 the model which arises from the new axioms is described and the centre of any system is shown to be a classical system in the sense of §3. In §5 the interaction between classical and non-classical systems is discussed briefly.

The set of <u>states</u> of a physical system is represented by a cone V^+ in a complete base norm space (2,3,4,5,6,10,11,12). Often V^+ is supposed to be closed in the base norm but this will not be assumed here. The base K for V^+ represents the set of <u>normalised states</u>. An <u>operation</u> on the system is represented by an element T of the convex set $\mathcal{L}(V)_1^+$ of positive norm non-increasing linear operators on V. Alternatively T can be thought of as an affine mapping from K into the cap $\mathrm{conv}(K \cup \{0\})$. The set of operations is therefore represented by some subset P of $\mathcal{L}(V)_1^+$, the minimum reasonable assumptions on P being that P is convex, $0,1 \varepsilon P, T_1, T_2 \varepsilon P$ implies that $T_1 T_2 \varepsilon P$ and $T \varepsilon P$, $T \lesssim 1$ implies that $1-T \varepsilon P$. The ordering of P is that defined by the cone $\mathcal{L}(V)^+$ of positive bounded linear operators on V.

The dual space V^* of V is a complete order unit space (11) (GM-space with unit (13)) with weak* closed cone V^{*+} and order unit e defined for $x_1, x_2 \varepsilon V^+$ by $e(x_1 - x_2) = \|x_1\| - \|x_2\|$. For $T \varepsilon P$, let T^* be the adjoint operator acting on V^*.

For $x \varepsilon V^+$, e(x) is the <u>strength</u> of the state represented by x and corresponds to a number recorded on some counting device. The normalised states are those of unit strength. For $T \varepsilon P$, $x \varepsilon V^+$

$$\frac{e(Tx)}{e(x)} = \frac{T^* e(x)}{e(x)}$$

is the <u>transmission probability</u> for the state represented by x under the operation represented by T. Hence $T_1, T_2 \varepsilon P$ represent operations indistinguishable by counting procedures, or <u>isotonic</u> operations, if and only if $T_1^* e = T_2^* e$. An isotony class of operations is called a <u>simple observable</u> (<u>effect</u>(10),<u>test</u>(7)). Hence if $\bar{T}$ is the isotony class containing $T \varepsilon P$, $\bar{T}$ is the simple observable measured by T. There exists a bijection $\bar{T} \to T^* e$ between the set $\bar{P}$ of simple observables and a convex subset L of the unit interval $[0,e]$ in V^*. In the sequel $\bar{P}$ and L are identified and therefore L is regarded as representing the set of simple observables. Notice that $0, e \varepsilon L$ and in the special case $P = \mathcal{L}(V)_1^+, L = [0,e]$. It will be assumed that P also has the property that if $T \varepsilon P$, there exists $T' \varepsilon P$ such that $T^* e + T'^* e = e$.

An element $T \varepsilon P$ represents a <u>strongly repeatable</u> operation provided that there exists $T' \varepsilon P$ such that (i) $T^* e + T'^* e = e$, (ii) $TT' = T'T = 0$, (iii) if $S \varepsilon P, T^* S^* e = e, T'^* S^* e = 0$, then $S^* e = T^* e$. In this case $T^* e \varepsilon E(L)$ the set of extreme points of L. E(L) is said to be the set of <u>propositions</u> (<u>decision effects</u>).

An operation represented by $T \varepsilon P$ is said to be <u>classical</u> provided that $T \lesssim 1$. Let P_c represent the set of classical simple observables and let $L_c = \{T^* e : T \varepsilon P_c\}$ represent the set of <u>classical simple</u>

observables. L_c is a convex subset of L and hence $E(L_c)$, the set of classical propositions, is contained in E(L). If $T\varepsilon P_c, T^2 = T$, then T is said to represent a superselection rule. Superselection rules are strongly repeatable and define classical propositions. Notice that for $x\varepsilon V^+$, $x = Tx+(1-T)x$ is a unique decomposition into elements Tx of the split face TV^+ of V^+ and x-Tx of the complementary split face $(1-T)V^+$. TV^+ represents the set of states for which the classical proposition represented by T^*e is true and $(1-T)V^+$ represents the set of states for which it is false. For $x_1,x_2\varepsilon V^+$ write $x_1 \sim x_2$ if and only if $T^*e(x_1) = T^*e(x_2)$, $\forall\ T\varepsilon P_c$ in which case the states represented by x_1,x_2 are indistinguishable under classical operations. The set V_c^+ of equivalence classes of elements of V^+ under the equivalence relation ~ represents the set of classical states of the system.

In general, the system the set of states of which is represented by V_c^+ and the set of simple observables of which is represented by L_c does not form a classical system in the usual sense of probability theory. However it will be shown below that P can be chosen in such a way that this is the case. Then the corresponding classical system is said to be the centre of the original system.

2. Preliminaries. Recall that the ideal centre $\mathcal{O}(W)$ of a real vector space W with generating cone W^+ consists of those linear operators S on W for which there exists $\lambda \geq 0$ such that $\lambda w \pm Sw\varepsilon W^+$, $\forall\ w\varepsilon W^+$ (13).

Let W be a complete order unit space with unit e. Then $\mathcal{O}(W)$ is algebraically and order isomorphic to the algebra $C(\Omega)$ of real-valued continuous functions on a compact Hausdorff space Ω. The mapping $S \to Se$ from $\mathcal{O}(W)$ into W is an order isomorphism onto a closed subspace Z(W) of W. Z(W) is said to be the centre of W. W is said to be monotone σ-complete if every uniformly bounded monotone increasing sequence in W has a least upper bound in W. A positive linear mapping S from W into another such space W' is said to be σ-normal if for every uniformly bounded monotone increasing sequence $\{w_n\}$ W, $\mathrm{lub}Sw_n = S(\mathrm{lub}w_n)$.

The proofs of the following results can be found in (5).

THEOREM 1. Let W be a monotone σ-complete order unit space with unit e and let the set C of σ-normal linear functionals on W satisfy the condition that $w\varepsilon W, x(w) \geq 0$, $\forall\ x\varepsilon C$ implies that $w \geq 0$. Then,

(i) $\mathcal{O}(W)$ is monotone σ-complete and the set C_1 of σ-normal linear functionals on $\mathcal{O}(W)$ satisfies the condition that $S \in \mathcal{O}(W)$, $g(S) \geq 0$, $\forall g \in C_1$ implies that $S \geq 0$.

(ii) The set $\mathcal{J}(W)$ of idempotents in $\mathcal{O}(W)$ forms a Boolean σ-algebra uniformly generating $\mathcal{O}(W)$ and the set C_1 of σ-additive positive measures on $\mathcal{J}(W)$ satisfies the condition that $S_1, S_2 \in \mathcal{J}(W)$, $g(S_1) = g(S_2)$, $\forall g \in C_1$ implies that $S_1 = S_2$.

(iii) If $S \in \mathcal{O}(W), S \geq 0$, then S is σ-normal.

(iv) The mapping $S \rightarrow Se$ from $\mathcal{O}(W)$ onto $Z(W)$ is σ-normal.

(v) $Z(W)$ is monotone σ-complete and the set C_c of σ-normal linear functionals on $Z(W)$ satisfies the condition that $z \in Z(W)$, $g(z) \geq 0$, $\forall g \in C_c$ implies that $z \geq 0$.

(vi) The set $E(L_c)$ of extreme points of the unit interval L_c in $Z(W)$ forms a Boolean σ-algebra uniformly generating $Z(W)$ and the set C_c of σ-additive positive measures on $E(L_c)$ satisfies the condition that $z_1, z_2 \in E(L_c), g(z_1) = g(z_2)$, $\forall g \in C_c$ implies that $z_1 = z_2$.

THEOREM 2. Let W satisfy the conditions of Theorem 1 and let $V^+ \subset C$ be a cone satisfying

(i) $w \in W, x(w) \geq 0$, $\forall x \in V^+$ implies that $w \geq 0$.

(ii) If $K = \{x : x \in V^+, x(e) = 1\}$, then $V = V^+ - V^+$ is a complete base norm space with base K.

Then the mapping $w \rightarrow w'$ defined for $w \in W$ by $w'(x) = x(w)$, $\forall x \in K$ is a σ-normal order isomorphism from W onto a monotone σ-complete order unit space which is weak* dense in the dual space V^* of V and which possesses the same order unit.

THEOREM 3. Let W, V^+ satisfy the conditions of Theorem 2. Then, the conditions of Theorem 2 are also satisfied when W is replaced by $Z(W)$ and V^+ is replaced by $V_c^+ = V^+/Z(W)^o$ where $Z(W)^o$ is the annihilator of $Z(W)$. Moreover, $Z(Z(W)) = Z(W), (V_c^+)_c = V_c^+$.

3. Classical Systems. In classical probability theory the set of propositions is represented by a Boolean σ-algebra $\mathcal{L}$ and the set of states is represented by a point separating family V^+ of σ-additive positive measures on $\mathcal{L}$. In order to describe the model in operational terms it is necessary to suppose that V^+ is a generating cone for a complete base norm space with base $K = \{x : x \in V^+, x(e) = 1\}$ where e is the largest element in $\mathcal{L}$. $\mathcal{L}$ can be regarded as a subset of the dual space V^* of V and the uniformly closed linear span W of $\mathcal{L}$ is then a monotone σ-complete order unit space and $\mathcal{L}$ is the set $E(L)$

of extreme points of the unit interval in W. The proofs of all these remarks follow from the observation that W can be identified with $C(\Omega)$ where Ω is the Stone space of $\mathcal{L}$. Classical systems therefore possess an adequate description in either the 'lattice' or 'operational' approach. Notice that W, V^+ satisfy the conditions of Theorem 2 above and that $Z(W) = W$.

4. Arbitrary systems. In (5) it is shown that plausible axioms can be introduced in such a way that an arbitrary statistical physical system can be described in the following manner.

POSTULATE. To each physical system there corresponds a complete base norm space V with generating cone V^+ and a monotone σ-complete order unit space W weak* dense in the dual space V^* of V and possessing the same order unit. V^+ is invariant under a set P of adjoints of elements of a subset P' of the set of norm non-increasing σ-normal linear operators on W. P' possesses the properties: P' is convex; $S \to Se$ maps P' onto the unit interval L in W; $1 \varepsilon P'$; $S_1, S_2 \varepsilon P'$ implies that $S_1 S_2 \varepsilon P'$; if $S \varepsilon P'$, $S \leq 1$, then $1 - S \varepsilon P'$. The sets of states and operations are represented respectively by V^+ and P.

It follows immediately that the set of simple observables is represented by L and the set of propositions by E(L). From Theorem 1 it follows that the set of classical operations is isomorphic to the unit interval in $\mathcal{O}(W)$. Hence the set of classical simple observables is represented by the unit interval L_c in Z(W) and the set of classical propositions by the Boolean σ-algebra $E(L_c)$. The set of classical states is represented by $V_c^+ = V^+/Z(W)^\circ$. The next result follows from Theorem 3.

THEOREM 4. Under the conditions of the postulate, let Z(W) be the centre of W, let $V_c^{\pm} = V^+/Z(W)^\circ$ and let $V_c = V_c^+ - V_c^+$. Let $\tilde{P}$ be the set of linear operators on V_c which are adjoints of elements of the set of restrictions to Z(W) of those elements of P' which leave Z(W) invariant. Then the postulate is satisfied with V,W,P replaced respectively by $V_c, Z(W), \tilde{P}$.

The corresponding physical system is said to be the centre of the original system. If follows from §3 that a system is classical if and only if it coincides with its own centre. In particular Theorem 3 shows that the centre of any system is classical.

5. Instruments. The notion of instrument as introduced in (1,2) fits particularly well into this setting. Suppose that a classical system is represented by the spaces V_1, W_1 and that an arbitrary system is represented by the spaces V_2, W_2. Then, an instrument is represented by a mapping $\mathcal{E}$ from L_1, the unit interval in W_1, into P_2, the set of linear operators on V_2 representing operations on the second system, satisfying,

(i) $\mathcal{E}$ is affine.

(ii) $\mathcal{E}(e_1)^*e_2 = e_2$.

(iii) If $\{a_{1n}\} \subset L_1$ is monotone increasing with least upper bound a_1, then

$$\text{lub}\, \mathcal{E}(a_{1n})^*a_2 = \mathcal{E}(a_1)^*a_2, \quad \forall\, a_2 \varepsilon L_2.$$

(iv) If $\{a_{2n}\} \subset L_2$ is monotone increasing with least upper bound a_2, then

$$\text{lub}\, \mathcal{E}(a_1)^*a_{2n} = \mathcal{E}(a_1)^*a_2, \quad \forall\, a_1 \varepsilon L_1.$$

Notice that if the classical system is that obtained by choosing $V_1 = R^2, V_1^+ = \{(x_1,x_2): x_1,x_2 \geqslant 0\}, K_1 = \{(x_1,x_2): x_1,x_2 \geqslant 0, x_1+x_2=1\}$, $W_1 = R^2$ and $P_1 = \mathcal{L}(R^2)_1^+$ an instrument consists of convex combinations of the operations $0,1,T,T' \varepsilon P_2$ where $T = \mathcal{E}((1,0)), T' = \mathcal{E}((0,1))$.

An instrument $\mathcal{E}$ gives rise to an observable (1,2) $\mathcal{A}$ by writing $\mathcal{A}(a_1) = \mathcal{E}(a_1)^*e_2, \forall\, a_1 \varepsilon L_1$. In particular the restriction of $\mathcal{A}$ to the Boolean σ-algebra $E(L_1)$ is a σ-homomorphism into L_2.

6. Example. Recall that a Baire* algebra $\mathfrak{A}$ is a C*-algebra the self-adjoint part W of which is monotone σ-closed and the cone V^+ of σ-normal linear functionals on which is point separating (9). Hence a certain class of physical systems is obtained by choosing the set of states to be represented by the cone of σ-normal linear functionals on a Baire* algebra $\mathfrak{A}$ with identity. Notice that any countably generated Baire* algebra possesses an identity. For such a system the set of states and simple observables of the centre of the system are represented respectively by the set of restrictions of elements of V^+ to the centre $\mathfrak{Z}(\mathfrak{A})$ of $\mathfrak{A}$ and the unit interval in $\mathfrak{Z}(\mathfrak{A})$. Notice that every W*-algebra is a Baire* algebra.

References

1. Davies, E.B.: On the repeated measurement of continuous observables in quantum mechanics. J.Funct.Anal. 6, 318-346 (1970)
2. Davies, E.B., Lewis, J.T.: An operational approach to quantum probability. Commun.math.Phys. 17, 277-304 (1970).
3. Edwards, C.M.: Classes of operations in quantum theory. Commun. math.Phys. 20, 26-56 (1971).
4. Edwards, C.M.: The theory of pure operations. Commun.math.Phys. 24, 260-288 (1972).
5. Edwards, C.M.: Alternative axioms for statistical physical theories. (preprint).
6. Edwards, C.M., Gerzon, M.A.: Monotone convergence in partially ordered vector spaces. Ann.Inst.Henri Poincaré A 12, 323-328 (1970).
7. Giles, R.: Foundations of quantum mechanics. J.Math.Phys. 11, 2139-2160 (1970).
8. Hartkämper, A.: Preparing and measuring procedures. Proceedings of the Advanced Study Institute on 'Foundations of quantum mechanics and ordered linear spaces', Marburg. Berlin: Springer 1973.
9. Kehlet, E.T.: On the monotone sequential closure of a C*-algebra. Math.Scand. 25, 59-70 (1969).
10. Ludwig, G.: The measuring and preparing process and macro theory. Proceedings of the Advanced Study Institute on 'Foundations of quantum mechanics and ordered linear spaces', Marburg. Berlin: Springer 1973.
11. Nagel, H.: Order unit and base norm spaces. Proceedings of the Advanced Study Institute on 'Foundations of quantum mechanics and ordered linear spaces', Marburg. Berlin: Springer 1973.
12. Neumann, H.: The structure of ordered Banach spaces in axiomatic quantum mechanics. Proceedings of the Advanced Study Institute on 'Foundations of quantum mechanics and ordered linear spaces', Marburg. Berlin: Springer 1973.
13. Wils, W.: The ideal center of partially ordered vector spaces. Acta Math. 127, 41-79 (1971).

OPERATIONS AND EFFECTS IN THE HILBERT SPACE FORMULATION OF QUANTUM THEORY

K. KRAUS

Physikalisches Institut der Universität Würzburg
Würzburg, Germany

1. INTRODUCTION

The notion of "effects" is a basic one in Ludwig's axiomatic approach to quantum theory [1]. Another approach, proposed by Davies, Edwards, Lewis and others [2], uses "operations" as a fundamental concept. In these lectures, I will investigate effects and operations in the Hilbert space model of quantum theory.

The intention of my lectures is twofold. First, they shall give a simple illustration of the axiomatic approaches just mentioned, thus facilitating the understanding of some other lectures contained in this Volume. Second, I want to show by means of examples how the notions of effects and operations may be applied to concrete physical problems. Quite apart from their significance for certain axiomatic formulations, these notions thus turn out to be useful even in ordinary Hilbert space quantum theory. As my intention is mainly pedagogical, I feel justified to keep the following discussion as self-contained and elementary as possible.

The investigations to be reported have been done mainly in collaboration with K.-E. Hellwig [3], and are reviewed comprehensively in Ref. [4].

2. A MODEL FOR EFFECTS AND OPERATIONS

The following considerations may also be found elsewhere in this Volume [1]. Nevertheless, I felt I should repeat them here, very shortly and in a somewhat simpler language, for the convenience of the reader.

Assume a quantum system, described in a Hilbert space $\mathcal{H}$, to be in a state corresponding to a normalized density operator W. It is convenient here to use the Heisenberg picture, such that W is independent of time unless there are external interventions of the type studied below. The expectation values of observables B in this state W are

$$\mathrm{Tr}(BW). \tag{2.1}$$

Now the system is coupled to an apparatus, described in another Hilbert space $\mathcal{H}'$ by an initial density operator W', such that some interaction between system and apparatus occurs during a finite time interval. After this interaction, the state of system plus apparatus is given by the density operator

$$\mathbf{S}(W \otimes W')\mathbf{S}^*$$

on $\mathcal{H} \otimes \mathcal{H}'$, with a unitary "scattering" operator $\mathbf{S}$. Thus the probability to observe after the interaction some property of the apparatus, described in $\mathcal{H}'$ by a projection operator E', is given by

$$\mathrm{Tr}\,[(1 \otimes E')\mathbf{S}(W \otimes W')\mathbf{S}^*]. \tag{2.2}$$

For fixed W', $\mathbf{S}$ and E', (2.2) defines a linear function of W which is continuous with respect to the trace norm and takes values between zero and one. Accordingly, (2.2) may be rewritten as

$$\mathrm{Tr}(FW) \tag{2.3}$$

with an uniquely defined operator F on $\mathcal{H}$ such that

$$F = F^*, \quad 0 \leq F \leq 1. \tag{2.4}$$

This operator F represents an effect in the sense of Ludwig. Any observation of an apparatus property E' after a "scattering" interaction of the above type thus corresponds to such an effect F, which is produced by the system in state W with the probability (2.3).

Usually one considers, instead of such general effects, the more particular ones corresponding to projection operators E only, which are called "properties", "propositions" or, in Ludwig's terminology, "decision effects". However, this further restriction implicitly requires a very particular choice of the initial apparatus state W', the apparatus property E' and the interaction $\mathcal{S}$. Consequently, the axiomatic approach of Ludwig starts with general effects F, and distinguishes the subclass of decision effects E at a later stage of the theory only.

The set of general effects F is also a very natural one since it is not further enlarged if we replace in Eq. (2.2) the projection operator E' on the apparatus state space $\mathcal{H}'$ by an F' which satisfies (2.4) only. On the other hand, as we shall see later on, any operator F on $\mathcal{H}$ which satisfies (2.4) may be obtained from a model of the above type with suitable $\mathcal{H}'$, W', $\mathcal{S}$ and E'.

In order to explain what is meant by an operation, we have to investigate the state of our system after it has triggered the effect F at the apparatus. Assume the above experiment to be done repeatedly with an ensemble consisting of a large number N of systems in the original state W. According to (2.3), N.Tr(FW) systems from this ensemble will trigger the effect F. We select this subensemble of systems for further measurements and ask for ensemble averages in this new ensemble.

The new ensemble average for B measurements obviously is

$$\mathrm{Tr}\,[(B \otimes E')\mathcal{S}(W \otimes W')\mathcal{S}^*] \,/\mathrm{Tr}\,[(1 \otimes E')\mathcal{S}(W \otimes W')\mathcal{S}^*] \qquad (2.5)$$

The numerator in this expression is, for fixed W, W', $\mathcal{S}$ and E', a positive linear functional of B, continuous with respect to the ultraweak operator topology.

Therefore we may rewrite it as

$$\mathrm{Tr}\ [(B \otimes E')\mathcal{S}(W \otimes W')\mathcal{S}^*] = \mathrm{Tr}(B\hat{W}) \qquad (2.6)$$

with a uniquely defined density operator $\hat{W}$. Moreover, $\mathrm{Tr}\hat{W}$ coincides with the denominator of (2.5), such that

$$\tilde{W} = \hat{W}/\mathrm{Tr}\hat{W} \qquad (2.7)$$

is a normalized density operator, in terms of which the expectation value (2.5) may be represented as

$$\mathrm{Tr}(B\tilde{W}). \qquad (2.8)$$

The overall effect of the procedure just described is thus the transformation of the original state W of the system into a new one, $\tilde{W}$.

This state change is essentially what we mean by an operation. It is more conveniently described as $W \to \hat{W}$, with the final state $\hat{W}$ defined by (2.6) which in general is not normalized, instead of $W \to \tilde{W}$ with a normalized final state. As a formal advantage, $\hat{W}$ is a linear function of W whereas $\tilde{W}$ in general is not. Moreover,

$$\mathrm{Tr}\hat{W} = \mathrm{Tr}\ [(1 \otimes E')\mathcal{S}(W \otimes W')\mathcal{S}^*] = \mathrm{Tr}(FW), \qquad (2.9)$$

thus $\mathrm{Tr}\hat{W}$ as a function of W also determines the effect F which is observed during the operation. Finally, $\mathrm{Tr}\hat{W}$ is the transition probability (selectivity) of the state change, since just this fraction of the original ensemble goes over into the new state W.

If $\mathrm{Tr}\hat{W} = 1$ for all W, the operation is called non-selective. An equivalent characterization is, by (2.9), F = 1. It may also happen that $\hat{W} = 0$ for some W. Such trivial state changes need not be considered further and have been excluded already, e.g., in Eq. (2.5).

3. EXPLICIT REPRESENTATION OF OPERATIONS AND EFFECTS

We are now going to derive formulae for the final state $\hat{W}$ of an operation and for the corresponding effect F which are explicit in the sense that they contain operators acting on the state space

$\mathcal{H}$ of the system only. For this purpose, it is convenient first to simplify the model somewhat. We will return to the general case later on. The simplifying assumption is that the apparatus initially is in a pure state, i.e.,

$$W' = |f'\rangle \langle f'|, \tag{3.1}$$

a projection operator onto some unit vector $f' \in \mathcal{H}'$.

In order to evaluate Eq. (2.6), we calculate the trace on the left hand side using a complete orthonormal system $f_i \otimes g_k'$ in $\mathcal{H} \otimes \mathcal{H}'$, with arbitrary complete orthonormal systems f_i in $\mathcal{H}$ and g_k' in $\mathcal{H}'$, respectively. After some rearrangement of the operators in the trace, we insert at appropriate places intermediate states $f_i \otimes f_k'$ with another orthonormal system f_k' in $\mathcal{H}'$ for which

$$f_1' = f' . \tag{3.2}$$

This leads to

$$\begin{aligned}
\mathrm{Tr}(B\hat{W}) &= \mathrm{Tr}\,[(B \otimes E')\mathbf{S}(W \otimes W')\mathbf{S}^*] \\
&= \mathrm{Tr}\,[(B \otimes 1')(1 \otimes E')\mathbf{S}(W \otimes W')\mathbf{S}^*\,(1 \otimes E')] \\
&= \sum_{\substack{ijk\\lmn}} ((Bf_i \otimes g_k'),(1 \otimes E')\mathbf{S}(f_j \otimes f_l'))((f_j \otimes f_l'),(W \otimes W')\cdot \\
&\qquad\qquad \cdot(f_m \otimes f_n'))((f_m \otimes f_n'),\, \mathbf{S}^*(1 \otimes E')(f_i \otimes g_k')) \\
&= \sum_{ikjm} ((Bf_i \otimes g_k'),(1 \otimes E')\mathbf{S}(f_j \otimes f'))(f_j,\, W f_m)\cdot \\
&\qquad\qquad \cdot((f_m \otimes f'),\, \mathbf{S}^*(1 \otimes E')(f_i \otimes g_k')).
\end{aligned} \tag{3.3}$$

Now we define operators A_k on $\mathcal{H}$ by

$$(g, A_k\, f) = ((g \otimes g_k'),\ (1 \otimes E')\mathbf{S}(f \otimes f')) \tag{3.4}$$

for arbitrary f and g in $\mathcal{H}$. These operators A_k obviously exist and satisfy $\| A_k \| \leqslant 1$, and Eq. (3.4) defines them uniquely. However, the choice of the orthonormal system g_k' in (3.4) is completely arbitrary, which may be used for simplifications. For instance, we may choose the g_k' such that a subset of them spans the subspace $E'\mathcal{H}'$ of $\mathcal{H}'$, which is particularly useful if the dimension of $E'\mathcal{H}'$ is finite. Eq. (3.4) implies also

$$(f, A_k^* g) = ((f \otimes f'), S^*(1 \otimes E')(g \otimes g_k')).$$

With this and (3.4), the calculation (3.3) may be continued as

$$\sum_{ikjm} (Bf_i, A_k f_j)(f_j, W f_m)(f_m, A_k^* f_i) = \sum_{ik} (f_i, BA_k WA_k^* f_i),$$

thus leading to

$$\mathrm{Tr}(B\hat{W}) = \mathrm{Tr}(B \cdot \sum_k A_k WA_k^*). \qquad (3.5)$$

This has to hold true for any observable B, and therefore

$$\hat{W} = \sum_k A_k WA_k^* . \qquad (3.6)$$

An analogous calculation, starting with a different rearrangement of the operators in Tr $[(B \otimes E')S(W \otimes W')S^*]$, leads to

$$\mathrm{Tr}(B\hat{W}) = \mathrm{Tr}(\sum_k A_k^* BA_k \cdot W), \qquad (3.7)$$

which of course follows trivially from (3.5) if the k sum is finite. If the k sum is infinite, one may show that $\sum_k A_k WA_k^*$ and $\sum_k A_k^* BA_k$ converge in the trace norm and ultraweak topology, respectively [4].

We may also derive explicit formulae like (3.6) and (3.7) for a generalized model, with the particular initial state (3.1) and the decision effect E' replaced by a general density matrix W' and a general effect F', respectively. It turns out that Eqs.(3.6) and (3.7) remain true, if the operators A_k are properly redefined. This may be shown either by elementary calculations, as described in Ref. [4]. Alternatively, one may exploit the fact, proved below in Section 6, that mappings of the form (3.6) are already the most general ones that have all the properties of operations as described in Section 2.

For the particular case $B = 1$, Eq. (3.7) together with (2.9) leads to the representation

$$F = \sum_k A_k^* A_k \tag{3.8}$$

of the effect F corresponding to the operation (3.6). This in turn implies that the operators A_k must satisfy

$$\sum_k A_k^* A_k \leq 1. \tag{3.9}$$

Eqs. (3.6) and (3.8) clearly indicate that the correspondence between effects and operations is very far from being one-to-one. As particular cases, Eq. (3.6) already contains the state changes $W \to EWE$ with a projection operator E and $W \to \sum_k E_k WE_k$ with mutually orthogonal projection operators E_k which are usually associated with so called "ideal measurements". It is again clear, however, that assumptions of this type for the operators A_k correspond to very stringent restrictions for W', $\mathcal{S}$ and E', which will not be satisfied in the vast majority of actual experiments.

4. FORMAL CONSTRUCTION OF AN APPARATUS FOR A GIVEN OPERATION

Our next task will be to show that any set of operators A_k satisfying the condition (3.9) belongs to a model of the type considered before with suitably chosen $\mathcal{H}'$, f', $\mathcal{S}$ and E'. An operation (3.6) with the given operators A_k may thus be considered as being realizable by means of the corresponding apparatus. In this sense, Eq. (3.9) is the only condition restricting possible operations (3.6) and effects (3.8). From a physical point of view, however, one has to expect additional conditions for a realizable apparatus which go beyond the mere existence of a formal description in terms of $\mathcal{H}'$, f', $\mathcal{S}$ and E'. In particular, the class of interactions $\mathcal{S}$ actually occurring in nature is severely restricted. Some of these restrictions have been discussed, although also in a more formal way, in Refs. [3], [4] and [5].

Assume we are given a set of operators $A_k \neq 0$ satisfying condition (3.9). Denote by K the number of these operators, K thus being finite or denumerably infinite [4]. We choose $\mathcal{H}'$ as a Hilbert space of K + 1 dimensions with a complete orthonormal system f'_k, k = 0 ...K. Any vector $\mathbf{f}$ in the product space $\mathcal{H} \otimes \mathcal{H}'$ may be decomposed as

$$\mathbf{f} = \sum_{k=0}^{K} f_k \otimes f'_k$$

with unique $f_k \in \mathcal{H}$. This establishes a canonical isomorphism of $\mathcal{H} \otimes \mathcal{H}'$ and the direct sum $\bigoplus_{k=0}^{K} \mathcal{H}_k$ of K+1 copies $\mathcal{H}_k$ of $\mathcal{H}$, if one identifies the above $\mathbf{f} \in \mathcal{H} \otimes \mathcal{H}'$ with the vector

$$\bigoplus_{k=0}^{K} f_k \in \bigoplus_{k=0}^{K} \mathcal{H}_k.$$

The latter may be written most conveniently as a column vector

$$\begin{pmatrix} f_0 \\ \cdot \\ \cdot \\ \cdot \\ f_K \end{pmatrix} .$$

The operator matrix

$$\mathbf{A} = \begin{pmatrix} A_1 \\ \cdot \\ \cdot \\ \cdot \\ A_K \end{pmatrix}$$

may be considered as a linear transformation of $\mathcal{H}$ into the direct sum $\bigoplus_{k=1}^{K} \mathcal{H}_k$ of K copies $\mathcal{H}_k$ of $\mathcal{H}$, if we define

$$\mathbf{A}f = \begin{pmatrix} A_1 f \\ \cdot \\ \cdot \\ \cdot \\ A_K f \end{pmatrix} \in \bigoplus_{k=1}^{K} \mathcal{H}_k.$$

Then, by (3.9),

$$\| \mathbf{A}f \|^2 = \sum_k \| A_k f \|^2 = \sum_k (f, A_k^* A_k f) \leq \| f \|^2,$$

thus

$$\|\mathbf{A}\| \leq 1. \tag{4.1}$$

The adjoint of $\mathbf{A}$ is given by

$$\mathbf{A}^* = (A_1^* \dots A_K^*),$$

which maps $\bigoplus_{k=1}^{K} \mathcal{H}_k$ into $\mathcal{H}$ according to

$$\mathbf{A}^* \begin{pmatrix} f_1 \\ \cdot \\ \cdot \\ \cdot \\ f_K \end{pmatrix} = \sum_{k=1}^{K} A_k^* f_k .$$

The operator

$$\mathbf{A}^* \mathbf{A} = \sum_{k=1}^{K} A_k^* A_k$$

maps $\mathcal{H}$ into itself and coincides with F, whereas the K x K operator matrix

$$\mathbf{A}\,\mathbf{A}^* = \begin{pmatrix} A_1 A_1^* & \dots & A_1 A_K^* \\ \cdot & & \\ \cdot & & \\ \cdot & & \\ A_K A_1^* & \dots & A_K A_K^* \end{pmatrix}$$

maps $\bigoplus_{k=1}^{K} \mathcal{H}_k$ into itself. By (4.1), the operators $(1-\mathbf{A}^*\mathbf{A})^{\frac{1}{2}}$ and $(1-\mathbf{A}\,\mathbf{A}^*)^{\frac{1}{2}}$ are bounded self-adjoint operators on $\mathcal{H}$ and $\bigoplus_{k=1}^{K} \mathcal{H}_k$, respectively. Since they may be approximated by polynomials in $\mathbf{A}^*\mathbf{A}$ and $\mathbf{A}\,\mathbf{A}^*$, respectively, the relations

$$\begin{aligned} \mathbf{A}(1 - \mathbf{A}^*\mathbf{A})^{\frac{1}{2}} &= (1 - \mathbf{A}\mathbf{A}^*)^{\frac{1}{2}}\, \mathbf{A}, \\ \mathbf{A}^*(1 - \mathbf{A}\mathbf{A}^*)^{\frac{1}{2}} &= (1 - \mathbf{A}^*\mathbf{A})^{\frac{1}{2}}\, \mathbf{A}^* \end{aligned} \tag{4.2}$$

follow easily [6].

After these preparations, we are now ready to give the explicit construction of $\mathbf{S}$, f' and E'. For $\mathbf{S}$ we take, in a self-explanatory matrix notation,

$$\mathbf{S} = \begin{pmatrix} (1 - \mathbf{A}^*\mathbf{A})^{\frac{1}{2}} & \mathbf{A}^* \\ \mathbf{A} & -(1 - \mathbf{A}\mathbf{A}^*)^{\frac{1}{2}} \end{pmatrix} \qquad (4.3)$$

as acting on arbitrary column vectors

$$\begin{pmatrix} f_o \\ \cdot \\ \cdot \\ \cdot \\ f_K \end{pmatrix} \in \bigoplus_{k=0}^{K} \mathcal{H}_k \equiv \mathcal{H} \otimes \mathcal{H}' .$$

The unitarity relation $\mathbf{S}^* \mathbf{S} = \mathbf{S}\mathbf{S}^* = 1$ follows trivially by matrix multiplication using (4.2). (In fact, $\mathbf{S}$ is also selfadjoint). The initial state vector f' and the observed property E' of the apparatus are chosen to be

$$f' = f_o'$$

and

$$E' = \sum_{k=1}^{K} |f_k' >< f_k'| .$$

Finally, the orthonormal system g_k' in $\mathcal{H}'$ entering Eq. (3.4) is identified with f_k', k = O...k.
The relations

$$((g \otimes f_k'),(1 \otimes E')\mathbf{S}(f \otimes f_o')) \qquad (4.4)$$

for k = 1...K, corresponding to Eqs. (3.4), can now be verified immediately. For k = O, (4.4) gives zero. This completes our explicit construction of a model for an arbitrary operation (3.6).

With this result, the set $\mathcal{G}$ of operations may be identified with the set of mappings $W \to \hat{W}$ described by (3.6) and (3.9). This implies the following statements concerning the structure of $\mathcal{G}$.

1. The set $\mathcal{G}$ of operations is a semigroup, if the product of two operations $O_1 : W \to \hat{W}$ and $O_2 : \hat{W} \to \hat{\hat{W}}$ is defined in the natural way as $O_2 \cdot O_1 : W \to \hat{\hat{W}}$.
2. The semigroup of operations $\mathcal{G}$ acts transitively on the set of density matrices, in the sense that for any two W and $\hat{W}$ with $\mathrm{Tr}\hat{W} \leq \mathrm{Tr}W$ there exists an operation $O: W \to \hat{W}$.
3. For two operations $O_1 : W \to \hat{W}$ and $O_2 : W \to \hat{\hat{W}}$ and $0 < \lambda < 1$, the mapping $\lambda O_1 + (1-\lambda) O_2 : W \to \lambda\hat{W} + (1-\lambda)\hat{\hat{W}}$ is an operation. Thus $\mathcal{G}$ is convex.
4. Given a normalized final state $\hat{W}$, there exists an operation O which transforms all normalized initial states W into $\hat{W}$. Such an operation O may thus be considered as a preparing procedure for the state $\hat{W}$.

For the simple proof of statements 1 to 3, we refer to [4]. Statement 4 may be easily verified by similar methods.

5. MUTUALLY EXCLUSIVE EFFECTS AND OPERATIONS

Two effects F_1 and F_2 are called mutually exclusive if

$$F_1 + F_2 \leq 1. \tag{5.1}$$

If there exists a normalized density matrix W with $\mathrm{Tr}(F_1 W) = 1$, then (5.1) implies $\mathrm{Tr}(F_2 W) = 0$, i.e., if F_1 occurs with certainty in some state W, then F_2 cannot occur at all in this state, and vice versa. This does not mean, however, that F_1 and F_2 strictly exclude each other in the sense that they never might be triggered together by a single system. An example for this will be given later on.

More generally, a finite or countably infinite number of effects F_i, $i=1...I$ are called mutually exclusive if

$$\sum_{i=1}^{I} F_i \leq 1. \tag{5.2}$$

Any effect F_i may be considered to correspond to an operation

$$O_i : W \to \hat{W}_i = \sum_{k=1}^{K_i} A_{ik} W A_{ik}^* \tag{5.3}$$

such that

$$F_i = \sum_{k=1}^{K_i} A_{ik}^* A_{ik} . \qquad (5.4)$$

Then (5.3) implies

$$\sum_{i=1}^{I} \sum_{k=1}^{K_i} A_{ik}^* A_{ik} \leq 1. \qquad (5.5)$$

In this case, the operations O_i, i=1...I given by (5.3) are called mutually exclusive, too.

Since (5.5) is of the same form as condition (3.9) for a single operation, the construction of Section 4 may be repeated, with $\mathbf{A}$ now being an operator matrix which contains all operators A_{ik} entering (5.5). The apparatus state space $\mathcal{H}'$ now has dimension $\sum_{i=1}^{I} K_i + 1$, and the orthonormal basis f_k', k = 0...K of $\mathcal{H}'$ introduced in Section 4 has to be replaced by a basis consisting of f_o' and f_{ik}', $k=1...K_i$, i=1...I. The operator $\mathfrak{S}$ is again defined by (4.3), and the initial state of the apparatus is chosen to be

$$W' = |f_o'\rangle \langle f_o'|,$$

as before. Finally, we define mutually orthogonal projection operators E_i' on $\mathcal{H}'$ by

$$E_i' = \sum_{k=1}^{K_i} |f_{ik}'\rangle \langle f_{ik}'| , \quad i=1...I.$$

This construction, as one easily realizes, leads to

$$Tr(F_i W) = Tr\,[(1 \otimes E_i')\mathfrak{S}(W \otimes W')\mathfrak{S}^*]$$

and

$$Tr(B\hat{W}_i) = Tr\,[(B \otimes E_i')\mathfrak{S}(W \otimes W')\mathfrak{S}^*]$$

for all states W and observables B of the system. Thus, in the sense explained in Section 2, the effect F_i and the operation O_i described by (5.4) and (5.3) correspond to the observation of the apparatus property E_i'. The following conclusions may be drawn immediately from this model.

1. Mutually exclusive effects F_i may be observed at a suitably constructed single apparatus, such that they correspond to mutually orthogonal projection operators E_i' on the apparatus state space. At such an apparatus the F_i mutually exclude each other in the strict sense, i.e., at most one of them can be triggered by a single system.

2. Mutually exclusive operations O_i may be performed by means of a single apparatus. Thereby the original ensemble of systems, corresponding to the initial state W, is subdivided according to the occurrence of mutually exclusive properties E_i' at the apparatus after its interaction with the system.

3. Any operation

$$W \rightarrow \hat{W} = \sum_k A_k \; W A_k^* \tag{5.6}$$

may be decomposed into mutually exclusive operations of the simple form

$$W \rightarrow \hat{W}_k = A_k \; W A_k^* . \tag{5.7}$$

Such operations are called pure since they transform the set of pure states into itself. A general operation may thus be considered as a mixture of pure operations. A suitable apparatus will be able to perform all operations (5.7) together, and this apparatus may be used to perform the operation (5.6), if one does not separate from each other the subensembles produced by the operations (5.7).
Note that the state change

$$W \rightarrow \hat{W} = \sum_k E_k \; W \; E_k$$

corresponding to an ideal measurement of mutually orthogonal projection operators E_k is usually interpreted in exactly this way.

4. Finally, we would like to comment on the notion of coexistent effects in Ludwig's axiomatic scheme. For simplicity we consider the case of two effects only. Operationally, effects F_1 and F_2 are defined to be coexistent if they may be observed both at a single apparatus. According to 1., two mutually exclusive effects are coexistent in this sense. In order to discuss the general case, we will again start with a model.

Consider an apparatus with state space $\mathcal{H}'$ and initial state W' and an interaction operator $\mathcal{S}$ on $\mathcal{H} \otimes \mathcal{H}'$, and take F_1 and F_2 to correspond to commuting projection operators E_1' and E_2' on $\mathcal{H}'$, such that

$$\mathrm{Tr}(F_i W) = \mathrm{Tr}\,[(1 \otimes E_i')\mathcal{S}(W \otimes W')\mathcal{S}^*], \quad i=1,2.$$

There exist three mutually orthogonal projection operators $\bar{E}_1'$, $\bar{E}_2'$ and E_3' on $\mathcal{H}'$ such that

$$E_1' = \bar{E}_1' + E_3' \;, \quad E_2' = \bar{E}_2' + E_3' \;,$$

and the corresponding effects $\bar{F}_1$, $\bar{F}_2$ and F_3 thus satisfy

$$F_1 = \bar{F}_1 + F_3, \quad F_2 = \bar{F}_2 + F_3 \qquad (5.8)$$

and

$$\bar{F}_1 + \bar{F}_2 + F_3 \leqslant 1. \qquad (5.9)$$

Vice versa, assume that there are given two effects F_1 and F_2 which satisfy (5.8) and (5.9) with suitable effects $\bar{F}_1$, $\bar{F}_2$ and F_3. Since, by (5.9), the latter are mutually exclusive, there exists an apparatus on which they correspond to mutually orthogonal projection operators $\bar{E}_1'$, $\bar{E}_2'$ and E_3'. This apparatus measures both F_1 and F_2, because they correspond to the commuting projection operators $\bar{E}_1' + E_3'$ and $\bar{E}_2' + E_3'$, respectively. Thus (5.8) and (5.9) are necessary and sufficient conditions for the coexistence of F_1 and F_2. (See also Ref. [7].) The conditions (5.8) and (5.9) remain unchanged if we replace in our model the commuting projection operators E_1' and E_2' by arbitrary coexistent effects F_1' and F_2' of the apparatus. This shows the internal consistency of these conditions.

The apparatus for the effects F_1 and F_2 considered above responds in a strictly exclusive way to the effects $\overline{F}_1$, $\overline{F}_2$ and F_3. In terms of the usual "logic" for the occurrence or non-occurrence of F_1 and F_2, F_3 means "F_1 and F_2", $\overline{F}_1$ means "F_1 but not F_2", $\overline{F}_1 + \overline{F}_2 + F_3$ means "F_1 or F_2", ect. In this way, our apparatus leads to an embedding of F_1 and F_2 into a Boolean algebra of effects which can also be measured at the given apparatus.

In general, however, the decomposition (5.8) of coexistent effects F_1 and F_2,and thus the apparatus and the Boolean algebra based on this decomposition, are not unique. Consider the example of mutually exclusive effects

$$F_1 = \frac{1}{2} E \ , \quad F_2 = 1 - \frac{1}{2} E$$

with some projection operator E.
Here we may take, e.g.,

$$\overline{F}_1 = \frac{1}{2}E, \ \overline{F}_2 = \frac{1}{2}E + (1-E), \ F_3 = 0,$$

but also

$$\overline{F}_1 = 0, \quad \overline{F}_2 = 1-E, \quad F_3 = \frac{1}{2}E.$$

The latter decomposition corresponds to an apparatus on which F_1 and F_2 do not strictly exclude each other, but occur together in state W with probability $\frac{1}{2}$ Tr(EW).

Ambiguities of this kind are absent for coexistent decision effects E_1 and E_2. In this case, E_1 and E_2 are commuting projection operators, and the decomposition (5.8) is unique. Moreover, the unique Boolean algebra thus generated by E_1 and E_2 consists of decision effects only and satisfies the familiar rules of "quantum logic" for commuting projection operators. This is easily verified by an elementary calculation. For a similar deduction in the axiomatic framework of Ludwig, compare Ref. [8].

6. COMPLETELY POSITIVE MAPPINGS OF $\mathcal{L}(\mathcal{H})$

Whereas up to now we have been concerned mainly with a rather concrete description of operations, we will now study them from a more abstract point of view. The general properties of an operation $O: W \to \hat{W}$ as a mapping of the set of density matrices will be derived from the model of Section 2. For the sake of generality, the projection operator E' occuring there is now replaced by a general effect F'. Eq. (2.6), thus modified, leads to

$$\mathrm{Tr}(B\hat{W}) = \mathrm{Tr}[(B \otimes 1')(1 \otimes F'^{\frac{1}{2}})S(W \otimes W')S^*(1 \otimes F'^{\frac{1}{2}})] \qquad (6.1)$$

for all observables B on $\mathcal{H}$. This implies, as well-known,

$$\hat{W} = \mathrm{Tr}'[(1 \otimes F'^{\frac{1}{2}})S(W \otimes W')S^*(1 \otimes F'^{\frac{1}{2}})], \qquad (6.2)$$

where Tr' denotes the partial trace with respect to $\mathcal{H}'$. Now we replace W by a general operator T from the trace class $\mathcal{L}_1(\mathcal{H})$ and B by an operator X, not necessarily self-adjoint, from the algebra $\mathcal{L}(\mathcal{H})$ of all bounded operators. By this, (6.2) leads to a linear mapping

$$O : T \to \hat{T} = \mathrm{Tr}'[(1 \otimes F'^{\frac{1}{2}})S(T \otimes W')S^*(1 \otimes F'^{\frac{1}{2}})] \qquad (6.3)$$

of $\mathcal{L}_1(\mathcal{H})$ which extends the operation $O : W \to \hat{W}$ and is therefore denoted by the same symbol. Likewise, (6.1) yields, for fixed $X \in \mathcal{L}(\mathcal{H})$, a linear functional

$$\hat{X}(T) = \mathrm{Tr}[(X \otimes F')S(T \otimes W')S^*] \qquad (6.4)$$

of $T \in \mathcal{L}_1(\mathcal{H})$ which turns out to be continuous with respect to the trace norm. Since $\mathcal{L}(\mathcal{H})$ is the dual of $\mathcal{L}_1(\mathcal{H})$ with respect to the functionals $\mathrm{Tr}(XT)$, $X \in \mathcal{L}(\mathcal{H})$, $T \in \mathcal{L}_1(\mathcal{H})$, the functional $\hat{X}(T)$ defined by (6.4) is of the form

$$\hat{X}(T) = \mathrm{Tr}(\hat{X}T) \qquad (6.5)$$

with a unique $\hat{X} \in \mathcal{L}(\mathcal{H})$. Eqs. (6.4) and (6.5) thus define a mapping

$$O^* \quad : X \to \hat{X} \tag{6.6}$$

of $\mathcal{L}(\mathcal{H})$ which, obviously, is also linear. Moreover, since

$$\mathrm{Tr}(\hat{X}T) = \mathrm{Tr}[(X \otimes F')\mathcal{S}(T \otimes W')\mathcal{S}^*] = \mathrm{Tr}(X\hat{T}),$$

the mapping O^* is the adjoint of the mapping O, as already indicated by the notation. Both mappings are easily seen to be positive, i.e.,

$$\begin{aligned} T = T^* \geqslant 0 \text{ implies } \hat{T} = \hat{T}^* \geqslant 0, \\ X = X^* \geqslant 0 \text{ implies } \hat{X} = \hat{X}^* \geqslant 0. \end{aligned} \tag{6.7}$$

This already implies the continuity of O and O^* with respect to the trace norm $\|\cdot\|_1$ and operator norm $\|\cdot\|$, respectively. By explicit calculation we get

$$\begin{aligned} \|\hat{X}\| &\leqslant \|X\| , \\ \|\hat{T}\|_1 &\leqslant \|T\|_1 . \end{aligned} \tag{6.8}$$

We will now prove that O^* is completely positive in the sense explained in Størmer's lectures [9]. Consider an n x n operator matrix

$$\underline{X} = (X_{\alpha\beta}) \ , \ X_{\alpha\beta} \in \mathcal{L}(\mathcal{H}), \ \alpha,\beta = 1\ldots n, \tag{6.9}$$

with n arbitrary but finite. Then $\underline{X}$ represents a bounded operator on the direct sum $\bigoplus_{\alpha=1}^{n} \mathcal{H}_\alpha$ of n copies $\mathcal{H}_\alpha$ of $\mathcal{H}$ or, equivalently, on $\mathcal{H} \otimes \mathcal{H}_n$ with a Hilbert space $\mathcal{H}_n$ of dimension n. Vice versa, any bounded operator $\underline{X}$ on $\mathcal{H} \otimes \mathcal{H}_n$ may be represented in this form. Therefore O^* may be used to define a mapping

$$\underline{X} \to \hat{\underline{X}} = (O^* X_{\alpha\beta}) = (\hat{X}_{\alpha\beta})$$

of $\mathcal{L}(\mathcal{H} \otimes \mathcal{H}_n)$. It turns out that this mapping is again positive, i.e.,

$$\underline{X} = \underline{X}^* \geqslant 0 \text{ implies } \hat{\underline{X}} = \hat{\underline{X}}^* \geqslant 0, \tag{6.10}$$

which means complete positivity of O^*.

In order to prove (6.10), we denote by $\underline{T}$ an arbitrary operator from $\mathcal{L}_1(\mathcal{H} \otimes \mathcal{H}_n)$ and by 1_n the unit operator on $\mathcal{H}_n$. We will show then that

$$\mathrm{Tr}(\hat{\underline{X}}\, \underline{T}) = \mathrm{Tr}[(\underline{X} \otimes F')(\mathcal{S} \otimes 1_n)(\underline{T} \otimes W')(\mathcal{S} \otimes 1_n)^*]. \tag{6.11}$$

Once we have shown this, (6.10) follows immediately because (6.11) means that the mapping $\underline{X} \to \hat{\underline{X}}$ is of the same type as the mapping $X \to \hat{X}$ studied before. Indeed, (6.11) may be interpreted physically as describing an operation performed on a compound system with state space $\mathcal{H} \otimes \mathcal{H}_n$. The interaction operator $\mathcal{S} \otimes 1_n$ is unitary on $\mathcal{H} \otimes \mathcal{H}_n \otimes \mathcal{H}'$, and its particular form implies that the subsystem described in $\mathcal{H}_n$ does not interact with the apparatus.

<u>Proof of Eq. (6.11):</u> Consider first the particular case $\underline{X} = X \otimes X_n$, $\underline{T} = T \otimes T_n$. Then the right hand side of (6.11) yields

$$\begin{aligned} &\mathrm{Tr}[(X \otimes F' \otimes X_n)(\mathcal{S} \otimes 1_n)(T \otimes W' \otimes T_n)(\mathcal{S} \otimes 1_n)^*] \\ &\quad = \mathrm{Tr}[(X \otimes F')\mathcal{S}(T \otimes W')\mathcal{S}^*]\mathrm{Tr}(X_n T_n) \\ &\quad = \mathrm{Tr}(\hat{X}T)\mathrm{Tr}(X_n T_n) = \mathrm{Tr}[(\hat{X} \otimes X_n)(T \otimes T_n)] = \mathrm{Tr}(\hat{\underline{X}}\underline{T}). \end{aligned}$$

Since arbitrary $\underline{X} \in \mathcal{L}(\mathcal{H} \otimes \mathcal{H}_n)$ and $\underline{T} \in \mathcal{L}_1(\mathcal{H} \otimes \mathcal{H}_n)$ are finite linear combinations of operators of the form just considered and both sides of (6.11) depend linearly on $\underline{X}$ and $\underline{T}$, the desired result now follows trivially. ∎

Finally, the mapping O^* is continuous with respect to the ultraweak operator topology. This is generally true for any mapping of $\mathcal{L}(\mathcal{H})$ which is the adjoint of a mapping of $\mathcal{L}_1(\mathcal{H})$, since the ultraweak topology on $\mathcal{L}(\mathcal{H})$ is just the weak * topology generated by the functionals $\mathrm{Tr}(XT)$, $T \in \mathcal{L}_1(\mathcal{H})$.

Of course, all properties of the mappings O and O^* mentioned so far would also follow easily [4] if we knew already that they have the form

$$O \quad : \quad T \to \hat{T} \ = \ \sum_k A_k T A_k^* \ , \qquad (6.12)$$

$$O^* \quad : \quad X \to \hat{X} \ = \ \sum_k A_k^* X A_k \ . \qquad (6.13)$$

However, since we have not yet proved this for the general case considered now, we will instead derive it as a by-product of the following consideration.

Now we exploit Theorem 2.1 of Ref. [9] in a slightly modified form.

Theorem: Any completely positive linear mapping ϕ of a C* - algebra $\mathfrak{A}$ of operators X on a Hilbert space $\mathcal{H}$ into the algebra $\mathcal{L}(\mathcal{H})$ may be represented as

$$\phi : X \to \mathcal{A}^* \pi(x) \mathcal{A}$$

with a representation π of $\mathfrak{A}$ on a Hilbert space $\tilde{\mathcal{H}}$ and a bounded linear operator $\mathcal{A}$ from $\mathcal{H}$ into $\tilde{\mathcal{H}}$.

If applied to our particular case $\mathfrak{A} = \mathcal{L}(\mathcal{H})$, $\phi = O^*$, we obtain the

Corollary: The adjoint mapping O* of an operation O is of the form

$$O^* : X \to \sum_k A_k^* X A_k$$

with an at most countably infinite number of operators $A_k \neq O$ on $\mathcal{H}$ such that

$$\sum_k A_k^* A_k \leq 1.$$

Thus O* is indeed, quite generally, given by (6.13), which in turn immediately implies [4] the representation (6.12) for the corresponding operation O.

Proof of the Corollary: Any representation π of $\mathcal{L}(\mathcal{H})$ on a Hilbert space $\tilde{\mathcal{H}}$ is the direct sum of identical representations and a representation π_o on a subspace $\tilde{\mathcal{H}}_o$ of $\tilde{\mathcal{H}}$ of the quotient algebra $\mathcal{L}(\mathcal{H})/\mathcal{C}(\mathcal{H})$ with the closed two-sided ideal $\mathcal{C}(\mathcal{H})$ of compact operators. Accordingly, the operator $\mathcal{A}$ in our case may be represented as a column matrix

$$\mathcal{A} = \begin{pmatrix} \mathcal{A}_o \\ A_k \end{pmatrix} \qquad (6.14)$$

with bounded linear operators A_k on $\mathcal{H}$, $k \in \mathcal{K}$ (some index set) and a bounded linear operator $\mathcal{A}_o$ from $\mathcal{H}$ into $\tilde{\mathcal{H}}_o$, such that

$$\mathcal{A}^*\pi(X)\mathcal{A} = \mathcal{A}_o{}^*\pi_o(X)\mathcal{A}_o + \sum_{k\in\mathcal{K}} A_k^* XA_k \qquad (6.15)$$

for all $X \in \mathcal{L}(\mathcal{H})$. Boundedness of $\mathcal{A}$ is equivalent to

$$\mathcal{A}_o^* \mathcal{A}_o + \sum_{k\in\mathcal{K}} A_k{}^*A_k \leq \|\mathcal{A}\|^2 \, 1 ,$$

and this is easily shown [4] to restrict the number of $A_k \neq 0$ in (6.14) to be at most countably infinite. Moreover, since the mapping

$$X \to \sum_{k\in\mathcal{K}} A_k^* X A_k$$

already corresponds to an operation of the particular type studied in Sections 3 and 4, it is ultraweakly continuous. This, together with (6.15) and the ultraweak continuity of O^*, implies that the mapping

$$X \to \mathcal{A}_o^* \pi_o(X) \mathcal{A}_o \qquad (6.16)$$

has to be ultraweakly continuous, too. Now, because any $X \in \mathcal{L}(\mathcal{H})$ is the ultraweak limit of operators from $\mathcal{C}(\mathcal{H})$ which are annihilated by π_o, (6.16) must give identically zero, and thus

$$\mathcal{A}^*\pi(X)\mathcal{A} = \sum_{k\in\mathcal{K}} A_k^* X A_k .$$

Finally, the first one of the conditions (6.8) implies for $X = 1$ that $\sum_{k\in\mathcal{K}} A_k{}^*A_k \leq 1$. ∎

Although the above Corollary covers quite nicely the class of operations considered so far, we want to add two critical remarks which might indicate how one should try to improve it.

1. Even in the Hilbert space formulation of quantum theory one may be tempted to generalize the notion of operations such that it also includes, e.g., antilinear mappings like

$$O : T \to T^* \;,\; O^* : X \to X^*, \tag{6.17}$$

   or

$$O : T \to ATA^* \;,\; O^* : X \to A^* XA \tag{6.18}$$

   with an antilinear operator A on $\mathcal{H}$. Since only convex linear combinations of density operators W and effects F have direct physical interpretations and antilinear mappings of $\mathcal{L}_1(\mathcal{H})$ and $\mathcal{L}(\mathcal{H})$ preserve this structure, such mappings may also be physically interesting. In fact, as well-known, (6.18) with an antiunitary A describes time inversion. One would like to have a representation theorem covering such more general operations too. Note, however, that operations like (6.17) or (6.18) can not be associated with models of the type considered above, since for such models the linearity of the scattering operator $\mathcal{S}$ already implies the linearity of O and O*.

2. In an axiomatic framework like Ludwig's which does not start from the Hilbert space formalism, a condition like complete positivity cannot yet be incorporated. Therefore one would like to replace it by an equivalent condition which fits more naturally into a purely axiomatic approach.

We conclude this Section with some remarks concerning the physical interpretation of the mapping O*. If applied to an arbitrary effect F, O* yields another effect $\hat{F}$ such that, for each density operator W,

$$\mathrm{Tr}(\hat{F}W) = \mathrm{Tr}(F\hat{W}) \tag{6.19}$$

if O maps W into $\hat{W}$. Consider an apparatus consisting of two parts, one performing the operation O, the second one measuring F afterwards, and define a new effect by the occurrence of F at the second apparatus. Then, by (6.19), this new effect is just $\hat{F}$. In most cases, this mapping will transform decision effects E into effects $\hat{E}$ which are not of this particular type. Such mappings therefore have no counterpart in theories which restrict their attention to decision effects.

More generally, O^* transforms the set of self-adjoint operators into itself. However, although this could be considered formally as a transformation of observables, it can not be interpreted physically in the same way as the mapping $F \rightarrow \hat{F}$ of effects. A measuring apparatus for an observable B with spectral representation

$$B = \int \lambda \, dE(\lambda)$$

measures together all decision effects $E(\lambda)$. If, as above, this apparatus is combined with a second one which first performs the operation O, the new apparatus now responds to the effects $\hat{E}(\lambda) = O^* E(\lambda)$ which need not have to do anything with the spectral decomposition of $\hat{B} = O^*B$. The composite apparatus is thus not a measuring apparatus for $\hat{B}$ in the usual sense, although it may be used to determine the expectation values $Tr(\hat{B}W)$ of $\hat{B}$ in arbitrary states W.

7. APPLICATIONS

As promised in the Introduction, I will now illustrate by means of examples that the notions of effects and operations may help solving physical problems. Since I cannot go into technical details anyway, I will give references to the literature only.

The notion of effects becomes useful if one wants to localize particles like photons with restmass zero. One knows that there is no self-adjoint operator $\vec{X}$ which has all properties of a position observable for such particles. On the other hand, photons in fact are localizable in the sense that they trigger suitable counters. A description of such counters in terms of effects is sketched in Ref. [4].

Operations are particularly useful for a discussion of joint probabilities for the triggering of several particle counters in scattering theory. Each counter may be described mathematically by the operation O which it performs, and the corresponding effect F. In the simplest case, O may be taken to be a pure operation $W \to \hat{W} = C\, W\, C^*$ with a single operator C such that $F = C^* C$. The probability that, in a given state W, n counters $C_1, C_2 \ldots C_n$ are triggered at times $t_1 < t_2 < \ldots < t_n$, is easily seen to be

$$\mathrm{Tr}(C_n \ldots C_1 W C_1{}^* \ldots C_n{}^*). \qquad (7.1)$$

In quantum field theory, operators C_i describing particle counters have to satisfy some simple and physically motivated conditions. For such C_i, Eq. (7.1) may be evaluated if the counters are sufficiently far from each other, and leads directly to a space-time description of the propagation and scattering of particles (O. Steinmann, Ref. [10].) Recently Steinmann applied his method also to the decay of unstable particles [11] and the scattering of infraparticles [12]. In these cases the conventional S-matrix description is much less convenient or even not applicable at all.

REFERENCES

[1] G. Ludwig: The Measuring and Preparing Process and Macro Theory. This Volume, p. 125

[2] A recent paper containing a quite complete list of references is: C.M. Edwards, Commun. Math. Phys. 24, 260 (1972). To my knowledge, the notion of operation first appears in: R. Haag and D. Kastler, J. Math. Phys. 5, 848 (1964).

[3] K.-E. Hellwig and K. Kraus, Commun. Math. Phys. 11, 214 (1969), and 16, 142 (1970).

[4] K. Kraus, Ann. Phys. (N.Y.) 64, 311 (1971).

[5] K.-E. Hellwig, Proc. Intern. School of Physics Enrico Fermi, Course IL: Foundations of Quantum Mechanics, p. 338. Academic Press 1971.

[6] F. Riesz and B. Sz.-Nagy: Functional Analysis. Third Ed., Appendix. F. Ungar, New York 1960.

[7] K.-E. Hellwig, Intern. J. Theor. Phys. 2, 147 (1969).

[8] G. Ludwig: Deutung des Begriffs "physikalische Theorie" und axiomatische Grundlegung der Hilbertraumstruktur der Quantenmechanik durch Hauptsätze des Messens (Lecture Notes in Physics, Vol. 4), p. 339. Springer, Berlin - Heidelberg - New York 1970.

[9] E. Størmer: Positive Linear Maps of C*-Algebras. This Volume, p. 88

[10] O. Steinmann, Commun. Math. Phys. 7, 112 (1968).

[11] O. Steinmann, Helv. Phys. Acta 44, 618 (1971).

[12] O. Steinmann, Scattering of Infraparticles (Preprint, 1973).

THE EMPIRICAL LOGIC APPROACH TO THE PHYSICAL SCIENCES

D.J. Foulis and C.H. Randall

Department of Mathematics & Statistics

University of Massachusetts

Amherst, Massachusetts, U. S. A.

1. Introduction. The aim of an experimental science is to order, explain, evaluate and predict the observable consequences associated with the execution of certain physical operations, procedures or experiments. In any such well conceived experimental program one is concerned not with a single physical operation, but rather with a coherent collection of such operations. In recognition of these facts, the primary object of study in the empirical logic approach to the experimental sciences is a coherent collection, or manual, of physical operations. Thus we shall begin by developing a flexible but formal language capable of describing, with an adequate precision, the physical occurrences affiliated with such a manual of physical operations. The work reported here, for the most part, is concerned with establishing the syntax, or as we shall say, the logic, for such a language.

It should be appreciated at the outset that a symbolic logic is not a collection of physical laws; it is not even a language (an instrument of communication). A symbolic logic is only a syntax (or grammar) of a language. A symbolic logic cannot become a language until its symbols are assigned physical significance. A language can express physical laws (and thus form physical theories) only when it in some sense predicts the observable consequences of executing actual physical operations. Thus for the most part we shall not be concerned with physical laws but with the pristine grammar of a language used to describe physical experience.

This point of view has the virtue of making the physical operations the unanalyzable primitives of any physical theory expressed in the language and consequently of making all other concepts (including the notions of outcomes, events, propositions, test procedures, states, observables, parameters, time, space, dynamics and physical systems) operationally accessible. This approach, moreover, tends to be constructive rather than analytic in that new physical theories can easily be synthesized by imposing conditions on the manual of operations under consideration without reference to any existing theories. This, of course, is not to say that existing physical theories cannot be incorporated in this formalism.

2. Manuals of Operations. By a physical operation we shall mean instructions that describe a well defined, physically realizable, reproducible procedure and furthermore that specify what must be observed and what can be recorded as a consequence of an execution of this procedure. In particular, a physical operation must require that as a consequence of each execution of the instructions one and only one symbol from a specified set E be recorded as the outcome of this realization of the physical operation. It must be emphasized at this point that it is the symbol that is the outcome of this realization of the physical operation rather than any real or imaginary occurrence in a "real world out there". Also note carefully, if we delete or add details to the instructions for any physical operation, especially if we modify the outcome set E in any way, we thereby define a new physical operation.

It is important to realize that the subjective judgment of the observer is implicit in every realization of every physical operation, not only in regard to the interpretation of the instructions but also in connection with the decision as to which symbol to record as the outcome. Thus if a competent observer is of the opinion that he has executed a specified physical operation and has obtained a particular outcome, then in fact the operation has been executed and the outcome in question has been secured. Each realization of a physical operation is to be understood here as a "Ding an sich"; physical history as it were begins and ends with each execution of a physical operation. If physical procedures are to be carried out in a "connected sequence", then the instructions for such a compound operation must say so. When a compound physical operation is built up from more primitive operations by concatenation, it is understood that each constituent operation thereby loses its identity since it may have temporal antecedents and consequences. Notice in regard to this that the recording of the passage of time is often an important constituent operation of a compound operation.

Once we have assembled those physical operations of concern to us in a particular discourse, we are almost inevitably moved either by custom or by a particular intent to identify certain outcomes of different operations. For example, it is common practice to identify those outcomes that are believed to signal the presence or successful preparation of a particular type of physical system such as an electron or an elephant. Similarly we often prefer to regard a number of outcomes of distinct physical operations as registering the same property, or if yo prefer as representing the same measurement. If a voltage is measured using different instruments, or even different methods, identical numerical results are ordinarily taken to be equivalent. Finally, the instructions for carrying out the admissible procedures may all but dictate the identification of certain outcomes on purely syntactic grounds We surely wish to avoid the necessity of taking a stand on the "acceptability" of these identifications, since we hope to keep our language as free as possible from such ad hoc decisions. On the other hand, we wish our language to be able to handle this common practice of outcome identification. Clearly, permitting an unrestricted identification of outcomes would lead to "grammatical chaos"; thus we shall subject the outcome identification process to certain mild constraints.

In order to make the nature of these constraints precise, let us suppose that we are considering a particular manual of admissible physical operations and that we have already made what we consider to be the appropriate outcome identifications. (Naturally, these identifications will often be motivated by some "picture of the real world".) As our first condition, which we shall refer to as the determination condition, we require that given any outcome of an admissible physical operation, the set of all admissible physical operations capable of yield ing this outcome can be discerned from the symbolic pattern of the outcome itself. Thus when we assert that such an outcome has been obtained, we understand that what we are really asserting is that this outcome was obtained as a consequence of an execution of one of these admissible physical operations, although it is not necessary to specify whic one.

The second condition on such an identification process, which we shall call irredundancy, requires that given any physical operation $\mathcal{P}$, with the outcome set E, there is only one physical operation, namely $\mathcal{P}$ itself, capable of yielding each and every outcome in E. An immediate consequence of this condition is that each physical operation not only determines, but is uniquely determined by, the set of all of its outcomes. As a result we can and will identify each physical operation with its set of outcomes.

In view of the above, we propose to introduce as a formal mathemat ical representation for a manual of physical operations a non-empty set $\mathcal{A}$ of non-empty sets. Each set $E \in \mathcal{A}$ is supposed to correspond uniquely to the set of all outcomes of one of the admissible physical operations and each admissible physical operation is supposed to be represented ir this way. The condition of irredundancy now translates into the assertion that if E,F $\in \mathcal{A}$ and $E \subseteq F$, then E = F. The elements of $\mathcal{A}$ will henceforth will be referred to simply as $\mathcal{A}$-operations, while the set theoretic union of all of the $\mathcal{A}$-operations will be referred to simply as the set of all $\mathcal{A}$-outcomes.

If x and y are two $\mathcal{A}$-outcomes, with $x \neq y$, and if there exists an $\mathcal{A}$-operation E with x,y $\in$ E, then we shall say that x is orthogonal to y, and we shall write $x \perp y$. Notice when $x \perp y$, as above, then x and y reject each other operationally in the sense that an execution of E which yields the outcome x cannot yield the outcome y and visa versa. A set D of $\mathcal{A}$-outcomes will be called an orthogonal set provided that $x \perp y$ holds for all x,y $\in$ D for which $x \neq y$. Obviously every $\mathcal{A}$-operation is an orthogonal set and every subset of an orthogonal set is again an orthogonal set.

If E is an $\mathcal{A}$-operation, then a subset D of E will be called an event for E. As usual, if the physical operation corresponding to E is executed and an outcome d $\in$ D is obtained as a consequence, then we shall say that the event D has occurred. Naturally an $\mathcal{A}$-event is defined to be a set of $\mathcal{A}$-outcomes which is an event for at least one $\mathcal{A}$-operation. When we assert that an $\mathcal{A}$-event D has occurred, we of course are asserting that it has occurred as a consequence of an execution of a physical operation corresponding to some E $\in \mathcal{A}$ for which $D \subseteq E$.

If A and B are two $\mathcal{A}$-events, then we shall say that A is orthogonal to B, and write $A \perp B$, provided that $a \perp b$ holds for all a $\in$ A and all b $\in$ B. Our third condition, named the coherence condition, in effect permits us to extend the above operational interpretation for the orthogonality relation $\perp$ from $\mathcal{A}$-outcomes to $\mathcal{A}$-events. Specifically, the coherence condition requires that if A and B are $\mathcal{A}$-events with $A \perp B$, then there must exist an $\mathcal{A}$-operation E such that both A and B are events for E.

Let us summarize by making the following formal definitions: a premanual is defined to be a non-empty set $\mathcal{A}$ of non-empty sets. An element E $\in \mathcal{A}$ is called an $\mathcal{A}$-operation and the set theoretic union $X = \bigcup \mathcal{A}$ is called the set of $\mathcal{A}$-outcomes. We shall call $\mathcal{A}$ an irredundant premanual provided that E,F $\in \mathcal{A}$ and $E \subseteq F$ implies that E = F. Two $\mathcal{A}$-outcomes x,y $\in$ X are said to be orthogonal, in symbols $x \perp y$, provided that $x \neq y$ and there exists an E $\in \mathcal{A}$ with x,y $\in$ E. A subset D of X is called an orthogonal set if $x \perp y$ holds for all x,y $\in$ D with $x \neq y$. A subset D of X is called an $\mathcal{A}$-event if there exists an E $\in \mathcal{A}$ with $D \subseteq E$. If A and B are subsets of X, we say that A and B are orthogonal, and we write $A \perp B$, provided that $a \perp b$ holds whenever a $\in$ A and b $\in$ B. We shall say that the premanual $\mathcal{A}$ is coherent provided that the set theoretic union of any two orthogonal $\mathcal{A}$-events is again an $\mathcal{A}$-event. We define a manual to be an irredundant coherent premanual.

3. The Operational Logic of a Manual. Let $\mathcal{A}$ be a manual corresponding, as above, to some collection of admissible physical operations. Let us consider here only those propositions that are operationally well defined in the sense that they are confirmed or refuted strictly in terms of evidence secured as a consequence of the execution of $\mathcal{A}$-operations. Thus we define an operational proposition (for $\mathcal{A}$) to be an ordered pair (A,B) of subsets A and B of $X = \bigcup \mathcal{A}$. If an operation $E \in \mathcal{A}$ is executed and the outcome $e \in E$ is obtained as a consequence, we shall say that the operational proposition (A,B) is confirmed precisely when $e \in A$ and that it is refuted precisely when $e \in B$. Thus A will be called the confirmation set and B the refutation set for (A,B). If $A \perp B$, that is if every outcome that could confirm (A,B) operationally rejects every outcome that could refute (A,B), we shall say that (A,B) is orthoconsistent.

If A is a subset of X, we define $A^{\perp} = \{x \in X \mid x \perp a \text{ for all } a \in A\}$ and we define $A^{\perp\perp} = (A^{\perp})^{\perp}$. Evidently $A \subseteq A^{\perp\perp}$, $A \cap A^{\perp} = \emptyset$ and if $A \subseteq B \subseteq X$ then $B^{\perp} \subseteq A^{\perp}$. If (A,B) is an operational proposition, then any outcome $x \in A^{\perp}$ will be said to virtually refute (A,B) since it operationally rejects every outcome which could confirm (A,B). Similarly an outcome $y \in B^{\perp}$ will be said to virtually confirm (A,B). An operational proposition (A,B) is said to be closed if it is orthoconsistent, $B^{\perp} \subseteq A$ and $A^{\perp} \subseteq B$. Notice, for instance, that the condition $B^{\perp} \subseteq A$ simply requires that any outcome that virtually confirms (A,B) actually confirms (A,B). It is easy to see that an operational proposition is closed if and only if it has the form $(C^{\perp\perp}, C^{\perp})$ for some subset $C \subseteq X$.

An $\mathcal{A}$-operation E is said to test the operational proposition (A,B) in case $E \subseteq A \cup B$; that is, in case every outcome of E either confirms or refutes (A,B). A collection of operational propositions is said to be simultaneously testable if there exists a single $\mathcal{A}$-operation that is a test for every operational proposition in the collection. It is natural to associate with every $\mathcal{A}$-event D the operational proposition $(D,D^{\perp})$, since its test operations are precisely those $\mathcal{A}$-operations for which D is an event. Thus if a test operation for $(D,D^{\perp})$ is executed, $(D,D^{\perp})$ is confirmed precisely when the event D occurs and is refuted precisely when the event D does not occur. However, there may be operations in $\mathcal{A}$ for which D is not an event that could confirm or refute $(D,D^{\perp})$. Thus the assertion that the $\mathcal{A}$-event D occurred is not quite the same as the assertion that the proposition $(D,D^{\perp})$ was confirmed. Also, in general, $(D,D^{\perp})$ is not closed; however, there exists a unique closed operational proposition $p(D)$ which is confirmed if

$(D,D^{\perp})$ is confirmed and refuted if $(D,D^{\perp})$ is refuted, namely $p(D) = (D^{\perp\perp},D^{\perp})$.

Thus let us define $\Pi(\mathcal{Q}) = \Pi$ to be the set of all operational propositions of the form $p(D) = (D^{\perp\perp},D^{\perp})$ as D runs through all $\mathcal{Q}$-events. If C and D are $\mathcal{Q}$-events, we shall say that p(C) <u>implies</u> p(D), and we write $p(C) \leq p(D)$, in case every $\mathcal{Q}$-outcome that confirms p(C) also confirms p(D), that is $C^{\perp\perp} \subseteq D^{\perp\perp}$. Notice that $p(C) \leq p(D)$ if and only if every $\mathcal{Q}$-outcome that refutes p(D) also refutes p(C), that is $D^{\perp} \subseteq C^{\perp}$. We shall say that p(C) is <u>orthogonal</u> to p(D), and write $p(C) \perp p(D)$, provided that every $\mathcal{Q}$-outcome that confirms p(C) also refutes p(D), that is $C^{\perp\perp} \subseteq D^{\perp}$. Observe that this is so if and only if $C \perp D$. The system $(\Pi(\mathcal{Q}), \leq, \perp)$ is called the <u>operational</u> <u>logic</u> of $\mathcal{Q}$.

In most of the manuals that have been considered, for every event D there exists at least one other event B for which $B^{\perp\perp} = D^{\perp}$. When such an event B exists, then $E = B \cup D$ is an operation with respect to which B and D are relative complements of one another. Given the existence of at least one of these relative complements B for each event D, we can define the mapping ′ on the operational logic by $p(D)' = p(B) = p(E \setminus D) = (D^{\perp},D^{\perp\perp})$. This mapping is easily seen to be an <u>orthocomplementation</u> for the operational logic and naturally the proposition $p(D)'$ is regarded as the <u>negation</u> of the proposition p(D). In most of the interesting manuals the operational logic is not only orthocomplemented in this way, but actually forms an orthomodular poset. This turns out to be equivalent to the requirement that if E and F are two operations both containing an event D then $p(E \setminus D) = p(F \setminus D)$. In brief, in the orthomodular case, $p(D)'$ corresponds to <u>all</u> of the relative complements of D.

The operational logic $\Pi(\mathcal{Q})$ of the manual $\mathcal{Q}$ will be an orthomodular poset, as above, if and only if $\mathcal{Q}$ satisfies the so-called Dacey condition: If x and y are $\mathcal{Q}$-outcomes and if there exists an $\mathcal{Q}$-operation E such that every outcome $e \in E$ which is not orthogonal to x is orthogonal to y, then x is orthogonal to y. The Dacey condition is similar to the irredundancy and coherence conditions and could be given a similar heuristic motivation. As we have observed, this condition requires that all of the relative complements of an event correspond to the same operational proposition.

Since any $\mathcal{Q}$-operation for which D is an event is a test operation for p(D), every operational proposition in $\Pi(\mathcal{Q})$ is testable. For a Dacey manual (that is, a manual satisfying the Dacey condition) the converse is true in the sense that the operational logic $\Pi(\mathcal{Q})$ consists precisely of the closed and testable operational propositions for $\mathcal{Q}$.

In a Dacey manual, an operation E tests the operational proposition p(D) if and only if there exists an event $C \subseteq E$ such that p(C) = p(D).

If $\mathcal{Q}$ is any manual and if A and B are $\mathcal{Q}$-events, we say that p(A) <u>commutes with</u> p(B) if there exists pairwise orthogonoal $\mathcal{Q}$-events A_1, B_1 and D such that $p(A) = p(A_1 \cup D)$ and $p(B) = p(B_1 \cup D)$. In the case in which $\mathcal{Q}$ is a Dacey manual this definition coincides with the customary notion of commutativity in the orthomodular poset $\Pi(\mathcal{Q})$ and here it can be shown that p(A) commutes with p(B) if and only if p(A) and p(B) are simultaneously testable. In any case we shall say that an operational proposition belongs to the <u>center</u> of the operational logic $\Pi(\mathcal{Q})$ if it commutes with every other proposition in this logic. It was shown by B. Jeffcott [7] that the center of any operational logic $\Pi(\mathcal{Q})$ is always a Boolean algebra. We shall say that $\mathcal{Q}$ is a Boolean manual if and only if $\Pi(\mathcal{Q})$ is its own center. Evidently, if $\mathcal{Q}$ is a Boolean manual it is automatically a Dacey manual and any <u>finite</u> collection of propositions is simultaneously testable. Thus we are inclined to regard the experimental situations described by Boolean manuals as being classical.

For a manual $\mathcal{Q}$ let $(D_i \mid i \in I)$ be a family of $\mathcal{Q}$-events indexed by the set I. It is reasonable to ask when there will exist an $\mathcal{Q}$-event C such that p(C) is effective as the infimum (greatest lower bound) in the operational logic $\Pi(\mathcal{Q})$ of the family $(p(D_i) \mid i \in I)$. The necessar and sufficient condition can be seen to be that $C^{\perp\perp} = \bigcap_{i \in I} (D_i)^{\perp\perp}$; that is, p(C) is confirmed precisely by those $\mathcal{Q}$-outcomes that confirm every one of the propositions $p(D_i)$, $i \in I$. By analogy with classical logic, we feel entitled to refer to such a p(C), when it exists, as the <u>conjunction</u> of the propositions $p(D_i)$, $i \in I$, and to write p(C) = $\bigwedge_{i \in I} p(D_i)$. We shall call $\mathcal{Q}$ a <u>conjunctive</u> manual if every finite family of operational propositions in $\Pi(\mathcal{Q})$ has a conjunction, noting that although every Boolean manual is conjunctive, there do exist non-conjunctive Dacey manuals. In fact, $\Pi(\mathcal{Q})$ is an orthomodular lattice if and only if $\mathcal{Q}$ is a conjunctive Dacey manual. In classical logic, it is a fact that the conjunction of a collection of propositions is false if and only if at least one of the propositions in the collection is false. In $\Pi(\mathcal{Q})$, if $p(C) = \bigwedge_{i \in I} p(D_i)$, then any $\mathcal{Q}$-outcome that refutes at least one of the propositions $p(D_i)$, $i \in I$, will refute p(C); however, there may be $\mathcal{Q}$-outcomes that refute the conjunction p(C) but fail to refute any of the $p(D_i)$. This departure from the canons of classical reasoning is a characteristic feature of the non-classical operational logics affiliated with manuals that are not Boolean.

Given a family $(D_i \mid i \in I)$ as above, it is also natural to ask whether there exists an $\mathcal{Q}$-event D such that p(D) is the analog of the classical <u>disjunction</u> of the propositions $p(D_i)$, $i \in I$; that is, such that p(D) is confirmed precisely by those $\mathcal{A}$-outcomes which confirm at least one of the $p(D_i)$. It can be shown that if such a disjunction p(D) exists, it is effective as the supremum (least upper bound) in the operational logic $\Pi(\mathcal{A})$ of the family $(p(D_i) \mid i \in I)$. However, simple examples show that even if p(D) is the supremum of the propositions $p(D_i)$, $i \in I$, in the logic $\Pi(\mathcal{A})$, p(D) need not be effective as the disjunction of the propositions $p(D_i)$. In any case we write the supremum, when it exists, as $\bigvee_{i \in I} p(D_i)$ and note that p(D) does behave like the classical disjunction to the extent that any $\mathcal{A}$-outcome that refutes p(D) will refute every $p(D_i)$, while any $\mathcal{A}$-outcome that confirms any one of the $p(D_i)$ will confirm p(D). If the logic $\Pi(\mathcal{A})$ is orthocomplemented and $p(D) = \bigvee_{i \in I} p(D_i)$, then an $\mathcal{A}$-outcome will refute p(D) if and only if it refutes every $p(D_i)$, $i \in I$; however, there may be $\mathcal{A}$-outcomes that confirm p(D) but do not confirm any one of the propositions $p(D_i)$.

It is an immediate consequence of coherence that any finite family of mutually orthogonal operational propositions $(p(D_i) \mid i=1,2,\ldots,n)$ has a supremum, namely $p(\bigcup_{i=1}^{n} D_i)$. It is important to observe that even in this orthogonal finite case, the supremum need not be a disjunction in the above sense.

Let $(L, \leq, ')$ be any orthomodular poset, let X be the set of nonzero elements of L and let $\mathcal{A}$ be the set of all finite maximal orthogonal subsets of X. Then $\mathcal{A}$ is a manual and $(\Pi(\mathcal{A}), \leq, ')$ is isomorphic to $(L, \leq, ')$. Consequently every Boolean algebra can be realized as the operational logic of some (Boolean) manual of operations; furthermore, the manual can even be chosen such that every operation is finite. Likewise, since every proposed "quantum logic" turns out to be at least an orthomodular poset, it follows that every quantum logic can also be realized as the operational logic $\Pi(\mathcal{A})$ of some manual $\mathcal{A}$.

4. The Compounding of Manuals. In this section let $\mathcal{A}$ be a manual with outcome set X. We shall refer to $\mathcal{A}$ as the base manual and we shall regard $\mathcal{A}$ as a reservoir of primitive physical operations from which we intend to synthesize compound operations requiring the execution of the primitive operations in "connected sequences". Suppose, for instance, that $E_1, E_2, \ldots, E_n$ are operations in $\mathcal{A}$ and that the primitive operations are executed in a connected sequence so as to obtain the sequence x_1, x_2, ..., x_n of respective outcomes. Let us agree to record the formal product $x_1x_2x_3\ldots x_n$ to denote the acquisition of such a sequence of outcomes. Denote by S the set of all such formal products $x_1x_2\ldots x_n$ with $x_1, x_2, \ldots, x_n \in X$ where n runs through the positive integers, so that S is the free semigroup over X. The product in S of the "word" $x_1x_2\ldots x_n$ and the "word" $y_1y_2\ldots y_m$ is, of course, the "word" $x_1x_2\ldots x_ny_1y_2\ldots y_m$.

In the following it will be convenient to adjoin a formal identity 1 to the semigroup S so as to obtain a semigroup $X^c = S \cup \{1\}$ with an identity which we shall refer to as the <u>free monoid</u> over X. If $b \in X^c$ with $b \neq 1$, then b is uniquely expressible in the form $b = x_1 x_2 \ldots x_n$ with $x_i \in X$ for $i = 1,2,\ldots,n$ and we define the <u>length</u> of the word b to be $|b| = n$. Naturally, we define $|1| = 0$. The elements of X^c of length 1 are, of course, identified with the corresponding elements of X, so that $X \subseteq X^c$.

A subset A of X^c is said to be <u>bounded</u> if there is a non-negative integer n such that $|a| \leq n$ holds for all $a \in A$. If A is non-empty and bounded, we define $|A|$ to be the minimum of all such non-negative integers n and we define $|\emptyset| = -1$. If $A, B \subseteq X^c$, we define AB to be the set of all elements of X^c of the form ab with $a \in A$ and $b \in B$. When it is convenient, we disregard the distinction between an element $a \in X^c$ and the singleton set $\{a\}$; so, for instance , aB means $\{a\}B$.

In the following, $\{1\}$ will be interpreted as representing a trivial physical operation requiring that we do nothing other than to record the symbol 1 as the outcome. Of course, each basic operation $E \in \mathcal{Q}$ can be regarded as a one-stage compound operation. A two-stage compound operation is formed as follows: First select a basic operation $E \in \mathcal{Q}$, an $\mathcal{Q}$-event $D \neq \emptyset$ such that $D \subseteq E$ and a basic operation $F_d \in \mathcal{Q}$ for each outcome $d \in D$. The two-stage compound operation in question, let us call it G, is executed by first executing E to obtain (say) the outcome e; if $e \notin D$, we are done and we record the outcome of G as e, but if $e \in D$, we are obliged to execute F_e immediately to obtain (say) the outcome $x \in F_e$ and to record the outcome of this execution of G as ex $\in X^c$. Evidently, the outcome set for G is $(E \setminus D) \cup (\bigcup_{d \in D} dF_d)$. If we set $F_e = \{1\}$ for $e \in E \setminus D$, then the outcome set for G is simply the set $\bigcup_{e \in E} eF_e$. Multistage compound operations can now be built up inductively by iteration of the above procedure. As usual, we propose to identify such compound operations with their outcome sets.

The above construction of compound operations over $\mathcal{Q}$ can be formalized as follows: If $E, G \subseteq X^c$ with $E \neq G$, and if there exists for each $e \in E$ a set F_e such that either $F_e \in \mathcal{Q}$ or else $F_e = \{1\}$, and if $G = \bigcup_{e \in E} eF_e$, then we shall call G a <u>direct</u> <u>successor</u> of E. If there exists a finite sequence $G_1, G_2, \ldots, G_n$ such that $E = G_1$ and G_{i+1} is a direct successor of G_i for $i = 1,2,\ldots,n-1$, we shall say that G_n is a <u>successor</u> of E. We define $\mathcal{Q}^c$ to be the collection of subsets of X^c consisting of $\{1\}$ together with all of the successors of $\{1\}$.

It should be clear that if E is a non-empty bounded subset of X^c and G is a direct successor of E, then G is bounded and $|G| = |E| + 1$. It follows that every $G \in \mathcal{Q}^c$ is a non-empty bounded subset of X^c. A set $G \in \mathcal{Q}^c$ will be called a <u>compound</u> <u>operation</u> over $\mathcal{Q}$. Evidently, if G is such a compound operation, then $|G|$ represents the maximal number of sequential executions of basic operations that could be required to complete an execution of G. Clearly $\mathcal{Q} \subseteq \mathcal{Q}^c$; in fact, $\mathcal{Q}$ is exactly the subset of $\mathcal{Q}^c$ consisting of the compound operations G with $|G| = 1$.

Let $\perp$ be the orthogonality relation (of operational rejection) defined on X by $\mathcal{Q}$. We extend $\perp$ to a relation (also denoted by $\perp$) on the free monoid X^c as follows: For $a,b \in X^c$, $a \perp b$ holds if and only if there exist $c,d,e \in X^c$ and there exist $x,y \in X$ such that $a = cxd$, $b = cye$ and $x \perp y$ holds in X. For obvious reasons we refer to this extension of $\perp$ to X^c as the <u>lexicographic orthogonality</u> induced on X^c. A subset D of X^c is called an <u>orthogonal</u> set provided that $a \perp b$ holds whenever $a,b \in D$ with $a \neq b$. Clearly any singleton subset of X^c is an orthogonal set; in particular, $\{1\}$ is a maximal orthogonal set in X^c. It is not difficult to verify that any direct successor of a maximal orthogonal set is again a maximal orthogonal set; hence, any compound operation over $\mathcal{Q}$ is a maximal orthogonal subset of X^c. From this it follows that $\mathcal{Q}^c$ is an irredundant premanual.

In order to show that $\mathcal{Q}^c$ is coherent, it will be useful to develop some additional notation. If $a \in X^c$ and $B \subseteq X^c$, we define $a^{-1}B = \{c \in X^c \mid ac \in B\}$. Evidently, $a^{-1}B$ is the set of all "tail ends" of words in B that initiate with the word a. If $a,b \in X^c$ and $B \subseteq X^c$, we have the following obvious computational rules: (i) $a^{-1}(aB) = B$. (ii) $a(a^{-1}B) \subseteq B$. (iii) $a^{-1}(b^{-1}B) = (ab)^{-1}B$. (iv) If B is an orthogonal set, then so is $a^{-1}B$. (v) If B is a maximal orthogonal set and $a^{-1}B$ is not empty, then $a^{-1}B$ is again a maximal orthogonal set. If D is a subset of X^c, we define $i(D) = \{x \in X \mid x^{-1}D \neq \emptyset\}$, noting that i(D) is just the set of initial letters of words in D. We shall say that $D \subseteq X^c$ is <u>levelwise admissible</u> provided that $i(a^{-1}D)$ is an $\mathcal{Q}$-event for every $a \in X^c$.

By reasonably straightforward arguments one can verify the following facts: (i) Every compound operation $G \in \mathcal{Q}^c$ is levelwise admissible and, conversely, every bounded levelwise admissible maximal orthogonal subset G of X^c is a compound operation. (ii) Every subset D of a compound operation G is bounded, orthogonal and levelwise admissible; conversely, every bounded orthogonal and levelwise admissible subset D of X^c is contained in some compound operation $G \in \mathcal{Q}^c$. (iii) If G is a compound operation and if $a \in X^c$ with $a^{-1}G \neq \emptyset$, then $a^{-1}G$ is a compound operation. (iv) $\mathcal{Q}^c$ is coherent; hence a manual. (v) X^c is precisely the set of all $\mathcal{Q}^c$-outcomes. (vi) If D is an $\mathcal{Q}^c$-event and if $a \in X^c$, then aD as well as $a^{-1}D$ are $\mathcal{Q}^c$-events. (vii) If A and B are $\mathcal{Q}^c$-events, then AB is an $\mathcal{Q}^c$-event.

The logic $\Pi(\mathcal{Q}^c)$ is always totally non-atomic provided that there is an $\mathcal{Q}$-operation E consisting of at least two distinct outcomes. To see this, notice that an atom in $\Pi(\mathcal{A}^c)$ would necessarily be of the form $p(\{a\})$ for some $a \in X^c$. Let $E \in \mathcal{A}$ be a basic operation consisting of two or more outcomes and let $b = ax$ for some $x \in E$; it follows that $p(\emptyset) < p(\{b\}) < p(\{a\})$, a contradiction.

It is easy to show that $\mathcal{A}^c$ is a Dacey manual if and only if $\mathcal{A}$ is a Dacey manual and R.J. Weaver [15] has shown that if $\mathcal{A}$ is a Dacey manual, then $\mathcal{A}^c$ is conjunctive if and only if $\mathcal{A}$ is conjunctive. Thus, $\Pi(\mathcal{A})$ is an orthomodular poset (respectively, an orthomodular lattice) if and only if $\Pi(\mathcal{A}^c)$ is an orthomodular poset (respectively, an orthomodular lattice). If $\mathcal{A}$ consists of a single operation, then both $\mathcal{A}$ and $\mathcal{A}^c$ are Boolean manuals; however, if $\mathcal{A}$ is a Boolean manual, it need not follow that $\mathcal{A}^c$ is a Boolean manual.

5. Complete Stochastic Models (Weights and States). By a weight function for a manual $\mathcal{A}$ with the outcome set $X = \bigcup \mathcal{A}$, we mean a real valued function ω defined on X and taking its values in the closed unit interval for which the unordered sums $\sum_{e \in E} \omega(e)$ converge to 1 for every $E \in \mathcal{A}$. The set of all such weight functions is denoted by $\Omega = \Omega(\mathcal{A})$. It is natural to extend an $\omega \in \Omega$ to the $\mathcal{A}$-events by defining $\omega(D) = \sum_{d \in D} \omega(d)$ for any $\mathcal{A}$-event D. It then follows that $0 \leq \omega(D) \leq 1$ for all $\mathcal{A}$-events D and that ω is finitely additive in the sense that $\omega(\bigcup_{i=1}^{n} D_i) = \sum_{i=1}^{n} \omega(D_i)$ for any finite family $(D_i \mid i=1,2,\ldots,n)$ of pairwise orthogonal $\mathcal{A}$-events. In general, an $\omega \in \Omega$ need not be countably additive; such additional features as countable additivity will depend on the detailed structure of the manual $\mathcal{A}$.

A weight function $\omega \in \Omega(\mathcal{A})$ is to be regarded as a possible complete stochastic model for the empirical situation described by the manual $\mathcal{A}$. This is to be understood in the sense that for every $\mathcal{A}$-outcome x in X, $\omega(x)$ is interpreted as the "long run relative frequency" with which the outcome x is secured as a consequence of the execution of operations for which x is a possible outcome. Notice that the implicit decision to ignore the actual operations that are executed is already present in the outcome identification process used to form the manual $\mathcal{A}$.

Suppose now that C and D are $\mathcal{A}$-events and that $p(C) \leq p(D)$, so $D^\perp \subseteq C^\perp$. Let $\omega \in \Omega(\mathcal{A})$ and select any $\mathcal{A}$-operation E for which $D \subseteq E$. Define $B = E \setminus D$, noting that $B \subseteq D^\perp \subseteq C^\perp$. It follows from coherence that $B \cup C$ is an $\mathcal{A}$-event and by the finite additivity of ω that $\omega(B) + \omega(C) = \omega(B \cup C) \leq 1 = \omega(B) + \omega(D)$; hence, $\omega(C) \leq \omega(D)$. In particular, if $p(C) = p(D)$, then $\omega(C) = \omega(D)$. This fact permits us to lift ω to a function (still denoted by ω) defined on the operational logic $\Pi(\mathcal{A})$ simply by setting $\omega(p(D)) = \omega(D)$ for any $\mathcal{A}$-event D. This ω, defined on $\Pi(\mathcal{A})$, will be referred to as the regular state induced on $\Pi(\mathcal{A})$ by the original weight function $\omega \in \Omega(\mathcal{A})$.

As usual, a state is a function defined on the logic -- in this case $\Pi(\mathcal{A})$ -- taking on its values in the closed unit interval, finitely additive over orthogonal propositions and assuming the value 1 on the unit, which in this case is p(E) for any $E \in \mathcal{A}$. Notice that every regular state is a bona fide state; however, there may exist states that are not regular for the manual $\mathcal{A}$. If ω is a regular state, it can be shown that if $E \in \mathcal{A}$ is a test operation for a proposition p(D) in $\Pi(\mathcal{A})$, then $\omega(p(D)) = \omega(E \cap D^{\perp\perp})$. Thus, for a regular state ω, we can interpret $\omega(p(D))$ as the long run relative frequency with which p(D) will be confirmed , according to the stochastic model ω, when it is tested (by a test operation for p(D)).

In the customary approach to the study of the foundations of quantum mechanics, one begins with states and observables. Those observables that can assume only the values 0 and 1 are singled out and identified as quantum mechanical propositions (or questions) and are presumed to band together to form at least an orthomodular poset. The states are then reinterpreted as generalized probability measures defined on this poset of propositions (or "quantum logic", if you prefer) satisfying the above conditions for states. Thus, in the final analysis, our approach yields substantially the same mathematical system as does the customary approach, but provides it with a rich operational infrastructure which is at best tacit in the usual approach.

It is surely desirable to have a lavish supply of potential stochastic models (that is, weight functions) for a manual $\mathcal{A}$ and the physical circumstances that it describes. Any ad hoc assumption assuring such a supply of models would necessarily be a non-trivial constraint on the manual since there are large classes of orthomodular lattices that admit only one state, or no states at all [5]. Nevertheless, in any realistic situation there always appears to be a generous supply of weight functions in one of the following senses: Let Δ be a set of states defined on the operational logic $\Pi(\mathcal{A})$ of the manual $\mathcal{A}$. As usual, we shall say that Δ is a full set of states for the logic $\Pi(\mathcal{A})$ provided that the condition $\alpha(p(C)) \leq \alpha(p(D))$ for all $\alpha \in \Delta$ implies the condition $p(C) \leq p(D)$, where p(C) and p(D) are any two propositions in $\Pi(\mathcal{A})$. Similarly, Δ is said to be a strong set of states for $\Pi(\mathcal{A})$ provided that whenever $\alpha(p(C)) = 1$ implies $\alpha(p(D)) = 1$ for all $\alpha \in \Delta$, then $p(C) \leq p(D)$, where again p(C) and p(D) are any two propositions in $\Pi(\mathcal{A})$. A set of weights $\Delta \subseteq \Omega(\mathcal{A})$ is said to be full (respectively, strong) provided that the corresponding set Δ of regular states is full (respectively, strong) for the logic $\Pi(\mathcal{A})$.

It can be shown that a set of weights $\Delta \subseteq \Omega(\mathcal{A})$ is full (respectively, strong) if and only if given any non-orthogonal pair of $\mathcal{A}$-outcomes x and y, there exists an $\omega \in \Delta$ such that $1 < \omega(x) + \omega(y)$ (respectively, $\omega(x) = 1$ and $\omega(y) \neq 0$). Evidently any strong set of weights or states is necessarily full. Moreover, any full set of states will separate the propositions of $\Pi(\mathcal{A})$, and the existence of a separating set of states for $\Pi(\mathcal{A})$ implies that $\Pi(\mathcal{A})$ is an orthomodular poset and hence that $\mathcal{A}$ is a Dacey manual.

6. Signed Weights. The example of quantum mechanics and the mathematical attractions of linear algebra encourage us to "linearize" the concepts introduced in the previous sections. Thus let $\mathcal{A}$ be a manual with outcome set $X = \bigcup \mathcal{A}$ and weight space $\Omega = \Omega(\mathcal{A})$. Assume that for each $\mathcal{A}$-outcome $x \in X$ there exists a weight $\omega \in \Omega$ such that $\omega(x) > 0$, noting that if Ω is full, then this condition is automatically met. The initial step in the linearization process is simply to recognize that the weight space Ω is a convex subset of $\mathbb{R}^X$, the locally convex topological vector space of all real valued functions defined on X (with the product topology). A function ν in $\mathbb{R}^X$ is called a signed weight for $\mathcal{A}$ provided that the unordered sum $\sum_{x \in E} \nu(x)$ converges for every $E \in \mathcal{A}$ to a finite numerical value independent of E. Denote by $W = W(\mathcal{A})$ the vector subspace of $\mathbb{R}^X$ consisting of all such signed weight functions. Define the linear functional e on W by $e(\nu) = \sum_{x \in E} \nu(x)$ for each $\nu \in W$, where by definition E can be any $\mathcal{A}$-operation. Each outcome $x \in X$ induces a linear functional f_x on $\mathbb{R}^X$ by evaluation, $f_x(\nu) = \nu(x)$, and the restriction of f_x to W is a linear functional on W which we shall also denote by f_x. Thus, Ω is a convex subset of W, the linear functionals f_x, $x \in X$, are non-negative on Ω and $e = 1$ on Ω. Define $K = K(\mathcal{A})$ to be the cone in W consisting of all non-negative scalar multiples of the weight functions $\omega \in \Omega$ and define $V = V(\mathcal{A}) = K - K$, so that V is a vector subspace of W and $\Omega \subseteq K \subseteq V$. The restriction of the linear functional e to V will again be denoted by e and similarly the restrictions of the linear functionals f_x, $x \in X$, to V will continue to be denoted by f_x.

Observe that $K = \{\nu \in W \mid f_x(\nu) \geq 0 \text{ for all } x \in X\} = \{\nu \in V \mid f_x(\nu) \geq 0 \text{ for all } x \in X\}$; while $\Omega = \{\nu \in K \mid e(\nu) = 1\}$, so that Ω is a base for the cone K. Furthermore, for $\nu \in K$ we have $f_x(\nu) \leq e(\nu)$ for every $x \in X$, that is $e - f_x$ is always a non-negative linear functional on W (or on V). The set of linear functionals $\{f_x \mid x \in X\}$ is total (separates points) for W as well as for V. Finally, given any $x \in X$ there exists at least one $\omega \in \Omega$ such that $f_x(\omega) > 0$. In the next section, these simple observations will provide the motivation for our definition of an abstract signed weight space.

Now let $\mathcal{A}^c$ be the compound manual over $\mathcal{A}$ and let us describe the relationship between $W(\mathcal{A})$ and $W(\mathcal{A}^c)$. To begin with, notice that each signed weight function $\nu \in W(\mathcal{A}^c)$ induces a mapping $\bar{\nu}: X^c \longrightarrow W(\mathcal{A})$ by the prescription $[\bar{\nu}(a)](x) = \nu(ax)$ for all $a \in X^c$ and all $x \in X$. As can be seen, $\bar{\nu}$ carries precisely the information that ν does, but is directly related to the base manual $\mathcal{A}$. For any $a \in X^c$ and any $\nu \in W(\mathcal{A}^c)$, we have $e(\bar{\nu}(a)) = \nu(a)$; hence, $e(\bar{\nu}(ax)) = f_x(\bar{\nu}(a))$ for all $x \in X$ and $a \in X^c$. This functional equation, satisfied by $\bar{\nu}$, is crucial here and consequently we shall refer to any function φ belonging to the vector space $W(\mathcal{A})^{X^c}$ which satisfies the equation $e(\varphi(ax)) = f_x(\varphi(a))$ for all $x \in X$ and all $a \in X^c$ as a <u>transition vector</u>. The set all such transition vectors φ, which we shall denote by $W^c(\mathcal{A})$, forms a vector subspace of $W(\mathcal{A})^{X^c}$ and the mapping $W(\mathcal{A}^c) \longrightarrow W^c(\mathcal{A})$ given by $\nu \rightsquigarrow \bar{\nu}$ is a vector space isomorphism. In fact, for $\varphi \in W^c(\mathcal{A})$, the unique $\nu \in W(\mathcal{A}^c)$ for which $\varphi = \bar{\nu}$ is given by $\nu(a) = e(\varphi(a))$ for every $a \in X^c$. Define $K^c(\mathcal{A})$ and $\Omega^c(\mathcal{A})$ to be the cone in $W^c(\mathcal{A})$ and its base corresponding to the cone $K(\mathcal{A}^c)$ and its base $\Omega(\mathcal{A}^c)$ in $W(\mathcal{A}^c)$ under this isomorphism. Direct calculation reveals that $K^c(\mathcal{A})$ is exactly the set of $K(\mathcal{A})$-valued transition vectors, while $\Omega^c(\mathcal{A})$ is exactly the set of $K(\mathcal{A})$-valued transition vectors φ for which $\varphi(1)$ belongs to $\Omega(\mathcal{A})$.

In order to provide sufficiently many weights in $\Omega(\mathcal{A})$, we assumed that for $x \in X$ there exists at least one $\omega \in \Omega(\mathcal{A})$ with $\omega(x)$ strictly positive. We shall now show that $\Omega(\mathcal{A}^c)$ inherits this property. Thus let $a \in X^c$ with (say) $a = x_1x_2\ldots x_n$ where $x_i \in X$ for $i = 1,\ldots,n$ and for each such i, select $\omega_i \in \Omega(\mathcal{A})$ with $\omega_i(x_i) > 0$. Define $\alpha = n^{-1}\sum_{i=1}^{n}\omega_i$, noting that α is in $\Omega(\mathcal{A})$ and that $\alpha(x_i) > 0$ for $i = 1, 2, \ldots, n$. Now define $\varphi \in W(\mathcal{A})^{X^c}$ recursively by $\varphi(1) = \alpha$ and $\varphi(bx) = \alpha(x)\,\varphi(b)$ for $b \in X^c$ and $x \in X$. A routine calculation will show that φ is indeed a transition vector and that, if ω is the corresponding signed weight function in $W(\mathcal{A}^c)$, then $\omega \in \Omega(\mathcal{A}^c)$ and $\omega(a) > 0$ as promised.

It is particularly convenient to compute with the transition vectors, rather than with the original signed weights, in those instances in which we wish to regard the weights as representing statistical ensembles or beams of particles. In this connection, the transition operations defined below can be interpreted directly as "filtering operators", in the sense, for example, of Mielnik [9], [1]. Thus for any $a \in X^c$, we define the <u>transition operation</u> $T_a: W^c(\mathcal{A}) \longrightarrow W^c(\mathcal{A})$ as the linear transformation on the vector space of all transition vectors φ according to $[T_a\varphi](b) = \varphi(ab)$ for all $b \in X^c$. These transition operations enjoy the following "filterlike" properties: (i) $T_a \geqslant 0$ in the sense that it maps the positive cone $K^c(\mathcal{A})$ into itself--actually, it maps the cone onto itself. (ii) T_1 is the identity transformation on $W^c(\mathcal{Q})$ and $T_1 - T_a \geqslant 0$. (iii) For any $a, b \in X^c$, $T_a T_b = T_{ba}$.

Now let $\omega \in \Omega(\mathcal{Q}^c)$ and suppose $a \in X^c$ with $\omega(a) \neq 0$. Then $T_a\overline{\omega}$ is a non-zero transition vector and if we define $\overline{\omega}_a = (\omega(a))^{-1}T_a\overline{\omega}$, then $\overline{\omega}_a$ belongs to $\Omega^c(\mathcal{Q})$; hence, there exists a unique weight function $\omega_a \in \Omega(\mathcal{Q}^c)$ such that $\overline{(\omega_a)} = \overline{\omega}_a$. A simple calculation shows that, for any $b \in X^c$, $\omega_a(b) = \omega(ab)/\omega(a)$. Evidently, $\omega_a(b)$ is the operationally conditioned long run relative frequency of b given a; hence, ω_a will be referred to as the weight function obtained by <u>operationally conditioning</u> the weight function ω by the outcome a. Note carefully the definite temporal order implied by such conditioning and the consequent lack of temporal symmetry as exemplified by the failure of the classical multiplication rule; that is, in general, $\omega_a(b)\,\omega(a) \neq \omega_b(a)\,\omega(b)$.

Conditioning, in the sense described above, is consistent with the orthodox view of conditioning in quantum physics. Indeed, suppose that the compound operations in $\mathcal{Q}^c$ represent physical operations that can be performed "on suitable physical systems". Then suppose that $a \in X^c$ and $\alpha, \omega \in \Omega(\mathcal{Q}^c)$ are such that, whenever the outcome a is secured as a consequence of executing an operation on a physical system in the state ω, then immediately after such a "measurement", the system will be in the state α. If states are interpreted as long run relative frequencie then $\alpha = \omega_a$ in the following sense: Let $b \in X^c$, $G \in \mathcal{Q}^c$ with $b \in G$, $E \in \mathcal{Q}^c$ with $a \in E$ and put $(E \setminus \{a\}) \cup aG = F \in \mathcal{Q}^c$, noting that $ab \in F$. Suppose F is executed N times on a system in state ω, so that an outcome of the form ag with $g \in G$ will be secured approximately $\omega(a)$ N times; whereas, the particular outcome ab will be secured approximately $\omega(ab)$ N times. Each time the outcome ab is secured, G will have been executed on a system in state α and outcome b will have been secured as a consequence, and this should occur approximately $\alpha(b)\,\omega(a)$ N times. Thus, $\alpha(b)\,\omega(a)$ N should approximately equal $\omega(ab)$ N, with better approximation as N becomes large. We conclude that $\alpha(b) = \omega(ab)/\omega(a) = \omega_a(b)$.

Ordinarily, when a physicist speaks of a physical system in the state ω, he presumes (often implicitly) that ω is a so-called <u>pure</u> state--that is, an extreme point of the convex set of admissible states For us, a weight function $\omega \in \Omega$ is said to be <u>pure</u> (or an extreme point of Ω) if it can not be written non-trivially as a convex combination of other weight functions in Ω. For $\alpha, \beta \in \Omega$, we define the real number $r(\alpha, \beta) = \inf \{\alpha(x)/\beta(x) \mid x \in X \text{ and } \beta(x) \neq 0\}$. The following convenient criterion for extreme points of Ω is a generalization of a result of Greechie and Miller [6]: An $\alpha \in \Omega$ is pure if and only if $r(\alpha,\beta) = 0$ for all $\beta \in \Omega$ with $\beta \neq \alpha$. Using this criterion, we have shown [4] that if $\omega \in \Omega(\mathcal{A}^c)$ and $a \in X^c$ with $\omega(a) \neq 0$, then the operationally conditioned weight ω_a is pure if ω is pure. This implies that the transition operations T_a map extreme rays of the cone $K^c(\mathcal{A})$ into extreme rays of this cone.

Just as it is possible to operational condition by outcomes in X^c, so it is also possible to condition by $\mathcal{A}^c$-events. Indeed, if D is such an event and $\omega \in \Omega(\mathcal{A}^c)$ with $\omega(D) \neq 0$, we define the operationally conditioned weight function $\omega_D \in \Omega(\mathcal{A}^c)$ by $\omega_D(b) = \omega(Db)/\omega(D)$. Evidently $\omega_D = \sum_{d \in C} t_d \omega_d$ where $C = \{d \in D \mid \omega(d) \neq 0\}$ and $t_d = \omega(d)/\omega(D)$; thus, in general, ω_D will be a non-trivial convex combination of the operationally conditioned weights ω_d. It is important to observe that even when $E \in \mathcal{A}^c$, ω_E need not be equal to ω; that is, the operation E need not be "gentle" for the weight ω. A great deal more might be said concerning this particular system; however, it is mathematically more efficient to abstract these notions, as we shall do in the next section.

7. Abstract Signed Weight Spaces. As a consequence of the preceding, let us define an (abstract) signed weight space (SWS) to be a triple (W,X,e) satisfying the following conditions:

(i) W is a real vector space and e is a linear functional on W.

(ii) Corresponding to each $x \in X$ is a linear functional f_x, and the set $\{f_x \mid x \in X\}$ is a total set of linear functionals on W.

(iii) Given $x \in X$, there exists at least one vector $\nu \in W$ such that $f_x(\nu) > 0$ and $f_y(\nu) \geq 0$ for all $y \in X$.

(iv) If $\nu \in W$ is such that $f_x(\nu) \geq 0$ for all $x \in X$, then $f_x(\nu) \leq e(\nu)$ for all $x \in X$.

Clearly, if $\mathcal{A}$ is a manual with outcome set X and if , for each $x \in X$, there exists an $\omega \in \Omega(\mathcal{A})$ with $\omega(x) > 0$, then both triples $(W(\mathcal{A}), X, e)$ and $(V(\mathcal{A}), X, e)$ as defined in the preceding section are signed weight spaces. We shall refer to $(V(\mathcal{A}), X, e)$ as the signed weight space for the manaual $\mathcal{A}$.

Let (W,X,) be an (abstract) SWS and define $K = K(W,X,e)$ to be the cone in W consisting of all of the vectors $\nu \in W$ such that $f_x(\nu) \geq 0$ for all $x \in X$. Define $\Omega = \Omega(W,X,e) = K \cap e^{-1}(1)$, so that Ω is a base for the cone K. Clearly, for any $x \in X$, there will exist at least one $\omega \in \Omega$ for which $f_x(\omega) > 0$; hence, for any finite sequence $x_1, x_2, \ldots, x_n \in X$, there will exist an $\omega \in \Omega$ such that $f_{x_i}(\omega) > 0$ for $i = 1, 2, \ldots, n$. If X is a finite set, it can be shown that K is a generating cone for W; however, in the general case this need not be so. If K does not generate W, we can always replace W by the subspace $V = K - K$

and cut all of the linear functionals down to V; the resulting triple (V,X,e) is again an SWS with the same positive cone and the same base.

Let (V,X,e) be an SWS with generating cone K and base Ω. Let B = B(V,X,e) be the convex hull of $\Omega \cup (-\Omega)$, noting that B = $\{t\alpha - (1-t)\beta \mid 0 \leq t \leq 1 \text{ and } \alpha, \beta \in \Omega\}$. Evidently, $-1 \leq f_x(\beta) \leq 1$ for all $x \in X$ and all $\beta \in B$; hence B is linearly bounded. It follows that (V, Ω) is a base norm space [1]; the norm $\|\cdot\|$, of course, is the Minkowski functional for B. For each $x \in X$, both f_x and $e - f_x$ are positive linear functionals on V; hence, f_x is continuous in the base norm topology and $\|f_x\| \leq 1$. Naturally, e is also continuous and $\|e\| = 1$.

As usual, the continuous dual (V*,e) is an order unit space and $0 \leq f_x \leq e$ for every $x \in X$. For purposes of motivation, even in the abstract case, the elements $x \in X$ can be thought of as outcomes of physical operations, the $\omega \in \Omega$ as stochastic models and $f_x(\omega)$ as the long run relative frequency with which x is secured. Again, an operationally suggestive structure is provided by X, under such an interpretation, and this permits--at least in principle--operational accessibility for all subsequently defined concepts.

Suppose that $(V(\mathcal{A}),X,e)$ is the signed weight space for a manual $\mathcal{A}$ possessing a full set of weights. If D is an $\mathcal{A}$-event, then D induces a linear functional f_D on V by $f_D(\nu) = \sum_{d \in D} \nu(d)$ for all $\nu \in V$. Then, if A,B are $\mathcal{A}$-events with p(A) = p(B) in $\Pi(\mathcal{A})$, it follows that $f_A = f_B$; hence, for any proposition $p(A) \in \Pi(\mathcal{A})$, we can define a linear functional corresponding to p(A) by $f_{p(A)} = f_A$. Since Ω is full, then, for any two propositions p(A) and p(B) in the logic $\Pi(\mathcal{A})$, we have $p(A) \leq p(B)$ if and only if $f_{p(A)} \leq f_{p(B)}$ in V*; furthermore, we have $0 \leq f_{p(A)} \leq e$ and $f_{p(A)'} = e - f_{p(A)}$. Consequently, the dual order interval $[0,e] \subseteq V^*$ contains a faithful copy of the operational logic $\Pi(\mathcal{A})$. In all of the concrete examples that we have considered in which Ω is a strong set of weights for $\mathcal{A}$, this faithful copy of $\Pi(\mathcal{A})$ consists precisely of all of the extreme points of the dual order interval $[0,e]$.

Just as we defined a transition vector for a manual, so we can define them for any (abstract) SWS (W,X,e). Thus, define such a <u>transition vector</u> to be a vector $\varphi \in W^{X^c}$ satisfying $e(\varphi(ax)) = f_x(\varphi(a))$ for all a belonging to the free monoid X^c and all $x \in X$. The vector space of all such transition vectors φ will be denoted by W^c and, for $a \in X^c$, we define the linear functional f_a on W^c by $f_a(\varphi) = e(\varphi(a))$ for all $\varphi \in W^c$. In particular, define $e^c = f_1$. Then (W^c,X^c,e^c) is a SWS, called the <u>compound</u> SWS over (W,X,e). Naturally, we define $K^c = K(W^c,X^c,e^c)$ and $\Omega^c = \Omega(W^c,X^c,e^c)$.

As before, each $a \in X^c$ will induce a linear transition operation T_a on the vector space W^c by $[T_a\varphi](b) = \varphi(ab)$ for all $\varphi \in W^c$ and all $b \in X^c$, and again, these will satisfy $0 \leq T_a \leq T_1$ and $T_a T_b = T_{ba}$. Also, every such T_a maps the cone K^c onto the cone K^c. If $\varphi \in W^c$, then $\varphi \in \Omega^c$ if and only if $\varphi(1) \in \Omega(W,X,e)$ and $\varphi(a) \in K(W,X,e)$ for every $a \in X^c$. Thus, for $\omega \in \Omega^c$ and for $a \in X^c$ with $e(\omega(a)) = f_a(\omega) \neq 0$, $\omega(a)/e(\omega(a))$ will belong to $\Omega(W,X,e)$. It can be shown that such an ω is an extreme point of Ω^c if and only if each such vector $\omega(a)/e(\omega(a))$ is an extreme point of $\Omega(W,X,e)$. Moreover, $e(\omega(a)) \neq 0$ implies that $0 \neq T_a\omega \in K^c$ and, consequently, $\omega_a = (T_a\omega)/e(\omega(a))$ will belong to Ω^c again. We refer to ω_a as the weight obtained by operationally conditioning ω by $a \in X^c$. Again, we can prove that if ω is an extreme point of Ω^c, then so is ω_a; hence, again, the transition operation T_a maps extreme rays in K^c back into extreme rays in K^c.

8. Conclusion. In the preceding sections, the principal ideas connected with the empirical logic approach to the physical sciences were developed (for the details see [3], [11], [14]) and then linearized with the introduction of the notion of the signed weight space $(V(\mathcal{Q}),X,e)$ and the consequent base norm space $(V(\mathcal{Q}),\Omega(\mathcal{Q}))$ and its dual order unit space $(V^*(\mathcal{Q}),e)$ for a manual of operations $\mathcal{Q}$. In the case where the weight space $\Omega(\mathcal{Q})$ was full, there even was a faithful copy of the operational logic $\Pi(\mathcal{Q})$ in the order interval $[0,e]$ of the dual $V^*(\mathcal{Q})$. This suggests the natural question: What are the necessary and sufficient conditions on the manual $\mathcal{Q}$ for this faithful copy of the operational logic $\Pi(\mathcal{Q})$ to be precisely the extreme points of the dual order interval $[0,e]$?

Thus, affiliated with every manual of operations $\mathcal{Q}$, there is an operational logic $\Pi(\mathcal{Q})$ and a signed weight space $V(\mathcal{Q})$. Since it is only the manual that may be tied directly to instructions for carrying out actual physical operations, it is important that we understand the manner in which the properties of the manual are reflected in the properties of its logic and its signed weight space. In the case where the outcome set $X = \bigcup \mathcal{Q}$ is finite, for example, it can be shown that the manual $\mathcal{Q}$ is a Boolean manual if and only if for every pair of operations $E,F \in \mathcal{Q}$ there exists a common "refinement" $G \in \mathcal{Q}$ in the sense that both E and F can be obtained as "coarsenings" of the single operation G (see [14] for definitions and details). It can also be shown, in this case, that the above is so if and only if the finite dimensional real ordered vector space $V(\mathcal{Q})$ is a lattice, and this is equivalent to requiring that the weight space $\Omega(\mathcal{Q})$ be a simplex. In [14] the connections between some of the properties of an operational logic $\Pi(\mathcal{Q})$ and its manual $\mathcal{Q}$ were studied; many of these results were reported above. Evidently similar studies concerning the signed weight spaces of manuals are in order.

The ability to formally describe repeated measurements is often an aim of an operational approach to physical theory; in this matter the empirical logic approach is no exception. The passage from a manual $\mathfrak{A}$ to a compound manual $\mathfrak{A}^c$ has been used here to represent the unrestricted compounding (by concatenation) of operations in $\mathfrak{A}$. Some of the facts concerning the operational logics $\Pi(\mathfrak{A}^c)$ of such compound manuals $\mathfrak{A}^c$ have been mentioned above; other facts can be found in [2], [11], [12] and [13]. When the $\mathfrak{A}^c$-outcomes are to be regarded as filters, as they often are in particle physics, it is convenient to recall that each outcome $a \in X^c$ is associated with a positive element in the unit ball of the space of all bounded linear operators from the compound signed weight space $W^c(\mathfrak{A})$ into itself, namely the transition operator T_a. Naturally, the transition vectors (and the corresponding weights) must be classified and studied according to how they transform under the transition operators.

In regard to the practical value of some of the programs mentioned above, and the many that suggest themselves and have not been mentioned, recall that the matter of how to interpret the infimum (or supremum) of a pair of propositions in a quantum logic has always been illusive; yet given the direct tie to a manual $\mathfrak{A}$, that one has in the empirical logic approach, we all have seen this mystery quickly and simply resolved. The question of the meaning of commuting propositions in the logic also enjoyed a similar fate. Thus we might hope that other outstanding problems of interpretation in the physical sciences, and in particular in quantum physics, might be resolved in this way.

References

1. C.M. Edwards, The operational approach to algebraic quantum theory I, Commun. Math. Phys. 16 (1970) 207-230.
2. D.J. Foulis and C.H. Randall, Conditioning maps on orthomodular lattices, Glasgow Math. J. 12 Part 1 (1971) 35-42.
3. ______________________________, Operational statistics I, basic concepts, J. Math. Phys. 13 NO. 11 (1972) 1667-1675.
4. ______________________________, The stability of pure weights under conditioning, to appear in Glasgow Math. J.
5. R.J. Greechie, Orthomodular lattices admitting no states, J. Combinatorial Theory 10 (1971) 119-132.
6. R.J. Greechie and F. Miller, On structures related to states on an empirical logic I, weights on finite spaces, Kansas State University mimeographed notes, 1969.
7. B. Jeffcott, The center of an orthologic, J. Symbolic Logic 37 NO. 4 (1972) 641-645.
8. G. Lüdders, Über die Zustandänderung durch den Messprozess, Ann. Physik 8 (1951) 322-328.
9. B. Mielnik, Theory of filters, Commun. Math. Phys. 15 (1969) 1-46.
10. J.C.T. Pool, Baer *-semigroups and the logic of quantum mechanics, Commun. Math. Phys. 9 (1968) 118-141.
11. C.H. Randall and D.J. Foulis, An approach to empirical logic, Amer. Math. Monthly 77 (1970) 363-374.
12. ______________________________, Lexicographic orthogonality, J. Combinatorial Theory Ser.A 11 (1971) 157-162.
13. ______________________________, States and the free orthogonality monoid, Math Systems Theory 6 No.3 (1972) 268-276.
14. ______________________________, Operational statistics II, manuals of operations and their logics, scheduled to appear J. Math. Phys. 14
15. R.J. Weaver, The conjunctive property in the free orthogonality monoid, Mount Holyoke College mimeographed notes, (1971).

THE STRUCTURE OF QUANTUM MECHANICS: SUGGESTIONS FOR A UNIFIED PHYSICS

Michael Drieschner
Max-Planck-Institut zur Erforschung der Lebensbedingungen der wissenschaftlich-technischen Welt, Starnberg

What we do in science depends on the questions we ask. In economics, for instance, we can ask for an easy way to increase the industrial power of a nation, or we can ask whether further economic growth yields any further advantages. In physics we could either look for the most elegant presentation of perturbation theory, or we could try to deduce physics from a pragmatic description of measurement. - It will depend on our goal what our science will look like.

I shall describe our program at Starnberg, which deviates from the usual physics, just as the program here at Marburg does. Therefore I will mainly try to show you that our questions are worth asking.

Let us first pose the problem of the justification of physical predictions or, if you like, of "Hume's challenge": David Hume saw the success of the physics of his time, but he also saw that it is impossible to deduce physical predictions logically from experiment. Let me say this in Hume's own words; his English is better than mine, anyway[1]

> "The contrary of every matter of fact is still possible; because it can never imply a contradiction, and is conceived by the mind with the same facility and distinctness, as if ever so conformable to reality. That the sun will not rise to-morrow is no less intelligible a proposition, and implies no more contradiction than the affirmation that it will rise. We should in vain, therefore, attempt to demonstrate the falsehood. Were it demonstratively false it would imply a contradiction, and could never be distinctly conceived by the mind." (21)

> "To say it is experimental, is begging the question. For all inferences from experience suppose, as their foundation, that the future will resemble the past, and that similar powers will be conjoined with similar sensible qualities. If there be any suspicion that the course of nature may change, and that the past may be no rule for the future, all experience becomes useless, and can give rise to no inference or conclusion. It is impossible, therefore, that any arguments from experience can prove this resemblance of the past to the future; since all these arguments are founded on the supposition of that resemblance." (32)

> "All inferences of experience, therefore, are effects of custom, not of reasoning". (36) "Belief is the true and proper name of this feeling." (40)

> "Here, then, is a kind of pre-established harmony between the course of nature and the succession of our ideas." (44) 2)

1) 'An Enquiry concerning Human Understanding', sections IV and V, London [2]1777.

2) The numbers refer to the paragraphs according to L.A.Selby-Bigge

It is impossible to refute this objection. But I suppose that your reaction is similar to mine: I don't feel quite easy. The argument is striking, but can it be the whole truth?

I am not going to speak immediately of Kant, who has given an answer to Hume's problem, but will ask how a physicist of today, who is not yet spoiled by philosophy of science, justifies a law of nature. Let us suppose that he devises an experiment to test a theory. He will never test an isolated law of nature, but always an aspect of a very complicated network of interrelated physical propositions. In every experiment a lot of theory has to be presupposed, at least the theory of the measuring apparatus.If the experiment does not confirm the prediction there is always the possibility left that "the experiment was wrong", namely that the theoretical interpretation of the experiment was not correct.

This complicated theoretical network is in part represented by a hierarchy of theories. In the progress of science a hypothetical law of nature can eventually be reduced to a more general one, i.e., one can strictly prove the special law when the general one is assumed valid. This more general law will eventually be reduced to a still more general one, and so on: If the physicists work successfully there will at last be only very general principles that must be supposed as premises for all laws of nature.

At which stage of this evolution is physics today?[3] There is quantum mechanics, a general theory of any physical object. There are complementary theories of special objects and forces, namely electrodynamics, gravitation theory, and the theories of strongly and weakly interacting elementary particles. Gravitation and electrodynamics are intimately connected with the relativistic theories of space and time, but there has not yet been a satisfactory unification of these theories with quantum mechanics, and still less has been achieved in the effort towards a unified relativistic and quantum mechanical theory of elementary particles and interactions. One has achieved, though, a deduction in principle of most of the "classical" physical theories and of chemistry from quantum mechanics and electrodynamics. For large systems and for irreversible processes, thermodynamics and information theory come in, which are not yet incorporated into the hierarchy of physical theories. There is, outside of the hierarchy, too, the theory of the one universe that really is, i.e. cosmology, including geology, and

[3] C.F. von Weizsäcker, 'Die Einheit der Natur' part II, especially pp.232 to 241.

biological evolution.

Thus we see a good deal of unification, but there are still many disconnected pieces and missing links. Empirically, therefore, the hypothesis of the unity of physics is neither proved nor refuted. But let us imagine a unified physics: What would it look like?

An axiomatic system of physics would apparently run into difficulties: It would have to presuppose logic (or a logic), but there is no justification of logic as yet. Perhaps it will only be possible to justify logic in connection with a justification of physics; such a program would be immense - but such an immense program might turn out to be the only one that has a chance of success. - I will not, however, talk about logic here.

Presupposing a logic, one would require that an axiomatic physics be consistent - naturally. But in view of the state of existing physical theories this requirement is far-reaching. We are going to extend it beyond the consistency of formalism to the requirement of <u>semantical consistency</u>. This means the following: Usually a physical theory has an "external semantics"; it deals with certain quantities that are supposed measurable by well known apparatus and techniques. But the claim of a unified physics would have to be universal; universal physics must include a possible description of the measuring apparatus and techniques - this is a truism. One should require then, that this theoretical description of measurement results in exactly those properties the measurement presupposed in its original, ordinary language definition. I have the impression that this is the deeper reason for the great interest taken in theories of measurement.

In order to give a more vivid picture of semantical consistency, I would like to talk about a cranky theory, the so-called 'hollow-world-theory': Imagine that the surface of the earth is not convex but concave, such that we live on the inside of a sphere that is surrounded by the solid ground and that contains the heaven with all its stars and planets in its inside. Our illusion of living on the outside of the sphere comes about because of the bending of the light rays. In reality all light rays are not straight lines but circles passing through the center of the world.

There were people who really believed in this theory and who could solve, I don't know which mysteries by the help of it. Apparently their belief was somewhat shaken by the appearance of Sputniks. This shows that they had not really understood their own theory, because if you do everything correctly, there is absolutely no empirical evidence that can jeopardize this theory. All you have to do is to reflect the usual theory on the sphere, that is to transform it according to the formula:

$$r \rightarrow r' = \frac{R^2}{r} ,$$

where R is the radius of the earth, and everything will be alright. Light rays will be, indeed, circles passing through the center of the world; and every body becomes smaller and smaller as it approaches the center of the world.

How about semantical consistency? According to its definition a measuring rod is a rigid body that has always the same length, wherever it is transported. The result of the 'hollow-world-theory' is, however, that <u>every</u> body "in reality" becomes smaller when it approaches the center of the world - and that makes the theory semantically inconsistent, although it is formally consistent.

A physicist, I suppose, will be reminded of relativity theory: Any body in motion or acceleration will, according to this theory, alter its dimensions as in the 'hollow world'. Hence semantical consistency cannot be as easy as it looked when we were dealing with cranks. Apparently the theory itself can result in necessary deviations from the original ideas about measurement; the theory may give a specification and an adjustment of the pre-theoretical definitions of the measurable quantities. This gives the theory a conventionalist element: If it is allowed to be semantically inconsistent with the original definitions of measurement, then there could be several such theories, and it is not at once clear which adjustment of the original definition is the right one.

In such a case, when it is impossible to have the theory semantically consistent with the original definition of measurement, it must at least be consistent with an adjusted definition. For relativity this means that for geometric measurements one would rather rely on light rays than on measuring rods; this is the way Einstein introduced his interpretation.

So much about derivation. But whence come the axioms for this derivation?

Let us first have a look at the possibility of such axioms. What can they look like? If <u>all</u> laws of nature can be derived from certain principles, those principles must be very general. They cannot contain any constants; constants rather have to follow from the theory. The axioms cannot be anything but general principles of any physics.

General principles of any physics: We should be able to name one or the other. What is it that we expect from physics? It ought to give an objective description of nature, a description that every one can, in principle, verify empirically. Therefore physics must at least allow unambiguous statements, empirically decideable as to yes or no. Empirically testable statements are predictions of the result of a test. To sum it up: physics ought to state <u>laws</u> for <u>predictions</u> of <u>unambiguously empirically decideable alternatives.</u> There is a lot packed into this formula. I can give here only a few explanatory key words.

1.) <u>Laws</u> are meant to hold for all times ("homogeneity of time"). Moreover they shall have other symmetries; I shall come back to this. There is an extensive literature in the Philosophy of Science that tries to say precisely what a "law-like proposition" is. (I should like to draw your attention to a paper by Andreas Kamlah[4] that deals especially with the relation between law-likeness and invariance under transformations.)

2.) <u>Predictions</u> may be stipulated as the central subject of physics:
   a) The concept of empirical or experimental science entails the concept of experience: Someone has experience when he has learned from the past for the future; experience arises in the past for application in the future, for predictions.
   b) An empirical proposition can only be checked <u>after</u> it is proposed, it must be a prediction.
   c) The generality of laws implies that they hold in particular also in the future.

3.) Predictions shall be <u>unambiguous</u> - this is self-evident to the extent that in general it is not even formulated. This postulate implies the strength but also the limits of physics. Hardly ever in daily life does one talk so precisely that one could not say afterwards: "Well, I didn't mean it <u>that</u> way"; - and the multiplicity of allusions, connotations and possible interpretations gives "ordinary" propositions their charm. Science, to the contrary has its very essence in the fact that there is no ambiguity, no allusion, no charm.

4.) Physics deals only with <u>empirically decideable alternatives</u>: Before every measurement a finite number of results is possible. Afterwards - "empirically" - one of them has turned out true, all

4) Zeitschrift für allgemeine Wissenschaftstheorie, Heft 2/1973

others have turned out wrong.
These, admittedly very short, indications will not impress you much; they are rather self-evident and hardly worth mentioning. They constitute the language of science, as C. Randall put it in his talk.

We have tried in Starnberg to formulate as much of what seems self-evident as possible, and to draw out as much physics from it as we could Out of this came an axiomatic structure of quantum mechanics that does not follow entirely from evident postulates, but that apparently does not need so very much more. In any case we have learned a lot about the structure of quantum mechanics and the concept of object.

I am not going to explain this here in any detail. A rather extensive description has been given at Penn State, but is not yet published. Let me just give a short description of the argument: The predictions postulated above can only be probabilities for measurements on objects. That probabilities are, in this framework, the most general predictions, is a rather long argument that I shall not repeat here. Physics then turns out to be an (objective) probability theory in a non-distributive lattice; the lattice is non-distributive, if one does not want to presuppose determinism to begin with. With a few more rather plausible assumptions this implies the abstract structure of quantum mechanics in Hilbert space without any reference as yet to definite quantities like position or momentum.

We would then have derived - supposing that the argument is convincing - the structure of quantum mechanics "a priori", namely by a reasoning of this sort: "What you mean by physics, I cannot prescribe to you. But _if_ you mean the same as I have formulated above, then only a theory in Hilbert space is physics, because it follows from the above postulates alone."

This is not the same usage of "a priori" as Kant's, but a more special one. According to Kant a judgement is true "a priori" if no experience would be possible at all without the truth of this judgement. This is Kant's answer to Hume's challenge: Certainly there is no implication from past to future, not even that there will be a future at all. But supposing that experience will be possible further on - this is only a hypothesis -, but _if_ experience will be possible, then certain conditions must be fulfilled, namely the preconditions of the possibility of any experience or, in Kant's technical terms, the _synthetical judgements "a priori"_.

We have started more modestly, not looking for conditions without which no experience would be possible at all, but only looking for conditions without which no physics would be possible. "Physics" indicates, though, any objectifying science, it would certainly comprise chemistry and biology, and also e.g. psychology and sociology in so far as they work under the conditions mentioned. This shows that these conditions are still far-reaching, even though they would not comprise all conceptual thinking - but perhaps it is not unthinkable that they might some day do even that.

SPACE

Until now we have only dealt with a very general framework of quantum mechanics. Space didnot yet occur, although it is a quite particular physical quantity: All measurements are, in the last analysis, spatial, and interaction always depends on spatial parameters. Perhaps these two particular features are really only one: Measurements are spatial because every measurement needs an interaction.
What is physical Space? (I shall capitalize it to distinguish it from the merely mathematical spaces.) Naturally I cannot here define Space, even briefly, but I will emphasize one aspect: Interaction always depends on the spatial relation between the interacting objects. Relativistically there should be only short-range interaction, i.e. every theory should be either strictly local such that interaction is only between fields at the same position, or interaction should be restricted to very small neighbourhoods.
Is this an accidental coincidence between interaction and Space? We want to examine the thesis that Space is defined as the one parameter on which interaction depends. This thesis seems to be justified by our every-day experience: In our childhood we have experienced the spatial distribution of things in a sometimes painful interaction; when we orient ourselves in Space and look around us we see the things where they interact with electromagnetic radiation from the sun.- The formulation leads us into the typical difficulties of studies of so fundamental a character. The truth we search for underlies all speech, such that every explicit mention seems ridiculous. Let us try, though: The objects interact where they are; they are where they interact. The position of an object is that one of its quantities which determines the possible interaction with other objects.

In order to make this Space physically more concrete, we introduce Weizsäcker's hypothesis of ur-objects: We can split every n-fold alternative into simple ("yes-no") alternatives[5]. Quantum mechanically

[5] cf., e.g., J.M. Jauch, Foundations of Quantum Mechanics, pp. 72 seq.

these simplest possible alternatives correspond to fundamental objects with a 2-dimensional state space; we call such a basic object, according to Weizsäcker's German usage, an "ur": "All physical objects, then, are composed of ur-objects (with a 2-dimensional state space)." This is the most radical atomism conceivable: There are no smaller objects. Ur-objects are atoms in the strict sense that one cannot even think their parts.

What does this hypothesis mean? One can construct any (n+1)-dimensional space as a symmetrical combination of n 2-dimensional spaces; so far the hypothesis is trivial. It can become non-trivial only if it in addition introduces or excludes certain symmetries or transformation properties. Let us investigate such symmetries.

A 2-dimensional state space, looked at by itself, has the symmetry group SU(2) or, to be precise, the projective group of the corresponding projective geometry. This is the group composed of $SU(2)Z_2$ and the corresponding anti-linear transformations. The relation of this group with the spatial rotations is well known: It gives the spin-representations; the anti-linear transformations are connected with the dotted indices.

This symmetry of the "urs" should play a special role, according to the hypothesis. Imagine an object composed of n urs. Its state space is a subspace of the product of n 2-dimensional spaces. If we transform all n urs with the same symmetry transformation the relation of the urs among each other is unchanged. If the total object is isolated, the above transformation does not change physics or dynamics at all.

The concept of total isolation of an object and of the corresponding symmetries is a rather difficult one. In transformations, we have to distinguish between (a.) simple changing of names for the states, and (b.) "physical" transformations:

a.) Names can be changed arbitrarily as long as the relations between the states remain unchanged. In quantum mechanics these relations are "transition"-probabilities; all unitary transformations leave these unchanged.

b.) In "physical" transformations not only relations between states are conserved, but individual states keep some of their properties and may keep their names; what is changed are (for the problem at hand) irrelevant properties. The energy of a hydrogen atom, e.g., is independent of its orientation in space. Thus the transformations of state space that correspond to rotations (and which must form a projective representation of the rotation group) are physical symmetry-transformations.

For an isolated "ur" these two types of transformations are merged: If the ur is really isolated there is no way to distinguish its states from one another; any labelling of individual states would create additional distinctions, i.e. additional urs. Therefore there is no individual property of a state that could be conserved in a "physical" transformation. Physical transformations and mere change of names are the same for urs.

Now we come to the point of the hypothesis: Imagine a part of the total object containing, say, $m (< n)$ urs. Transform all urs of this partial object by the same transformation. For the partial object this again is a symmetry transformation that does not change the "internal" properties of the partial object, but does change the relation of its urs to the urs of the "rest of the world". One should expect, therefore, that the interaction between our partial object and the rest of the world will be changed by such a transformation.

We can now express the contents of the "hypothesis of ur-objects" thus: The interaction between objects is changed exactly by those transformations that transform all urs of one of the objects by the same element of the 1-ur symmetry group.

Above we have defined as spatial transformations exactly those that determine interaction. We conclude then, in accordance with the hypothesis of ur-objects: <u>the spatial transformations of an object are those which transform all urs of that object by the same symmetry transformation</u>.

The state space of an object composed of n urs is the product of n 2-dimensional spaces. The action of the symmetry group in this space corresponds to a tensor representation. We are looking for a representation of all such objects in one space, which is then Space ("position-space"). Such a representation is furnished by the regular representation of the symmetry group. Let me explain this briefly:

Consider the group SU(2). (This is not exactly the symmetry group, but for the moment let us forget the difference.) We can construct the Hilbert space of the square-integrable functions of the group (i.e. on a S_3); this is our representation space. The action of the representation in this space is generated by the group multiplication: If r and s are elements of the group,and f(s) is an element of the representation space (function of the group), then the regular representation D(s) is defined by

(*)
$$D(s)\ f(r) = f(s \cdot r)$$

Every irreducible representation of the group is contained in the regular representation. Hence the state space of every object that is composed of urs is contained in the function space. This could be achieved just as easily with any other abstract Hilbert space. The point here is that the action of the symmetry transformation, expressed as in equation (*), is a point transformation of the underlying S_3. According to our argument these symmetry transformations are the spatial transformations: Then the underlying S_3 would be Space.[6)]

Thus from the structure of quantum mechanics together with the hypothesis of ur-objects we have derived 3-dimensional closed Space, provided the argument is stringent.

Actually the argument, as it stands, is more or less heuristic. Only when a unified theory of elementary particles has been submitted will we be able to give a rigorous account of our considerations. But we have the impression that this new line of - however heuristic - arguments already provides new insight, physically as well as philosophically. Space and the Lorentz group are fundamental in any physical theory; they are usually taken as empirical, and usually no further question is asked about them. Our considerations suggest reconsidering the issue and trying to derive these fundamentals from the very concepts of object, prediction and alternative.

---

6) There are some considerations by D.Finkelstein and L.Castell, aimed at extending this argument to the Lorentz group, or rather in this context to the de-Sitter-Group, or to the Conformal Group. (Phys.Rev.D 5 (1972) 320)

IRREVERSIBILITY AND DYNAMICAL MAPS OF STATISTICAL OPERATORS°

Vittorio Gorini+ and E.C.G. Sudarshan

Department of Physics, Center for Particle Theory, the University of Texas, Austin Texas, 78712

1. Introduction

Let $\mathfrak{C}$ denote the space generated by the states of a quantum mechanical system. We wish to discuss some problems and results, mainly in connection with N-level systems, regarding the structure of the convex set of linear maps $\mathfrak{C} \to \mathfrak{C}$ which map states to states. In the conventional Hilbert space formulation of quantum theory, $\mathfrak{C}$ is taken to be the Banach space $T(\mathfrak{h})$ of trace-class (t.c.) operators acting on the Hilbert space $\mathfrak{h}$ of the system, under the t.c. norm $\|\sigma\|_1 = \mathrm{Tr}(\sigma^* \sigma)^{\frac{1}{2}}$. The states of the system are the positive elements of $T(\mathfrak{h})$, with trace 1 (statistical operators). They form a convex set $K(\mathfrak{h})$ whose algebraic span is $T(\mathfrak{h})$. Linear maps $f:T(\mathfrak{h}) \to T(\mathfrak{h})$ such that $f(K(\mathfrak{h})) \subseteq \bigcup_{0\le\lambda\le 1} \lambda K(\mathfrak{h})$ are studied in relation with the measuring process and called <u>operations</u> [18,12,6] . Operations which map $K(\mathfrak{h})$ into itself are called <u>non-selective</u> [18]. Our interest is precisely in the convex set $F(\mathfrak{h})$ of non--selective operations. However, our motivation for studying these maps is not connected with the change undergone by an ensemble as a consequence of a measurement, but rather with the description of the dynamical evolution of a system undergoing an irreversible process, as we shall presently explain briefly. For this reason, we prefer to call the elements of $F(\mathfrak{h})$ <u>dynamical maps</u>.(°°) A more detailed discussion, together with the proofs of the results will appear in a forthcoming paper [10] .

The program of our investigation is twofold:

i) find out whether $F(\mathfrak{h})$ has sufficiently many extreme elements to make it possible to approximate in a suitable topology every dynamical map by means of a finite convex combination of extreme (pure) dynamical maps, in the same way that any state can be approximated by a finite convex combination of extreme (pure) states;

° Presented by V. Gorini.

+ Present Address: Istituto di Fisica dell'Università di Milano, Italy.

(°°) This terminology was introduced by one of us in ref. [28], where some properties of dynamical maps were studied. See also [14,15] .

ii) possibly classify all the extreme dynamical maps.

We briefly touch upon the first question and then state and discuss a theorem which, as a particular case, gives a classification of the extreme dynamical maps of a two--level system and clarifies their geometrical structure and their symmetry properties. This leads us to a conjecture as to how the result might be generalized to N-level systems, for arbitrary N. Størmer [26] has computed the normalized extreme positive linear maps of the 2x2 complex matrices. By taking adjoints, his result is equivalent to our classification above. However, we give a more general theorem. In addition, Størmer's proof does not allow for a straightforward geometrical interpretation and does not display the symmetry properties of the extreme maps, thus giving little hint as to how the problem could be solved for the NxN matrices.

The physical motivation for studying dynamical maps is discussed at length in [10]. Here it is sufficient to observe that the dynamical evolution of a quantum system Σ is described by a one-parameter family $t \to A(t)$, $t \geq 0$, of dynamical maps which is determined by the hamiltonian H of Σ and by the nature and degree of coupling of Σ to its surroundings. Only in the limiting case when Σ can be treated as isolated is the dynamics hamiltonian and reversible, thus having the form $A(t)=\exp(-i\mathcal{H}t)$, where $\mathcal{H} = [H,\ldots]$ is the Liouville-von Neumann operator. On the other hand, in general, the interaction of the system with the external world plays a definite role in producing an element of irreversibility in the dynamical evolution, which ceases to be hamiltonian. The essential difference between hamiltonian and non-hamiltonian evolution lies in the fact that the latter brings about a variation in time of the "purity" of the state, which depends on the particular dynamics and on the initial condition. For example, the state of a system which is coupled to a thermal reservoir, eventually ends up in the equilibrium canonical distribution, independently on the original preparation.

Models of irreversible non-hamiltonian evolutions, based on various types of "master equations" and in which the coupling of the system to the surroundings is treated either stochastically or mechanically, have been considered by several authors in general contexts and in specific physical situations [7,30,31,11,20,2,29, 23,21,13,8] . Some of these models are discussed in [10] . Here we only make a remark which we deem important. Concerning macroscopic systems which are adiabatically isolated, the hope that their macroscopic dynamics and in particular the features of the approach to equilibrium of the macroscopic observables can be explained starting from the Liouville-von Neumann equation which describes the detailed microscopic dynamics in the approximation of complete isolation is certainly justified, and much progress has recently been made in this direction [19,9,32] . However, as regards the problem of irreversibility, the small residual interaction of the system with the surroundings is still important in bringing about a progressive decrease of the purity of the statistical operator and thus a progressive loss of memory of the initial state [1,21] . In this connection, non-hamiltonian dynamics is again important.

We hope that a knowledge of the extreme dynamical maps and their possible physical interpretation might help to clarify the structure of various dynamical evolutions described by a one-parameter family of dynamical maps A(t), by looking at special convex decompositions $A(t)=\sum_i \alpha_i(t)A_i(t)$ in terms of extreme maps $A_i(t)$, provided there are enough extreme maps that decompositions of this type exist. For example, it is sometimes possible [22] to analyze the dynamics of an open system as $A(t)=\sum_n \alpha_n \exp(-i\mathcal{H}_n t)$ where the coefficients α_n of the convex combination do not depend on time and $\{\mathcal{H}_j\}_{j=1,2,..}$ is a sequence of Liouville-von Neumann operators (it can be seen that $\exp(-i\mathcal{H}_n t)$ is extreme since it maps pure states to pure states). Another example is provided by models of dynamical semigroups $t\to A(t)$ induced by stochastic processes on topological groups, for which a natural convex decomposition in terms of extreme maps is given and which seem to find application in the analysis of master equations of laser theory [17] .

2. Notations

M(N):=unitary algebra of the NxN complex matrices with inner product $(a,b)=\mathrm{Tr}(a^* b)$.

co Y:=convex hull of Y.

extr X:=set of the extreme elements of the convex set X.

$K(N):=\{w|w\in M(N);\ w\geq 0,\ \mathrm{Tr}\ w=1\}$ = set of the NxN density matrices.

$(\vec{b},T)$: $\vec{x}\to T\vec{x}+\vec{b}$ denotes an affine map $\mathbb{R}^n\to\mathbb{R}^n$.

O(n):=group of the orthogonal nxn real matrices.

$SO(n):=\{B|B\in O(n);\ \det B=1\}$.

SU(n):= group of the unitary nxn complex matrices with determinant one.

Ad:u $\to$ Ad u, $u\in SU(n)$, denotes the adjoint representation of SU(n).

B_n:= closed unit ball in $\mathbb{R}^n$.

S_n:= boundary of B_n.

$\mathrm{diag}\{\alpha_i\}_n$:= nxn diagonal matrix with diagonal elements $\alpha_1,\ldots,\alpha_n$.

E^* := topological dual of the topological vector space E.

3. Extreme Dynamical Maps

Let $\mathfrak{A}$ be the C^*-algebra generated by the bounded observables of a quantum mechanical system Σ and assume that $\mathfrak{A}$ has an identity. Then the set K of states on $\mathfrak{A}$ is $\sigma(\mathfrak{A}^*,\mathfrak{A})$-compact and its algebraic span is $\mathfrak{A}^*$ [5]. Let θ be the point-open topology [16] on the space $\mathcal{N}$ of linear maps $\mathfrak{A}^* \to \mathfrak{A}^*$, where $\mathfrak{A}^*$ is taken in the $\sigma(\mathfrak{A}^*,\mathfrak{A})$-topology. Define the set F of the (mathematical) dynamical maps of Σ as $F = \{ A | A \in \mathcal{N} ; A(K) \subseteq K \}$. Then the following theorem results as a corollary of the Krein-Milman theorem [24] and of a theorem of Kadison [16] .

Theorem 3.1. co(extr F) is θ-dense in F.

If the physical states of Σ were to be represented by the totality of the elements of K, the above theorem would give a positive answer to the question whether there are "sufficiently many" extreme dynamical maps. On the other hand, in the conventional formulation of quantum theory which applies to finitely extended systems and to which our philosophy about the explanation of irreversibility conforms, one identifies $\mathfrak{A}$ to $B(\mathfrak{h})$, the C^*-algebra of bounded operators on a separable Hilbert space $\mathfrak{h}$ and assumes the only physical states to be the normal ones. Via the correspondence $\omega(a)=\mathrm{Tr}(wa)$, these are identified to the set $K(\mathfrak{h})$ of statistical operators, which spans $T(\mathfrak{h})$. Since $T(\mathfrak{h})$ is the dual of the C^*-algebra of completely continuous operators [25] which does not have an identity, we cannot apply theorem 3.1 to $F(\mathfrak{h})$ and, to our knowledge, the problem whether $F(\mathfrak{h})$ has "sufficiently many" extreme elements is open. However, because of the properties of statistical operators ans since the elements of $F(\mathfrak{h})$ are bounded, we conjecture that an element of $F(\mathfrak{h})$ is the limit of a norm Cauchy sequence of elements of co(extr $F(\mathfrak{h})$). We also remark that an element of $F(\mathfrak{h})$ which maps pure states to pure states is extreme.

Now we consider an N-level system Σ whose Hilbert space (respectively, whose C^*-algebra of observables) is isomorphic to $\mathbb{C}^N$ (respectively, to M(N)). Let $\{v_\mu\}_{\mu=1,\dots,N^2}$ be a complete orthogonal set (c.o.s.) for M(N) with the normalization $(v_\mu, v_\nu) = (1/N)\delta_{\mu\nu}$. Choose the v_μ's to be hermitian and, in particular, $v_{N^2}=(1/N)\mathbb{1}_N$. The states of Σ are the density matrices, forming the set K(N). Expand a density matrix in terms of the v_μ's:

$$w = \frac{1}{N} \mathbb{1}_N + \sum_{i=1}^{N^2-1} \alpha_i v_i .$$

The map $\ell : w \to \{\alpha_1,\dots,\alpha_{N^2-1}\} = \vec{\alpha}$ is a bijection of K(N) onto a compact and convex neighbourhood of the origin in $\mathbb{R}^{N^2-1}$. We identify henceforth a density matrix w with the corresponding vector $\vec{\alpha}=\ell(w) \in \ell(K(N)) \overset{\text{def.}}{=} L(N)$. Since $\mathrm{Tr}\, w^2 \leq 1$ we have

$\vec{\alpha}^2 \leq N-1$, and $\vec{\alpha}^2=N-1$ iff w is a pure state. Hence L(N) is contained into the closed ball of radius $(N-1)^{\frac{1}{2}}$ and its intersection with the boundary of the ball is the set extrL(N) of the pure states. The set of dynamical maps of Σ is defined as

$$G(N) = \{A|A:M(N) \to M(N),\ A \text{ linear}; \quad w\in K(N) \Rightarrow Aw\in K(N)\}.$$

Let $\{A_{\mu\nu}\}_{\mu,\nu}$ be the matrix representing an element A of G(N) with respect to the c.o.s. $\{v_\mu\}_\mu$. Then $A_{N^2N^2}=1$ and $A_{N^2i}=0$ $(i=1,\dots,N^2-1)$. Writing $Aw=(1/N)\,\mathbb{1}_N+\sum_{i=1}^{N^2-1}\alpha_i' v_i$ and $A_{iN^2}=n_i$ $(i=1,\dots,N^2-1)$ we have $\alpha_i' = \sum_{j=1}^{N^2-1} A_{ij}\,\alpha_j+b_i$ $(i=1,\dots,N^2-1)$. Hence we can identify G(N) with the set of affine maps of $\mathbb{R}^{N^2-1}$ into itself which map L(N) into itself. The map $g:A\to\{b_i,A_{rs}\}_{i,r,s}$ is a bijection of G(N) onto a compact and convex neighbourhood of the origin in $\mathbb{R}^{N^2(N^2-1)}$, and we henceforth identify A with the corresponding set of matrix elements $\{b_i, A_{rs}\}_{i,r,s}\in g(G(N)) \overset{def}{=} F(N)$. The Krein-Milman theorem ensures that F(N) = co(extrF(N)) and the problem that we are interested in is the classification of the extreme elements of F(N).

Consider first the simplest case N=2. Then $L(2)=B_3$ and we look for the extreme elements of the set $F(2) \overset{def}{=} D_3$ of affine maps $\mathbb{R}^3 \to \mathbb{R}^3$ which map B_3 into itself or, more generally, for the extreme elements of the set D_n of affine maps $\mathbb{R}^n \to \mathbb{R}^n$ which map B_n into itself.

<u>Theorem 3.2</u>

$$D_n = \{(\vec{b},T)\,|\,(\vec{b},T) = (\vec{0},Q_1)(\vec{a},\Lambda)(\vec{0},Q_2) =$$

$$= (Q_1\vec{a},\ Q_1\Lambda Q_2);\ Q_1,\ Q_2\in O(n);$$

$$a_i = \beta\xi_i(1-\alpha\omega_i^2),\ i = 1,\dots,n;$$

$$\Lambda = \mathrm{diag}\{\alpha\beta\omega_\ell(\sum_{j=1}^{n} \xi_j^2\omega_j^2)^{\frac{1}{2}}\}_n\ ;$$

$$0\leq\alpha\leq 1,\ 0\leq\beta\leq 1,\ 0\leq\omega_n\leq\dots\leq\omega_2\leq\omega_1 = 1,$$

$$0\leq\xi_k\leq 1,\ k=1,\dots,n;\quad \sum_{\ell=1}^{n}\xi_\ell^2 = 1\}.$$

The boundary of D_n is obtained by taking $\beta=1$; extrD_n is obtained by taking $\beta=\alpha=\omega_1=\omega_2=\dots=\omega_{n-1}=1$ and $\xi_n>0$.

Remarks on theorem 3.2.

a) Using the polar decomposition of a real matrix, B=SQ, S symmetric and positive, Q orthogonal, we can split $(\vec{b},T)$ as a product $(\vec{0},Q_1)$ $(\vec{a},\Lambda)(0,Q_2)$, where $\Lambda=\mathrm{diag}\{\lambda_\ell\}_n$, $\lambda_1 \geqslant \lambda_2 \geqslant \ldots \lambda_n \geqslant 0$, $a_i \geqslant 0 (i=1,\ldots,n)$. $(\vec{a},\Lambda)$ maps B_n to an ellipsoid E_n whose axes have lengths $\lambda_1, \lambda_2, \ldots \lambda_n$ respectively, and whose center is $\vec{a}$.

b) extr D_n contains in particular the elements of D_n which map S_n into itself. There are two types of such maps: those of the form $(\vec{0},Q)$, $Q \in O(n)$, and those which map B_n onto a point of S_n.

c) In the physical case n=3, $(\vec{0},Q)$ is induced by a unitary transformation on $\mathbb{C}^2$ if $Q \in SO(3)$, by an antilinear unitary transformation if $Q \in O(3) \setminus SO(3)$.

d) The relation among the elements of D_n specified by $(\vec{b},T) \sim (\vec{b}',T')$ iff $\exists\, Q, \hat{Q} \in O(n)$ such that $(\vec{b}',T') = (\vec{0},Q)(\vec{b},T)(\vec{0},\hat{Q})$ is an equivalence relation and $(\vec{b},T) \in$ extr D_n iff the whole equivalence class of $(\vec{b},T)$ is contained in extr D_n.

e) The geometrical meaning of the parameters ω_i is clear from the relation $\omega_i = \lambda_i/\lambda_1$.

f) As to the meaning of the ξ_i's take $\beta=1$ and $\alpha < 1$. Then $E_n \cap S_n = \{\vec{\xi}\}$.

g) β and α are parameters of convex combinations. With the notation $(\vec{a},\Lambda)=\Delta(\alpha,\beta,\vec{\xi},\vec{\omega})$ we have: 1) $\Delta(\alpha,\beta,\vec{\xi},\vec{\omega}) = \beta\Delta(\alpha,1,\vec{\xi},\vec{\omega}) + (1-\beta)\Delta(\alpha,0,\vec{\xi},\vec{\omega})$ and we note that $\Delta(\alpha,0,\vec{\xi},\vec{\omega}) = (\vec{0},0)$; 2) $\Delta(\alpha,1,\vec{\xi},\vec{\omega}) = \alpha\Delta(1,1,\vec{\xi},\vec{\omega}) + (1-\alpha)\ (0,1,\vec{\xi},\vec{\omega})$ and we note that $\Delta(0,1,\vec{\xi},\vec{\omega})$ maps B_n onto $\vec{\xi}$.

h) Take $\alpha=\beta=1$ and $\xi_1 > 0$. Then, if $\omega_2 < 1$ we have $E_n \cap S_n = \{\vec{\xi},\vec{\xi}'\}$, where $\vec{\xi}'=(-\xi_1,\xi_2,\cdots\xi_n)$ If $\omega_2=1$ and $\omega_3 < 1$, $E_n \cap S_n$ is a circle. If $\omega_3=1$ and $\omega_4<1$, it is a three-dimensional sphere, and so on. If $\omega_{n-1}=1$, $\omega_n < 1$ and $\xi_n > 0$, $E_n \cap S_n$ is an (n-1)-dimensional sphere and the map is extreme (as well as when $\omega_n=1$, which gives the identity map). The remaining extreme maps are obtained in the limit case $\xi_n=1$, for which the (n-1)-dimensional sphere $E_n \cap S_n$ degenerates to the point $(0,0,\ldots,0,1)$.

i) Observe that the extreme elements have a high symmetry. Precisely, if $(\vec{b},T)$ is extreme, then $\exists\, C \in O(n)$ and a subgroup of O(n), say Γ, isomorphic to O(n-1), such that $QTC^{-1}\, Q^{-1}C = T$ and $\vec{Q}b = \vec{b}$, $\forall\, Q \in \Gamma$. However, this conditon is not sufficient for $(\vec{b},T)$ to be extreme, as the example $\beta=\alpha=\omega_1=\ldots=\omega_{n-1}=1, \omega_n<1, \xi_n=0$ shows.

j) Observe that an element of the form $(\vec{0},T)$ is extreme iff $T \in O(n)$ (in the physical case n=3, iff it is induced by either a unitary or an antilinear unitary map on $\mathbb{C}^2$).

Remarks on the case $N > 2$.

a) If $N > 2$, L(N) is a proper subset of B_{N^2-1}, and the problem of finding extrF(N) is more complicated. A partial classification is provided by theorem 3.3 below [27,4]. L(N) is no more mapped onto itself by arbitrary rotations, but by those rotations which are the elements of the adjoint representation of SU(N). Such rotations are induced by transformations $w \mapsto uwu^*$, u unitary, on the density matrices. Antilinear unitary transformations induce rotations if N=4k, 4k+1, reflections if N=4k-2, 4k-1 (k=1,2,3,......). All the above maps are extreme because they map pure states to pure states (and we conjecture that they are the only extreme maps among those for which $(1/N)\mathbb{1}_N$ is a fixed point). For the same reason, also the maps which map L(N) onto a given pure state are extreme.

b) We guess that the extreme elements of F(N) have a high symmetry also in the case $N > 2$. To be precise, we make the following

Conjecture. Let V(N) denote the subgroup of $O(N^2-1)$ generated by $[\mathrm{Ad}\ SU(N)] \cup A_o$, where A_o is a given dynamical map induced by an antilinear unitary transformation. Then, if $(\vec{b},T)$ is extreme, $\exists\, C \in V(N)$ and a subgroup of V(N), Γ say, isomorphic to the subgroup generated by $[\mathrm{Ad}\ SU(N-1)] \cup A_o$ such that $QTC^{-1}\ Q^{-1}C=T$ and $Q\vec{b}=\vec{b}$, $\forall\, Q \in \Gamma$.

c) Let 1_n denote the identity map $M(n) \to M(n)$. By definition [27,3] an element A of G(N) is completely positive iff, for all positive integers n, $A \otimes 1_n$ is a positive map of $M(N) \otimes M(n)$ into itself.

Theorem 3.3 [27,4] An element A of G(N) is extreme among the completely positive maps iff $Aa=\Sigma_i\ \ell_i a \ell_i^*$, where $\Sigma_i \ell_i^* \ell_i = \mathbb{1}_N$ and $\{\ell_i^* \ell_j\}_{i,j}$ is a linearly independent set in M(N).

Remark. The elements of G(N) which are induced by antilinear unitary transformations on $\mathbb{C}^N$ are not completely positive. To our knowledge, it is an open question which other elements of G(N), if any, are not completely positive.

REFERENCES

1. Blatt, J.M.: Progr. Theor. Phys. (Kyoto) 22, 745 (1959).
2. Bloch, F.: Phys. Rev. 70, 460 (1946).
3. Choi, M.D.: Canad. J. Math. 24, 520 (1972).
4. Choi, M.D.: Thesis, University of Toronto (1972).
5. Dixmier, J.: Les C*-algèbres et leur représentations, Gauthier-Villars, Paris (1969). 2.5.6 and 12.3.4.
6. Edwards, C.M.: Commun. Math. Phys. 24, 260 (1972).
7. Emch, G.G. and Sewell, G.L.: J. Math. Phys. 9, 946 (1968).
8. Fonda, L., Ghirardi, G.C. and Rimini, A.: Evolution of Quantum Systems Subject to Random Measurements, preprint IC/73/44, ICTP, Trieste (1973).
9. George, C., Henin, F., Prigogine, I. and Rosenfeld, L.: A Unified Formulation of Dynamics and Thermodynamics, preprint University of Brussels (1972) and references quoted therein.
10. Gorini, V. and Sudarshan, E.C.G.: to appear.
11. Gross, E.P. and Lebowitz, J.L.: Phys. Rev. 104, 1528 (1956).
12. Haag, R. and Kastler, D.: J. Math. Phys. 5, 848 (1964).
13. Haake, F.: Statistical Treatment of Open Systems by Generalized Master Equations, Springer-Verlag, Berlin (1973).
14. Jordan, T.F., Pinsky, A.M. and Sudarshan, E.C.G.: J. Math. Phys. 3, 848 (1962).
15. Jordan, T.F. and Sudarshan, E.C.G.: J. Math. Phys. 2, 772 (1961).
16. Kadison, R.V.: Proc. Am. Math. Soc. 12, 973 (1961).
17. Kossakowski, A.: Rep. Math. Phys. 3, 247 (1972).
18. Kraus, K.: Operations and Effects in the Hilbert Space Formulation of Quantum Theory, lectures at this Institute.
19. Lanz, L., Lugiato, L.A. and Ramella, G.: Physica 54, 94 (1971) and references quoted therein.
20. Mathews, P.M., Shapiro, I.I. and Falkoff, D.L.: Phys. Rev. 120, 1 (1960).
21. Mehra, J. and Sudarshan, E.C.G.: Nuovo Cimento 11 B, 215 (1972).
22. Okubo, S. and Isihara, A.: J. Math. Phys. 12, 2498 (1971).
23. Rau, J.: Thesis, Brandeis University (1962).
24. Schaefer, H.H.: Topological Vector Spaces, Macmillan, New York (1966). 10.4.
25. Schatten, R.: Norm Ideals of Completely Continuous Operators, Springer-Verlag, Berlin (1960). IV.1, theorem 1.

26. Størmer, E.: Acta Math. 110, 233 (1963). Section 8.

27. Størmer, E.: Positive Linear Maps of C^*-Algebras, lectures at this Institute.

28. Sudarshan, E.C.G., Mathews, P.M. and Rau, J.: Phys. Rev. 121, 920 (1961).

29. Wangsness, R.K. and Bloch, F.: Phys. Rev. 89, 728 (1953).

30. Willis, C.R.: Phys. Rev. 127, 1405 (1962) and J. Math. Phys. 5, 1241 (1964).

31. Willis, C.R. and Bergmann, P.G.: Phys. Rev. 128, 391 (1962).

32. Zwanzig, R.: Physica 30, 1109 (1964).

THE INNER ORTHOGONALITY OF CONVEX SETS IN AXIOMATIC QUANTUM MECHANICS

U. Krause

Fachbereich Mathematik der Universität Bremen

Bremen, West Germany

Introduction

There are at least three reasons for considering Axiomatic Quantum Mechanics:

1. Rediscovery of the Hilbert space structure of conventional Quantum Mechanics with the help of acceptable physical axioms.
2. A deeper understanding of the relationship between Classical Mechanics and Quantum Mechanics - especially of the measuring process.
3. Evaluation of more appropriate models between the two conventional extremes of Classical and Quantum Mechanics.

There is a variety of approaches to Axiomatic Quantum Mechanics. One of them seems to be particularly adapted for the three reasons given above, the so called "convex approach" with a convex set of physical states as fundamental notion (see[12]). For example the work of E.B. Davies and J.T.Lewis ([6]), C.M.Edwards ([7]), S.Gudder ([11]), B. Mielnik ([21]) and of G.Ludwig ([19],[20]) goes in this direction. (For point 1 mentioned above see[19], for 2 Ludwig in this volume and the almost convex formulation of minimal measurement of F.Herbut[14], and for 3[12],[21]).

This paper deals with structures in convex sets which are relevant for Axiomatic Quantum Mechanics, especially in the convex approach. First, an axiomatization of convex sets based on the concept of state-mixing is given; from this an inner distance for states comes out which generalizes the distance in Hilbert space (this concept was independently formulated by S.Gudder[11]; see also[19]); the inner distance again leads to a generalized version of orthogonality, formulating disjointness for physical states. The crucial affine property of the orthogonality relation is studied under the heading "orthoconvex sets", finally, as an example of orthoconvexity, a convex set with the most significant axioms of Ludwig's set up is exhibited.

Since in the convex approach linear spaces, especially base norm spaces and order unit spaces, are often used - therefore also "ordered linear space approach"(Pool) - let me make a remark on the relationship between convex sets and linear spaces: Usually one looks at convex sets as an appendage to linear spaces (recall the definition as a certain subset of a linear space in all books on this topic). But a more careful examination under categorical aspects shows that convexity has a theory in its own right and that the categorical behavior of (normed,ordered) linear spaces is less regular than it is for convex sets. (Therefore Z.Semadeni [26] has proposed that it would be better to look at the unit balls of Banach spaces for example than the whole Banach spaces itself.) Also from the standpoint of interpretation convexity seems more appealing than linear spaces, because convexity can be directly interpreted as a certain type of mixing whereas neither the multiplication with arbitrary big scalars nor the addition of arbitrary distant elements seems to have a testable sense.

Because of lack of space only the more basic notions and theorems are given here and the proofs are omitted (for this see [18]).

The author wishes to express his gratitude to S. Gudder for a preprint of his paper cited above, and is also greatly indebted to G.Ludwig and his coworkers A.Hartkämper and H.Neumann for stimulating discussions about how reasonable physics should look like. Last but not least thanks to F.R.FitzRoy for correcting the authors English.

1. Mixture spaces

A central procedure (not only theoretically but also practically) in physics is the combination of two entities x and y to form a new one $z = x\alpha y$, which contains α percent of x and $1-\alpha$ percent of y. This type of mixing can be axiomatized as follows.

Definition A nonempty set X together with a map $f: [0,1] \times X \times X \longrightarrow X$, where $f(\alpha,x,y)$ is also denoted by $x\alpha y$, is called a mixture space, if the following conditions are fulfilled

(a) $x\alpha y = y(1-\alpha)x$

(b) $x\alpha x = x$

(c) If $x\alpha z = y\alpha z$ for some $\alpha > 0$, then $x = y$

(d) $(x\alpha y)\beta z = x\alpha\beta\left(y \dfrac{\beta(1-\alpha)}{1-\alpha\beta} z\right) \qquad (\alpha\beta \neq 1)$

for all $x,y,z \in X, \beta,\alpha \in [0,1]$; $[0,1]$ is the closed intervall between 0 and 1 in the reals $\mathbb{R}$.

It can easily be seen that a convex set in a real linear space with the usual convex combination of elements is a mixture space; therefore both the state space of Classical Mechanics and of Quantum Mechanics are mixture spaces with the usual convex combination of states.

The following theorem says that the convex sets are the only mixture spaces (for a proof see [13]).

Theorem For a mixture space (X,f) there exists a real linear space E, a convex subset Y of E and a bijection $g:X \longrightarrow Y$ with $g(x\alpha y) = \alpha g(x)+(1-\alpha)g(y)$.

In view of this the definition of a mixture space can be taken as an axiomatization of a convex set, which is independent of the physically irrelevant surrounding linear space. It is this interpretation of a convex set as a mixture space, which makes the concept of convexity important in physics, especially in Axiomatic Quantum Mechanics. In many cases it is better to look at the mixture space itself instead of its convex representation in some linear space (the deeper reason for this can be seen by thinking in terms of categories).

Now, let X be a (nonempty) convex set in a real linear space F. From the idea of mixing the following concepts can be derived:

Definition For $x,y\in X$, $\alpha\in[0,1]$, denote

$x \underset{\alpha}{\prec} y$ if there exists a $z\in X$ with $y=\alpha x+(1-\alpha)z$

$x \underset{\alpha}{\sim} y$ if $x \underset{\alpha}{\prec} y$ and $y \underset{\alpha}{\prec} x$

$x \prec y$ if there exists $\alpha>0$ with $x \underset{\alpha}{\prec} y$

$x \sim y$ if $x \prec y$ and $y \prec x$.

($x \prec y$ means, that x is a "component" of y).

It can easily be seen that $\prec$ is a reflexive, transitive (and in general anti-symmetric) relation on X; therefore $\sim$ is an equivalence relation on X, called the part relation. The equivalence classes are called parts by H.Bauer and H.S.Bear ([4]), since the concept comes from the so called Gleason parts in the theory of function algebras. More informative than this relations is the greatest percentage for which x is a component of y; thus the

Definition The function $\beta:X \times X \longrightarrow [0,1]$, defined by $\beta(x,y) = \sup\{\alpha\in[0,1]: x \underset{\alpha}{\prec} y\}$ is called the structure function of X, and $\alpha(x,y) = \min(\beta(x,y),\beta(y,x))$ the symmetrized structure function of X.

A dual description of mixing, that is a description of the mixing in X by functions on X is given in the following

Theorem

For $x\in X$, $y \longmapsto \beta(x,y)$ is a concave function on X,

for $y\in X$, $x \longmapsto \dfrac{1}{\beta(x,y)}$ is a convex function on X.

For $x,y\in X$ $\beta(x,y) = \inf\{u(y): u \text{ concave on } X, u\geq 0 \text{ on } X, u(x)=1\}$.

(For the notions "convex","concave" see [1]).

2. The inner metric

Let X be normalized in the space F, i.e. from $\lambda x=\mu y$ ($\lambda,\mu\geq 0$, $x,y\in X$) it follows that $\lambda=\mu$; every convex set can be assumed to be normalized, e.g. as $X\times\{1\}$ in $F\times\mathbb{R}$. $K = \mathbb{R}_+X = \{\lambda x: \lambda\geq 0, x\in X\}$ is the convex cone generated by X,

$E = \text{lin } X = K - K$ is the linear space generated by X. The convex set $k(X\cup -X)$ generated by $X\cup -X$ is absorbing in E and the <u>Minkowski functional</u> p of this set, defined by $p(z) = \inf\{\lambda\geq o: z\in\lambda\cdot k(X\cup -X)\}$ $(z\in E)$, is therefore a seminorm on E ([24]).

Let E^* be the space of all linear functionals on E, E' the subspace of the functionals which are continous with respect to p. The dual semi norm p' defined by $p'(f) = \inf\{\lambda\geq o: f\leq\lambda p\}$ $(f\in E')$ is a norm on E'.

Let $A(X)$ be the space of all real affine functions on X, $A_b(X)$ the subspace of all bounded affine functions. The mapping

$\varphi: E^* \longrightarrow A(X),\ f \longmapsto \text{Res}_X f$ is a linear bijection, and

$\varphi: (E',p') \longrightarrow (A_b(X), \|.\|)$ is a norm preserving linear bijection,

where $\|l\| = \sup_{x\in X}|l(x)|$ for $l\in A_b(X)$. Note especially: There exists exactly one $f_o\in E'$, which is identically 1 on X; f_o is called the <u>support functional</u> of X.

The following theorem shows, how to calculate the Minkowskifunctional by means of the structure function.

<u>Theorem</u> For $x,y\in X$, $\alpha,\beta\geq o$

$$p(\alpha x-\beta y) = \sup_{l\in A_b(X), \|l\|\leq 1} \alpha l(x)-\beta l(y) = 2\cdot\inf_{z\in X} \max\left(\frac{\beta}{\beta(y,z)},\frac{\alpha}{\beta(x,z)}\right) - (\alpha+\beta)$$

For $L=L(X)=\{l\in A(X): o\leq l\leq 1 \text{ on } X\}$ d defined by $d(x,y) = \sup_{l\in L}(l(x)-l(y))$ satisfies the conditions $d(x,x)=o$, $d(x,y)=d(y,x)$, $d(x,y)\leq d(x,z)+d(z,y)$. Therefore d is a metric on X iff $A_b(X)$ separates the points of X.

<u>Definition</u> A convex set X is called (internally)<u>bounded</u>, if $A_b(X)$ separates the points of X. For a bounded convex set the metric d is called the <u>inner metric</u> of X.

<u>Corollary</u> $$d(x,y) = \frac{1}{2}\,p(x-y) = \frac{1}{\sup_{z\in X}\min(\beta(x,z),\beta(y,z))} - 1$$

<u>Remarks</u>

(1) A generalized version of the Hahn-Banach-Theorem done by H.König ([16]) yields the useful rule

$d(A,B) = \sup_{l\in L} \inf l(A) - \sup l(B)$, where A,B are nonempty convex sets,

$d(A,B) = \inf_{x\in A, y\in B} d(x,y)$ and $\inf f(M)$, $\sup f(M)$ denotes the infimum resp. supremum of a real function f on a set M.

For $d(A,B)\neq o$ there exists $l_o\in L$ such that $d(A,B) = \inf l_o(A) - \sup l_o(B)$ and $\sup l_o(X)=1$, $\inf l_o(X)=o$.

(2) From the Corollary it follows that

$\frac{1}{1+d(x,y)} = \sup\{\alpha\in[0,1]: \alpha x+(1-\alpha)x' = \alpha y+(1-\alpha)y' \text{ for certain } x',y'\in X\}$,

and therefore d turns out to be the "intrinsic metric" recently defined by S.Gudder ([11]).

(3) Connected with the part relation there is another metric on X, the so called <u>part metric</u> $\delta(x,y) = \log \frac{1}{\alpha(x,y)}$ of H.Bauer and H.S.Bear (more precisely δ is a quasi-metric, since the value $+\infty$ is allowed)([4]). Sometimes δ is also called a "interne Metrik" ([3]), but δ is less internal than the inner metric d, or more precisely: Even for the simple convex set [0,1] it can happen that a bijection of X onto itself which preserves δ need not be necessarily affine; in other words, the metric δ is not intimately enough connected with the underlying mixture structure, to distinguish different mixture structures on a set. However every bijection of [0,1] - more generally of a strictly convex set - onto itself which preserves the inner metric must be affine.

Two extreme examples of a mixture space are the classical state space (a simplex - see [12]) and the state space of conventional Quantum Mechanics.

<u>First example</u>

Let X be a simplex, i.e. the cone generated by the normalized X is a lattice cone, which is bounded. For $x,y\in X$, $d(x,y) = \frac{1}{2} f_o(|x-y|)$, where f_o is the support functional of X and $|.|$ the usual module in a vector lattice. Especially for the so called Bauer simplex, i.e. X is the set of all probability measures on a compact Hausdorff space ([1]), it follows $d(x,y) = \frac{1}{2}\|x-y\|$, where $\|.\|$ is the usual norm of a measure.

<u>Second example</u>

Let X be the set of all density operators on an infinite complex and separable Hilbertspace. The inner distance between two density operators x,y is given by $d(x,y) = \frac{1}{2}$ trace $(|x-y|)$, where $|.|$ is the "local module" given by $|z| = +\sqrt{z^2}$ (for a bounded self-adjoint operator z).
For pure states, i.e. for extreme points of X, the expression above can be calculated in terms of the underlying Hilbert space H: Let φ, ψ be two vectors of norm 1 in H and x,y be the projections on the subspaces generated by φ resp. by ψ; a lengthy, but elementary calculation gives
$d(x,y) = \sqrt{1 - |\langle\varphi,\psi\rangle|^2}$.(See also [15], [23]).
Therefore the metric structure of H can be rediscovered via the inner metric of the state space, especially orthogonality in H means maximal inner distance 1 in the state space.

<u>Remark</u> The different module operations in the two examples can be unified in the following concept of a local module: Let X be a normalized convex set; an element $u\in K=\mathbb{R}_+X$ is called a <u>local module</u> of $z\in E=K-K$, if $-z,z\leq u$

and from $-z, z \leq v \in K$ it follows that $f_o(u) \leq f_o(v)$ ($\leq$ the order relation defined by K).

If for two points $x,y \in X$ there exists a local module u of x-y, then the inner distance is given by $d(x,y) = \frac{1}{2} f_o(u)$.

Every element in E has a local module if and only if E posesses the minimal decomposition property of A.J.Ellis ([9]).

The following theorem characterizes the regular semi-norms on E as certain convex functions on X; where a semi-norm p on the ordered vector space (E,K) is called <u>regular</u>, if for every $z \in E$ $p(z) = \inf_{z=z_1-z_2, z_i \in K} p(z_1+z_2)$; it can easily be seen, that this is a short formulation of the original definition given by E.B.Davies ([5]); important examples of regular norms are base norms and order unit norms.

<u>Theorem</u> The map $p \longmapsto Res_X p$ is a bijection of the set of all regular semi-norms on E onto the set of all convex functions u on X for which $0 \leq u(x)\beta(x,y) \leq u(y)$ holds for all $x,y \in X$.

Topological notions with respect to the topology defined by the inner metric on X, the <u>inner topology</u>, are denoted by the additional term "inner" or "internally". The following theorem gives a test for internal completeness; for an analogous characterization of completeness in terms of cones see [8].

<u>Theorem</u> A nonempty convex set X is internally bounded and internally complete if and only if X is σ- convex and bounded and closed in some locally convex Hausdorff topological vector space.

Here a convex set X is called <u>σ- convex</u>, if for every sequence (x_n) in X and sequence (α_n) in $[0,1]$ with $\sum_1^\infty \alpha_n = 1$ there exists some x in X such that $l(x) = \sum_1^\infty \alpha_n l(x_n)$ for all bounded affine functions l on X.

3. Orthoconvex sets

In both the classical and the quantum mechanical state space, the orthogonality of two states x,y can be characterized by $d(x,y)=1$. This gives also the appropriate description of disjointness with respect to the mixture operation in the general case; therefore two points x,y in a bounded convex set are called <u>orthogonal</u>, $x \perp y$, if $d(x,y)=1$. Or equivalently $\beta(x,z)+\beta(y,z) \leq 1$ for all $z \in X$, which means that every mixture z contains x and y in complementary proportions (sum of the proportions ≤ 1).

For a subset A of X the set $A^\perp = \{x \in X: x \perp a \text{ for all } a \in A\}$ is always internally closed, but not necessarily convex, which is seen e.g. by the square in the plane.

It is easy to show, that for every bounded convex set X the family $\mathcal{T}=\{A^{\perp\perp}: A\subseteq X\}$ is an orthocomplemented complete lattice with respect to set-theoretic inclusion and orthocomplementation $A^{\perp\perp}\longmapsto A^{\perp\perp\perp}=A^{\perp}$; for this only the symmetry and anti-reflexivity of $\perp$ is needed (see [10]).

For a subset A of X let $\overline{1}_A = \inf\{l\in L: l=1 \text{ on } A\}$; then for two convex subsets A,B $A\subseteq B^{\perp}$ iff $\overline{1}_A+\overline{1}_B \leq 1$. The orthogonality relation $\perp$ is called <u>affine</u>, if for every tripel x,y,z in X and every $\alpha\in[0,1]$ from $x\perp z$, $y\perp z$ it follows that $\alpha x+(1-\alpha)y\perp z$.

A bounded convex set X is called <u>orthoconvex</u>, if $\perp$ is affine and if for every subset A of X the following holds:

(*) $\overline{1}_A + \overline{1}_{A^{\perp}} > 0$ on X $\Longrightarrow$ $\overline{1}_A + \overline{1}_{A^{\perp}} \geq 1$ on X.

A variation of (*) is

(**) $\overline{1}_A > 0$ on X $\Longrightarrow$ $\overline{1}_A = 1$ on X (for all $A\subseteq X$).

It can easily be seen, that X is orthoconvex iff $\overline{1}_A+\overline{1}_{A^{\perp}}=1$ on X for every subset A of X.

A general notion of superposition given by V.S.Varadarajan ([27]) which generalizes the usual one of conventional Quantum Mechanics, can be formulated with respect to mixture spaces in the following way:

A point $x\in X$ is called a <u>superposition</u> of points in $A\subseteq X$, if every $l\in L$ which is zero on every point of A is also zero at x.

The set of all superpositions of points in A is denoted by $\mathcal{S}$up A, and the set of all superposition sets in X by $\mathcal{S}$up, $\mathcal{S}\text{up} = \{\mathcal{S}\text{up } A: A\subseteq X\}$.

A useful description of $\mathcal{S}$up A is $\mathcal{S}\text{up } A = [\overline{1}_A=1]$ (for a real function f on a set M and $r\in\mathbb{R}$ denotes $[f=r]$ the set $\{x\in M: f(x)=r\}$).

The regular behavior of $\perp$, made precise by the notion of an orthoconvex set allows the identification of the families $\mathcal{S}$up, $\mathcal{T}$ and $\mathcal{E}=\{\overline{1}_A: A\subseteq X\}$, namely

<u>Theorem</u> For an orthoconvex X the families $\mathcal{S}$up, $\mathcal{T}$ and $\mathcal{E}$ are isomorphic complete orthomodular lattices.

The isomorphism $\mathcal{E}\longrightarrow\mathcal{S}$up is given by $\overline{1}_A \longmapsto [\overline{1}_A=1]$.

(A lattice with orthocomplementation $\perp$ is called <u>orthomodular</u>, if for two elements a,b of the lattice $a\leq b$ implies $b=a\vee(b\wedge a^{\perp})$.)

Now, if $\perp$ is affine, then the sets $A^{\perp}$ are internally closed faces of X; where a <u>face</u> S of X is a convex subset S for which $\alpha x+(1-\alpha)y\in S$ with $\alpha\in[0,1]$ implies $x,y\in S$.

Let $\mathcal{S}$ denote the family of all internally closed faces of X (including the empty set $\emptyset$). $\mathcal{S}$ is stable under arbitrary intersections and therefore for $A\subseteq X$ the smallest closed face S(A) containing A exists.

In general $\mathcal{S}$ and $\mathcal{S}$up do not coincide, but

<u>Corollary</u> For an orthoconvex set X with $\mathcal{S}\text{up } A\subseteq S(A)$ for all $A\subseteq X$ we have $\mathcal{S}=\mathcal{T}=\mathcal{S}\text{up}$.

Especially in this case $\mathcal{S}$ is a complete orthomodular lattice (with respect to $\subseteq$ and $\perp$).
Since L is a weakly compact convex set there are by the Krein-Milman-Theorem many extreme points in L. For an orthoconvex set $\mathcal{S}$up is "almost" the set of extreme points L_e of X, more precisely

Theorem The following propositions are equivalent:
(a) X is orthoconvex and every element of L_e attains its supremum on X.
(b) $L_e = \mathcal{E}$ and (**) holds.

Corollary Let X be an orthoconvex set which is compact with respect to the finest topology on X for which all internally continous affine functions on X are continous. Then $L_e = \mathcal{E}$.

Especially $L_e = \mathcal{E}$ for an orthoconvex set which is closed in some finite dimensional real vector space.

Remarks
(1) The orthogonality relation in a bounded normalized convex set X can be described by the orderrelation $\leq$ of the cone $K = \mathbb{R}_+ X$ in the following manner:
$x \perp y$ and $x \vee y$ exists in $K \Longleftrightarrow x \vee y = x + y$
($x, y \in X$ and $\vee$ denotes the supremum with respect to $\leq$).
From this it follows that for a simplex $x \perp y$ iff $x \wedge y = o$ in the lattice $E = K - K$ - this is the usual orthogonality relation in vector lattices ([24]).
(2) W.Wils ([28], [29]; see also [1]) has proposed and analyzed another orthogonality relation for convex sets. Two faces S,T of a convex set X are called orthogonal in the sense of Wils, $S \between T$, if their convex hull is a face and every point in the convex hull can be uniquely represented as a convex combination of elements in S and T; two points $x, y \in X$ are called Wils-orthogonal, $x \between y$, if $S(x) \between S(y)$.
For a (normalized) simplex $x \between y$ holds iff $x \wedge y = o$ in the lattice E, and therefore in this (classical) case the notions $\between$ and $\perp$ coincide.
But in general there is no simple implication between $\between$ and $\perp$, which can be seen by looking at the two following examples:
(a) Let X be the following trapezoid in $\mathbb{R}^2$

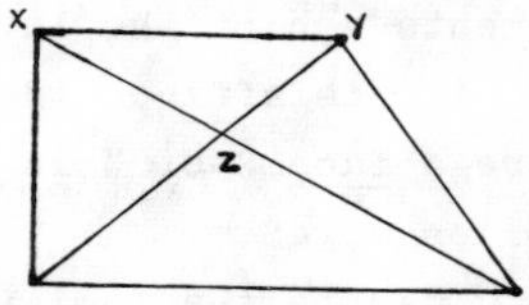

Here $x \between y$, but $\beta(x,z), \beta(y,z) \geq \frac{2}{3}$ and therefore $\beta(x,z) + \beta(y,z) > 1$, which means not $x \perp y$.
(b) Let X be the unit ball in $\mathbb{R}^n$ ($n \geq 2$). Here $d(x,y) = \frac{1}{2}\|x-y\|$ ($\|\cdot\|$ the euclidean norm), therefore $x \perp y$ iff x and y are diametrically opposite boundary points of X. But the convex hull of such points is never a face of X and

therefore they are not orthogonal in the sense of Wils. A direct calculation shows, that the quantum mechanical state space with an underlying two-dimensional complex Hilbert space can be identified (as a convex set) with the unit ball in $\mathbb{R}^3$; in the quantum mechanical case therefore $x \perp y \Longrightarrow x ⫰ y$ is in general false.

More generally one can prove the following theorem:

<u>Theorem</u> Let X be an orthoconvex set with $S(x)=\mathcal{S}up\{x\}$ for all $x \in X$.

(a) Two extremepoints $x,y \in X$ are Wils-orthogonal if and only if $\mathcal{S}up\{x,y\}=k\{x,y\}$

(b) $x ⫰ y \Longrightarrow x \perp y$ for different extreme points of X.

Because the quantum mechanical state space satisfies the conditions of the theorem (for orthoconvex sets $S(x)=\mathcal{S}up\{x\}$ is equivalent with (L4) of the next section), two pure states are Wils-orthogonal iff there are only classical superpositions of the two states. Therefore in conventional Quantum Mechanics ⫰ is not a useful relation , because $x ⫰ y$ never occurs for pure states. In the case of direct convex sums of quantum mechanical state spaces $x ⫰ y$ may occur and is always stronger than $x \perp y$; this shows that Wils-orthogonality is strongly connected with the existence of superselection rules. The following convex set

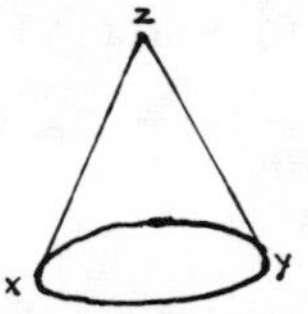

x,y,z are pairwise orthogonal with respect to $\perp$; $x ⫰ z$, $y ⫰ z$, but not $x ⫰ y$.

which is the direct convex sum of a point and the unit ball in $\mathbb{R}^2$ (and therefore orthoconvex with $S(.)=\mathcal{S}up\{.\}$) gives a simple example of a convex set for which implication (b) of the theorem is true(but not trivial) and for which the reversal of (b) is false.

4. The axioms of Ludwig

In the approach of G.Ludwig ([19]) to Quantum Mechanics the state space of ensembles is a bounded internally complete convex set X.

Ludwig places certain axioms on X, which can be formulated in the following way (for a detailed discussion see[19]; see also A.Hartkämper and H.Neumann in this volume and[25]).

(L1) Finiteness axiom: X is seperable (with respect to the inner metric).

(L2) Sensitivity increase of two effects:
For $l_1, l_2 \in L$ there exists $l \in L$ with $l_1, l_2 \leq l$ and $[l_1=o] \cap [l_2=o] \subseteq [l=o]$.

(L3) Sensitivity increase of one effect (axiom 4bz in [19]):
$\vec{1}_A > o$ on $X \Longrightarrow \vec{1}_A \geq 1$ on X (for all $A \subseteq X$).

(L4) Two related ensembles have the same components.
Thereby two ensembles x,y are called related, if for every effect $l \in L$ $l(x)=o$ iff $l(y)=o$, and they have the same components iff $S(x)=S(y)$.

The following lemma gives a translation into terms employed here:

Lemma

(a) (L2) $\Longleftrightarrow$ $\bar{1}_A$ is affine for all $A \subseteq X$.

(b) (L2) and (L3) $\Longleftrightarrow$ X is orthoconvex.

(c) If X is orthoconvex and (L1) holds, then
(L4) $\Longleftrightarrow$ $\mathcal{S}$up A = S(A) for all $A \subseteq X$.

From the theorem on orthoconvex sets of the last section and this translation the following theorem follows directly

Theorem For a bounded, internally complete convex set X which satisfies (L2) and (L3) the families $\mathcal{T}$, $\mathcal{S}$up, $\mathcal{E}$ are isomorphic complete orthomodular lattices.

If X satifies in addition (L1) and (L4) then also $\mathcal{S}$ is a complete orthomodular lattice, isomorphic to $\mathcal{T}$, $\mathcal{S}$up, $\mathcal{E}$.

The first corollary of the last section gives the

Theorem For a bounded, internally complete convex set X in some finite dimensional real vector space which satisfies (L2) and (L3) one has $L_e = \mathcal{E}$.
If X satisfies in addition (L1) and (L4) then the families $\mathcal{T}$, $\mathcal{S}$up, $\mathcal{S}$, $\mathcal{E}$, L_e are isomorphic complete orthomodular lattices.

Remarks

(1) For an orthoconvex X $\mathcal{E} \subseteq L$ holds, therefore the elements of $\mathcal{E}$ are effects which are called decision effects by Ludwig.
The two theorems mentioned above were proved by Ludwig ([19]); for the second theorem he uses some type of a finite dimensional spectral theorem.
(2) Compact convex sets with the property (L2) are studied by A.Ancona ([2]) with the aim of generalizing the finite results of Ludwig. But the strong assumption of compacticity leads one back nearly to the finite dimensional case, since e.g. a finite spectral theorem holds.
(3) The operation $\bar{1}_A$ is an analogue to the $\hat{1}_A$ - notion common in Choquet-theory ([1]). I conjecture, that the two notions coincide exactly in the classical case of a simplex under suitable topological assumptions, which would suggest, that the usual Choquet-theory is something like a classical case of some more general "Choquet-theory"(see for this the spectral formulation of Choquet-theory in [17]); recall, that Choquet-theory is based on Radon measures which are classical in the sense that they are measures on a distributive lattice.

References

1. Alfsen, E.M.: Compact convex sets and boundary integrals
Ergebnisse der Mathematik 57, Springer Verlag 1971

2. Ancona, A.: Sur les convexes de Ludwig
Ann.Inst.Fourier 2o,2, 21-44 (197o)

3. Bauer, H.: Intern vollständige konvexe Mengen
Aarhus Universitet Preprint Series No. 3o 197o/71

4. - , Bear,H.S.: The part metric in convex sets
Pac.J.Math. 3o, 15-33 (1969)

5. Davies, E.B.: The structure and ideal theory of the predual of a
Banach lattice. Trans.Amer.Math.Soc. 131, 544-555 (1968)

6. - , Lewis, J.T.: An operational approach to quantum probability
Commun.Math.Phys. 17, 139-26o (197o)

7. Edwards, C.M.: Classes of operations in quantum theory
Commun.Math.Phys. 2o, 26-56 (1971)

8. - , Gerzon, M.A.: Monotone convergence in partially ordered vector
spaces. Ann.Inst.Henri Poincaré, 12(4), 323-328 (197o)

9. Ellis, A.J.: Minimal decompositions in base normed spaces
This volume

1o.Foulis, D.J., Randall, C.H.: Lexicographic orthogonality
J.Combinatorial Theory 11, 157-162 (1971)

11.Gudder, S.: Convex structures and operational quantum mechanics
Commun.Math.Phys. 29(3), 249-264(1973)

12.Haag, R.: Bemerkungen zum Begriffsbild der Quantenphysik
Z.Physik 229, 384-391 (1969)

13.Hausner, M.: Multidimensional utilities. In:
Thrall, R.M., Coombs, C.H., Davis, R.L.(ed.): Decision
processes. Wiley 1954

14.Herbut, F.: Derivation of the change of state in measurement from the
concept of minimal measurement
Annals of Physics 55, 271-3oo (1969)

15. Jauch, J.M., Misra, B., Gibson, A.G.: On the asymptotic condition of
scattering theory. Helv.Phys.Acta 41, 513-527 (1968)

16.König, H.: Über das von Neumannsche Minimax - Theorem
Arch.Math. 19, 482-487 (1968)

17.Krause, U.: Der Satz von Choquet als ein abstrakter Spektralsatz und
vice versa. Math.Ann. 184, 275-296 (197o)

18. - : Strukturen in unendlichdimensionalen konvexen Mengen
Forthcoming

19. Ludwig, G.: Deutung des Begriffs "physikalische Theorie" und axiomatische Grundlegung der Hilbertraumstruktur der Quantenmechanik durch Hauptsätze des Messens.
Lecture Notes in Physics 4. Springer Verlag 197o

2o. - : The measuring and preparing process and macro theory
This volume

21. Mielnik, B.: Theory of filters
Commun.Math.Phys. 15, 1-46 (1969)

23. Robinson, D.W.: Normal and locally normal states
Commun.Math.Phys. 19, 219-234 (197o)

24. Schaefer, H.H.: Topological vector spaces. Macmillan 1966

25. Schmidt, H.J.: Die Kategorie der Operationen in der Axiomatischen Quantenmechanik.Typoscript. Marburg 1972

26. Semadeni, Z.: Categorical methods in convexity. In:
Proc.Coll.Convexity , 281-3o7, Copenhagen 1967

27. Varadarajan, V.S.: Geometry of quantum theory. Vol.I. Van Nostrand 1968

28. Wils, W.: The ideal center of partially ordered vector spaces
Acta mathematica 127, 41-77 (1971)

29. - : Centers and central measures
This volume

REDUCED DYNAMICS IN QUANTUM MECHANICS *

L. Lanz[+], L.A. Lugiato[+], G. Ramella.

Istituto di Fisica dell'Università, Milano, Italy

1. Introduction

A macrosystem M is usually described as having an N-body structure, such that all the physics of the system can be obtained by the quantum mechanics of the N-body system. We shall discuss the general problem of extracting from this description the macroscopic one.

Looking at M as a N-body system, the statistics of the experiments on M is given by the expression

(1.1) $\mu(V, F) = \mathrm{Tr}(V F)$, $V \in K$, $F \in L$,

where K is the base of $\tau c(\mathcal{H})$ (trace class operators on a Hilbert space $\mathcal{H}$), L is the $[0,1]$ order interval of $\mathbb{B}(\mathcal{H})$ (bounded operators on $\mathcal{H}$). $\mathcal{H}$ is given by

(1.2) $\mathcal{H} = \sum_{n=0}^{\infty \oplus} (\mathcal{H}_1^{\,n})^{s,a}$,

$(\mathcal{H}_1^{\,n})^{s,a}$ being the symmetric or the antisymmetric subspace of n times the direct product of the one-particle Hilbert space $\mathcal{H}_1$. On $\mathcal{H}$ a unitary representation $U(g)$, of the Galilei group $\mathcal{G}$ is defined, which is up to a factor in each superselection space $(\mathcal{H}_1^{\,n})^{s,a}$; (-i) x the restriction of the generator of time translations to $(\mathcal{H}_1^{\,n})^{s,a}$ is the N-body Hamiltonian H_N. Then one has a representation $\mathcal{U}(g)$ of $\mathcal{G}$ on $\mathbb{B}(\mathcal{H})$ mapping L onto L

(1.3) $\mathcal{U}(g)Y = U(g)\,Y\,U^{-1}(g)$, $g \in \mathcal{G}$, $Y \in \mathbb{B}(\mathcal{H})$.

Now the general problem arises of connecting concrete experiments on M with pairs F, V; particular aims could be to derive from (1.1) the equilibrium thermodynamics for M, the hydrodynamics for a fluid, etc. Such a problem has been generally considered as a problem of good guessing for F and V; no general characterization of the relevant part of the definition domain of the function $\mu(V, F)$ has been given so far.

Recently Ludwig and collaborators have developed a new axiomatic approach to the description of physical systems in which real experi-

---

* Presented by L. Lanz.

\+ Also Istituto Nazionale di Fisica Nucleare, Sezione di Milano, Italy.

ments are the starting point [6,7,8]. In such a way general aspects of a new theory of macrosystems have been established [7,8]. A pure macro system is characterized by the statistics of the registrations of the trajectories of the system in a state space $\mathcal{Z}$ during the time interval $[0,\infty)$. The frequencies of such registrations are given by a function $\mu_m(u,f)$ with

(1.4) $\quad 0 \le \mu_m(u,f) \le 1 \quad , u \in K_m \, , f \in L_m ,$

where K_m is the set of "macrocollections" and L_m the set of trajectory registration effects. K_m is the base of a base-normed space B_m, L_m is the $[0,1]$ order interval in the dual space B'_m. The restriction of the canonical bilinear form $\langle x|y\rangle$, $x \in B_m$, $y \in B'_m$ to $K_m \times L_m$ is the function $\mu_m(x,y)$, $x \in K_m$, $y \in L_m$. B'_m is linked in a suitable way [7,8] to the space $\mathcal{Y}$ of trajectories at positive times. Since the transformations $g(\tau, \underline{a}, \underline{v}, R)$ with $\tau \ge 0$ of the Galilei semigroup $\mathcal{G}^{(+)}$ can be defined on $\mathcal{Y}$, one can define such transformations also on L_m and obtain a representation $U_m(g)$ of $\mathcal{G}^{(+)}$ on B'_m. In particular positive time translations are represented by a norm continous semigroup $U_m(\tau)$ on B'_m. Such a semigroup character of $U_m(\tau)$ is a very important feature of the theory, which provides a far reaching generalization of the "markoffian" character of the time evolution, which was assumed as typical for macrosystems in many previous attempts to characterize macroobservables.

2. Definition of the embedding.

Now the problem arises of embedding the axiomatic description of M into the general N-body description. One then expects that subsets $S_1 \subset K$ and $S_2 \subset L$ can be found so that $S_1 \times S_2$ is μ-isomorphic to $K_m \times L_m$, i.e. two bijective affine mappings γ_1, γ_2

(2.1) $K_m \xrightarrow{\gamma_1} S_1 ,$

(2.1') $L_m \xrightarrow{\gamma_2} S_2 ,$

exist such that

(2.2) $\mu_m(u,f) = \mu(\gamma_1 u, \gamma_2 f) \quad , \forall u \in K_m \, , f \in L_m .$

We shall call then S_2 the set of macroeffects and S_1 the set of macrocollections. By (2.1') the representation $U_m(g)$ induces a representation of $\mathcal{G}^{(+)}$ on S_2 given by

(2.3) $\mathcal{U}_m(g) =: \gamma_2 U_m(g) \gamma_2^{-1}$,

so that by eq. (2.2)

(2.4) $\mu_m(u, U_m(g)f) = \mu(V, \mathcal{U}_m(g)F)$,

with $V = \gamma_1 u$, $F = \gamma_2 f$, $\forall u \in K_m$, $f \in L_m$. We want to look for a relation between $\mathcal{U}_m(g)$ and $\mathcal{U}(g)$: such a relation cannot be trivial in the case of time translations since an irreversible and a reversible time evolution must be connected. One expects that the effects $\mathcal{U}_m(g)F$ and $\mathcal{U}(g)F$, $F \in S_2$, are "equivalent" with respect to all preparations of M, i.e.

(2.5) $\mu(V, \mathcal{U}_m(g)F) = \mu(V, \mathcal{U}(g)F)$
$\forall V \in S_1$, $g \in \mathcal{G}^{(+)}$.

The embedding condition (2.5) has been proposed by Ludwig [7] and some consequences have been studied with particular reference to time translations $\mathcal{U}(\tau)$. It turns out that (2.5) is equivalent to the assumption that a projection $\mathcal{P}$ on $\tau\mathfrak{c}(\mathcal{H})$ exists such that the one-parameter family $\mathcal{P}\mathcal{U}(\tau)\mathcal{P}$, $\tau \geq 0$,has still the semigroup property. Such a requirement is very restrictive and there is no evidence that it can be fulfilled in an exact way. An approximate treatment of the same problem is met in the deduction of a markoffian master equation from the generalized master equation.

Our attitude at this point is the following: condition (2.5) should be replaced by a less restrictive one; on the other hand the effects $\mathcal{U}(\tau)F \neq \mathcal{U}_m(\tau)F$ involved in (2.5) have a somewhat pathological character. In fact they are microeffects, which are equivalent to macroeffects if preparations of macrosystems are considered; they must arise therefore in measurements on all the particles of a macrosystem. Such measurements are practically impossible; their presence in the N-body theory is an unsatisfactory feature of the theory itself; in fact N-body theory for very large N is only a formal extrapolation of a well established theory for small N. Therefore we want to formulate the equivalence of $\mathcal{U}_m(\tau)F$ and $\mathcal{U}(\tau)F$, $F \in S_2$, without considering $\mathcal{U}(\tau)F$ realistic effects with a statistics given by $\mu(V, \mathcal{U}(\tau)F)$. We replace eq. (2.5) by the assumption that for any $V \in S_1$, the dependence of the l.h.s. of eq.(2.5) on $g \in \mathcal{G}^{(+)}$ and $F \in S_2$ can be reformulated as an affine dependence on $\mathcal{U}(g)F$. More specifically, indicating by $\mathcal{V}_2$ the linear manifold spanned by the elements $\mathcal{U}(g)F, g \in \mathcal{G}^{(+)}, F \in S_2$, we require that for every $V \in S_1$, exists $\tilde{V} \in \mathcal{V}_2^*$ such that

$$(2.6) \quad \mu\left(V, \mathcal{U}_m(g)F\right) = \tilde{V}\left(\mathcal{U}(g)F\right), \quad \forall F \in S_2, \; g \in \mathcal{G}^{(+)},$$

where $\mathcal{V}_2^*$ is the algebraic dual of $\mathcal{V}_2$. The mapping λ_1: $S_1 \xrightarrow{\lambda_1} \mathcal{V}_2^* : \lambda_1 V = \tilde{V}$ is then injective and affine. The set $\tilde{S}_1 =: \{\tilde{V} = \lambda_1 V, V \in S_1\}$ can be looked upon as the set of generalized macrocollections in terms of which the embedding relations (2.2), (2.6) can be summarized as follows

$$(2.7) \quad K_m \xrightarrow{\lambda_1 \gamma_1} \tilde{S}_1, \quad L_m \xrightarrow{\gamma_2} S_2,$$

$$\mu_m\left(u, U_m(g)f\right) = \tilde{\mu}\left(\mathcal{U}(g)\gamma_2 f, \lambda_1\gamma_1 u\right), \; \forall u \in K_m, f \in L_m$$

where $\tilde{\mu}$ is the canonical bilinear form of the duality $\langle \mathcal{V}_2, \mathcal{V}_2^* \rangle$.

3. Embedding: scheme for a concrete realization.

Let us sketch how the definitions given in the previous section can be translated into a constructive scheme. We shall treat only time translations, for which one meets the main difficulties. A more complete treatment will be found in [5] ; an analogous treatment in the framework of the Liouville space is given in [2].

Let $\mathcal{P}$ be a projection onto a subspace of $\tau\mathbb{C}(\mathcal{H})$; let $-i\mathcal{H}$ be the generator of the strongly continous, unitary one-parameter group $\underline{\mathcal{U}}(t)$ on $\tau\mathbb{C}(\mathcal{H})$,

$$\underline{\mathcal{U}}(t)X =: U^{-1}(t)\, X\, U(t), \quad \forall X \in \tau\mathbb{C}(\mathcal{H}).$$

By (1.3) $\mathcal{U}(t) = \underline{\mathcal{U}}'(t)$. Under very general assumptions on $\mathcal{P}$, the following decomposition of the resolvent $[z - \mathcal{H}]^{-1}$ has been proven in [1] * :

(*) Eq. (3.1) is fundamental also in the theory of subdynamics [9] formulated by Prigogine and coworkers; the analysis in the present section has many features in common with the technique developed in [9] and has been partially inspired by it.

(3.1)
$$[z-\mathcal{H}]^{-1} = [\mathcal{P}-\overline{\mathcal{N}_2}(z)] \frac{\mathbb{1}}{z+\overline{\mathcal{M}}(z)} [\mathcal{P}-\mathcal{N}_1(z)] + $$
$$+ (\mathbb{1}-\mathcal{P}) \frac{\mathbb{1}}{z-(\mathbb{1}-\mathcal{P})\mathcal{H}(\mathbb{1}-\mathcal{P})} (\mathbb{1}-\mathcal{P}) \ , \ \mathrm{Im}\, z \neq 0,$$

where

(3.2) $$\mathcal{M}(z) =: \mathcal{P}\mathcal{H}(\mathbb{1}-\mathcal{P}) \frac{\mathbb{1}}{(\mathbb{1}-\mathcal{P})\mathcal{H}(\mathbb{1}-\mathcal{P})-z} (\mathbb{1}-\mathcal{P})\mathcal{H}\mathcal{P} - \mathcal{P}\mathcal{H}\mathcal{P},$$

(3.3) $$\mathcal{N}_1(z) =: \mathcal{P}\mathcal{H}(\mathbb{1}-\mathcal{P}) \frac{\mathbb{1}}{(\mathbb{1}-\mathcal{P})\mathcal{H}(\mathbb{1}-\mathcal{P})-z} (\mathbb{1}-\mathcal{P}),$$

(3.4) $$\mathcal{N}_2(z) =: (\mathbb{1}-\mathcal{P}) \frac{\mathbb{1}}{(\mathbb{1}-\mathcal{P})\mathcal{H}(\mathbb{1}-\mathcal{P})-z} (\mathbb{1}-\mathcal{P})\mathcal{H}\mathcal{P},$$

$\overline{\mathcal{N}_2}(z)$ is the closure of $\mathcal{N}_2(z)$ etc. One gets the following properties

1) $\mathcal{N}_1(z)$ is bounded - holomorphic for $\mathrm{Im}\, z \neq 0$,
2) the domain $\mathcal{D}$ of $\overline{\mathcal{M}}(z)$, $\mathrm{Im}\, z \neq 0$, is independent of z ,
3) $\overline{\mathcal{M}}(z)A$, $A \in \mathcal{D}$, $\mathrm{Im}\, z \neq 0$, is a holomorphic vector-valued function of z ,
4)

(3.5) $$\mathcal{M}(z)-\mathcal{M}(z') = (z-z')\,\mathcal{N}_1(z)\,\mathcal{N}_2(z').$$

Let us make the following additional assumptions:

i) for any $A \in \mathcal{D}$ the vector $\overline{\mathcal{M}}(z)A$ can be analytically continued in z from the upper to the lower halfplane, which defines a linear operator $\mathcal{M}_+(z)$, with domain $\mathcal{D}$, such that $\mathcal{M}_+(z) = \overline{\mathcal{M}}(z)$ when $\mathrm{Im}\, z > 0$. Let D be the (simply connected) region of the z-plane on which $\mathcal{M}_+(z)A$, $\forall A \in \mathcal{D}$, is holomorphic.

ii) A subspace $S \subset \mathcal{P}\tau\mathfrak{C}(\mathfrak{H}) \cap \mathcal{D}$ and a linear closed operator $\mathcal{Q}_S$ mapping S into S exist such that

a) the spectrum of $\mathcal{Q}_S$ is contained in a regular closed path $\gamma \subset D$,

b) for every $V \in S$ the vector-valued function

(3.6) $\left[\mathcal{M}_+(z)+z\right]\frac{\mathbb{1}}{z-Q_S}V$ is holomorphic in D.

The subspace S can be constructed e.g. as the direct sum of linearly independent null eigenspaces of $\mathcal{M}_+(z)+z$, and the corresponding operator Q_S is defined as follows: if $\left[\mathcal{M}_+(\bar{z})+\bar{z}\right]U=0$, then $Q_S U=\bar{z}U$. Condition (3.6) is then trivially satisfied for V = U. We stress that the null space of $\mathcal{M}_+(z)+z$ can be nontrivial only for Im $z\leq 0$ [2].

On the basis of such assumptions, we can show what follows. For any $V\in S$, let us consider the linear functional $\mathcal{F}$ on the linear manifold spanned by the elements $\mathcal{U}(t)F$, $F\in\mathcal{B}'\mathbb{B}(\mathcal{S})$, $t\geq 0$,

(3.7) $$\mathcal{F}(\mathcal{U}(t)F)=\frac{1}{2\pi i}\oint_\gamma dz\left\{\mu\left(\left[\mathcal{S}-\mathcal{N}_2(z)\right]\frac{\mathbb{1}}{z-Q_S}V,\mathcal{U}(t)F\right)\right\}_{a.c.},$$

where the suffix a.c. means "analytically continued in z from the upper to the lower halfplane". By (3.1) and (3.5) one has

(3.8) $$\mathcal{F}(\mathcal{U}(t)F)=\frac{-1}{(2\pi i)^2}\oint_\gamma dz\int_{-\infty+i\varepsilon}^{+\infty+i\varepsilon}dz'\,e^{-iz't}\cdot\mu\left(\frac{\mathbb{1}}{z'+\overline{\mathcal{M}}(z')}\,\frac{\overline{\mathcal{M}}(z')+z'-\left[\mathcal{M}_+(z)+z\right]}{z'-z}\,\frac{\mathbb{1}}{z-Q_S}V,F\right).$$

Taking into account (3.6) and the resolvent identity

(3.9) $$\frac{1}{z'-z}\,\frac{\mathbb{1}}{z-Q_S}=\frac{1}{z'-z}\,\frac{\mathbb{1}}{z'-Q_S}+\frac{\mathbb{1}}{z-Q_S}\,\frac{\mathbb{1}}{z'-Q_S},$$

one obtains the result

(3.10) $$\mathcal{F}(\mathcal{U}(t)F)=\mu\left(\exp(-iQ_S t)V,F\right).$$

Assuming that $\exp(-iQ_S t)$ can be extended to $\tau_{\mathbb{C}}(\mathcal{S})$ to yield a one-parameter semigroup $\underline{\mathcal{W}}(t)$ (e.g. if S admits a projection $\mathcal{P}_S$, one can define $\underline{\mathcal{W}}(t)=\exp(-iQ_S t)\mathcal{P}_S$), we have

(3.11) $$\mathcal{F}(\mathcal{U}(t)F)=\mu\left(V,\mathcal{W}(t)F\right),\qquad \mathcal{W}(t)=\underline{\mathcal{W}}'(t)\quad t\geq 0.$$

One sees therefore that (3.11) is a realization of (2.6) for $g = (t, 0, 0, 1)$, with $S_1 \subset S$, $S_2 \subset \mathcal{S}'\mathcal{B}(\mathcal{S})$ and $\mathcal{U}_m(t) = \mathcal{W}(t)$.
The problem to interpret S_2 as a trajectory effect set and $\mathcal{U}_m(t)$ as the time translation operator on this set remains open.
We stress finally that a problem similar to the embedding is met in the treatment of unstable particles in the framework of quantum field theory. For a discussion of such a problem, we refer to [3 , 4].

References

1 Alberti, A., Cotta, P. Ramella, G., On a Banach space formulation of subdynamics in quantum statistics, preprint IFUM(Milano), March 1973.

2 Lanz, L., Lugiato, L.A., Ramella, G., Physica 54, 94 (1971)

3 Lanz, L., Lugiato, L.A., Ramella, G., On the quantum mechanical treatment of unstable nonrelativistic systems, in press on Int. Journ. Theor. Phys.

4 Lanz., L., Lugiato, L.A., Ramella, G., Sabbadini, A., The embedding of unstable nonrelativistic particles into Galilean Quantum Field Theories, in press on Int.Journ. Theor. Phys.

5 Lanz, L., Lugiato, L.A., Ramella, G., in preparation

6 Ludwig, G., Lecture Notes in Physics n. 4, Springer Verlag, Berlin 1970.

7 Ludwig, G., Makroskopische Systeme und Quantenmechanik, Notes in Mathematical Physics 5, Universität Marburg 1972

8 Ludwig, G., Lectures in the Proceedings of this Institute.

9 Prigogine, I., George, C., Henin, F., Physica 45, 418 (1969).

THE QUANTUM MECHANICAL HILBERT SPACE FORMALISM AND THE QUANTUM MECHANICAL PROBABILITY SPACE OF THE OUTCOMES OF MEASUREMENTS

Mioara MUGUR-SCHACHTER
Faculté des Sciences de l'Université de REIMS (France)

ABSTRACT :

The relation between the quantum mechanical descriptive basis - in its Hilbert space formulation - and the quantum mechanical probability space is analysed. The EVERETT - WHEELER - GRAHAM - DE WITT interpretation of the quantum mechanical formalism, GLEASON's theorem on the measure of the closed subspaces of a Hilbert space, and the theory of probabilities are used as a combined system of reference. The conclusion is reached that the quantum mechanical probability law is not obtainable in a purely deductive way from the basic descriptive system of quantum mechanics : the link between the quantum mechanical basis and the quantum mechanical probability law is insured by an independant postulate, of which the complex semantic contents and syntactic structure are established in detail. The performed analysis brings forth new perspectives and an efficient methodologic fact.

I. INTRODUCTION

1° The problem

In its present stage of development the orthodox quantum mechanics (QM) exists in several different forms. Each one of these forms proposes a mathematical descriptive structure (DS) by aid of which it represents micro-phenomena, and it asserts a probability law π concerning the outcomes of measurements performable on microsystems, included in a probability space $(\mathcal{E}, \mathcal{R}, \pi)$ associated to these outcomes. Whereas the descriptive structure varies rather strongly from one formulation of QM to another one, the asserted probability space is a fixed element. Most physicists consider that the quantum mechanical probability law has syntactically the nature of a postulate inside the quantum theory. However, it has been recently claimed by several physicists that this law can be derived inside the formal structure generated by a certain minimal system of basic descriptive elements. Anyhow it is obvious that the quantum mechanical probability law on the outcomes of measurements has to be expressed in terms of certain descriptive elements belonging to an admitted formal descriptive structure and thereby it acquires inside this structure certain syntactic valencies and a certain semantic charge. I shall try, in the particular case of the Hilbert space variant of QM, to analyze the composition of the link between the descriptive structure used and the quantum mechanical probability space, with its probability law. This analysis has a double aim : to gain a better insight into the physical implications of the quantum mechanical probability law, and to obtain a clearer knowledge of the configuration of syntactical constraints and liberties into which one has to insert coherently any eventual attempt to improve the description of microphenomena, while having chosen to continue to make use of the descriptive elements of QM.

2° The method

EVERETT[1], WHEELER[2], De WITT[3,4] and GRAHAM[5] (E.W.G.D.) have proposed an internal interpretation of the reduction problem, which includes a derivation of the orthodox quantum mechanical probability law, from the basic quantum mechanical descriptive structure, in its

Hilbert space variant. In a different line of thought, GLEASON[6] has shown that in a separable Hilbert space $\mathfrak{H}$ of dimension at least three, real or complex, every measure on the closed subspaces A has the form

$$\mu(A) = \text{trace}\ (TP_A)$$

where μ is a regular "frame function" in $\mathfrak{H}$, T is a positive semi-definite self-adjoint operator of the trace class, and P_A denotes the orthogonal projection of $\mathfrak{H}$ on A. Finally, QM is rooted into the conceptual ground of the general theory of probabilities. In what follows I shall use the E.W.G.D. development, Gleason's result, and the general theory of probability, as a combined system of reference appropriate for the analysis of the relation between the Hilbert space variant of the quantum mechanical descriptive structure, and the quantum mechanical probability space associated to the outcomes of measurements performable on microsystems.

3° The conclusion

The main conclusion will be that the quantum mechanical descriptive structure, in the Hilbert space variant - if stripped of any additional element, if considered only as a mathematical being - does not entail the quantum mechanical probability space by syntactical necessity, but is formally compatible with a whole class of probability spaces conceivable a priori, among which the quantum mechanical probability space realizes only one possibility. If physical reasons were established which would justify the acceptance of a non-orthodox probability space contained in this class, no logical reason would interdict the conservation of the orthodox mathematical descriptive elements, and their free utilisation, eventually in coherent association with additional descriptive elements.

From a methodological point of view, it will appear that a probability law has to be always conceived as an element organically integrated into a given probability space, if one wants to become aware of all the physical implications associated with the assertion of this probability law.

II. THE ESSENCE OF THE EWGD THEORY

1° The fundamental claim

Consider the quantum mechanical operators O_Q which describe the dynamical quantities Q_i (satisfying the known commutation equations) and the quantum mechanical Hilbert vectors $|\Psi\rangle$ which describe any isolated system, which are acted upon by the O_{qi}, and which are admitted to be everywhere and at all times solution of the linear Schrödinger equation of evolution. The EWGD fundamental claim is the following :

($\mathcal{C}$) : the basic descriptive structure $(O_Q\ ,\ \Psi)$ is able to generate in a purely deductive way, without any additional assumption, at the same time, the whole orthodox quantum mechanical descriptive structure in its Hilbert space variant, $(DS)^{\mathcal{H}}$, the orthodox quantum mechanical probability law on the outcomes of measurements, and the physical interpretation of all the mathematical expressions related to these laws.

2° Definition of a "good observation"

A "good" observation of a quantity Q corresponding to the operator O_Q with eigenfunctions Φ_{q_i}, performed on an object system S, by an observer system O (apparatus) having the initial state Ψ^O, is such that if the initial object system state is an eigenstate $\Phi^S_{q_i}$ (uncoupled with Ψ^O), then the total (isolated) system (S + O), considered as the world, undergoes by the observation process a Schrödinger evolution.

(1) $$\Phi^S_{q_i}\ \Psi^O \ \longrightarrow\ \Phi^S_{q_i}\ \Psi^O\ (q_i^O\ \ldots)$$

which relates the initial total state $\Phi^S_{q_i}\ \Psi^O$ to a final total state $\Phi^S_{q_i}\ \Psi^O\ (q_i^O\ \ldots)$ in which the object system state is unchanged, and the observer system state "has learned it" (the dots indicate the various apparatus variables, and q_i^O is a value of such a variable characterizing the eigenvalue q_i of Q which corresponds to the object system state Φ_{q_i}).

3° The good Schrödinger observation - evolution for one object system S in an arbitrary initial state of S

(1) and the superposition principle implied by ($\mathcal{C}$) entail that, if the initial object system state is $\Psi^S = \Sigma_i\ c_i\ \Phi^S_{q_i}$, the total state of (S + O) indergoes by observation the Schrödinger evolution symbolizable by

(2) $$\Big(\sum_i c_i\ \Phi^S_{q_i}\Big)\ \Psi^O \ \longrightarrow\ \sum_i c_i\ \Phi^S_{q_i}\ \Psi^O\ (\ldots\ q_i^O\ \ldots)$$

4° The Schrödinger observation - evolution for a statistical ensemble of replicas of S in arbitrary identical initial states

Consider now the "world" as constituted by a statistical ensemble of N non-interacting replicas S_1, S_2 ... S_N of a given object system S, having all a same initial state $\Psi^{S_1} \equiv \Psi^{S_2} \equiv \dots \equiv \Psi^{S_N}$. The observer system O with initial state Ψ^O performs successively on S_1.. .. S_N, in this order, a measurement of the quantity Q. EVERETT[1] shows by a sequence of propositions where the only assumptions used are (1), (2) and the superposition principle implied by ($\mathcal{C}$), that after $r \leqslant N$ measurements, the Schrödinger evolution of the total system (S_1 + ... + S_N + O) has operated a transformation which can be indicated by

$$(3) \quad \Psi^{S_1} . \Psi^{S_2} \dots \Psi^{S_N} . \Psi^O \longrightarrow \sum_{i,j..k} c_i \dots c_k \, \Phi^{S_1}_{q_i} \dots \Phi^{S_r}_{q_k} \Psi^{S_{r+1}} .. \Psi^{S_N} \times$$

$$\times \; \Psi^O \left(\dots q_i^{O,1}, q_j^{O,2}, \dots q_k^{O,r} \dots \right)$$

(obvious significance of the $q_l^{O,m}$).

5° The interpretation of the final total states in (2) and (3) and the qualitative formulation of the internal interpretation of the reduction problem

The final total states in (2) and (3) can be regarded respectively as superpositions of "elements" of the forms

$$(2') \qquad \Phi^S_{q_i} \; \Psi^O \left(\dots q_i^O \dots \right)$$

$$(3') \qquad \Phi^{S_1}_{q_i} \Phi^{S_2}_{q_j} \dots \Phi^{S_r}_{q_k} . \Psi^{S_{r+1}} \dots \Psi^{S_N} . \Psi^O \left(\dots q_i^{O,1}, q_j^{O,2} \dots q_k^{O,r} .. \right)$$

Now, EWGD assert that in each element (2'), while the object system state is a particular eigenstate of the observed quantity, the corresponding observer state "describes the observer as definitely perceiving that particular system state"[1], and that each element (3') "describes the observer with a definite memory sequence $(\dots q_i^{O,1}, q_j^{O,2}, \dots q_k^{O,r} \dots)$", constituted by the "perceived" values $q_i^{O,1} \dots q_k^{O,r}$. As a consequence of this interpretation of the elements (2') and (3'), EWGD obtain, by considerations on the "branching memories" structure of the final state (3), a first(qualitative)formulation of their well known "internal interpretation" of the reduction problem.

6° The quantitative formulation of the internal interpretation of the reduction problem (the quantum mechanical probability law)

The quantitative formulation of the EWGD internal interpretation of the reduction problem is realized by assignation of a measure to the elements of a superposition of orthogonal states :

... "in order to establish quantitative results, we must put some sort of measure (weighting) on the elements of a final superposition (...). We wish to make quantitative statements about the relative frequency of the different possible results of observation - which are recorded in the memory - for a typical observer state ; but to accomplish this we must have a method for selecting a typical element from a superposition of orthogonal states. We therefore seek a general scheme to assign a measure to the elements of a superposition of orthogonal states $\Sigma_i\ c_i\ \Phi_{q_i}$. We require a positive function μ of the complex coefficients of the elements of the superposition, so that $\mu(c_i)$ shall be the measure assigned to element Φ_{q_i} "1(*).

It is then noted that in order to prevent the normalization condition and the phase factor in c_i to introduce ambiguities, μ has in fact to be required as a function of the amplitude $|c_i|$ alone.

(4) $$\mu = \mu\ (|c_i|)$$

Then it suffices to impose also the additivity requirement

(5) $$\mu\ (|\gamma|) = \sum_{i=1}^{N} \mu(c_i)$$

where γ is defined by $\gamma\ \Phi' = \sum_{i=1}^{N} c_i\ \Phi_{q_i}$, in order to be able to demonstrate without further assumptions that, if the total measure is assigned the value 1,

(6) $$\mu\ (|c_i|) = |c_i|^2 .$$

which is precisely the value of the quantum mechanical probability for the outcome of the eigenvalue q_i of Q.

III. THE PROBABILITY SPACE OF QM

Throughout what follows I shall keep explicitely in mind the fact that every probability law has to be conceived as organically integrated into a given probability space, that it is rooted into a specifiable basic set of elementary events and concerns the elements of the ring of subsets of this basic set. Thereby it will be possible to bring into evidence all the syntactic relations between the quantum mechanical probability law and the descriptive system generated by $(\Psi,\ O_Q)$ and to make perceptible the whole semantic content with which the probability law is progressively charged while is achieved its integration in this descriptive system. This will show that it is fruitful to treat the concept of probability space as an entity more fundamental than the concept of probability law.

(*)p. 460 : we made renotations : $a_i \rightarrow c_i$, $\Phi_{Q_i} \rightarrow \Phi_{q_i}$, $m \rightarrow \mu$

1° Examination of the EWGD theory

A. Concerning (1)

[a(1)] - The definition (1) of a "good observation" (Wigner's "first type of measurement")[7] is not a logical consequence of the acceptance of (O_Q , Ψ) as a system of convenient basic descriptive elements for microphenomena. In EWGD, (1) is an independent additional proposition.

[b(1)] - For the particular case that the initial state of the object system S is an eigenstate $\Phi^S_{q_i}$ of O_Q, (1) constitutes (in consequence of [a(1)]) an independent - though implicit - postulation $\mathcal{P}_1(\pi)$ of a particular probability law π for the outcomes of "good observations" of Q :

$$(\mathcal{P}_1 (\pi)) \qquad \pi (\Phi^S_{q_i} , q^O_i) = 1 , \quad \pi (\Phi^S_{q_i}, q^O_j, j \neq i) = 0$$

Whatever may be established concerning the physical realizability of the distribution $\mathcal{P}_1(\pi)$, from a purely syntactical point of view, this implicit postulation - which introduces an a priori restriction on the probability law $\pi \in (\mathcal{E}, \mathcal{R}, \pi)$, in a theory aimed towards a deduction of this probability law - is obviously objectionable, since it imparts to the reasoning a partially circular character.

[c(1)] - But (1) does not act independently only on $\pi \in (\mathcal{E}, \mathcal{R}, \pi)$, it also acts independently on $\mathcal{E} \in (\mathcal{E}, \mathcal{R}, \pi)$:

The definition (1) is formulated for the measurement of one quantity only. What are its consequences concerning joint measurements of several quantities ? It is obvious that (1) admits an immediate reformulation for the case of a joint measurement of several commuting quantities. But if non-commuting quantities are envisaged, the situation becomes complex and requires a systematic distinction between the concepts which concern the statistical level of observation and those which concern the individual level of observation (see ref. 8, 9,10,11). For the moment I enter upon this subject no more than is strictly necessary for the limited purpose of establishing the action of (1) on $\mathcal{E} \in (\mathcal{E}, \mathcal{R} , \pi)$.

In QM a measurement of one quantity Q is called "rigorous" if $\pi \in (\mathcal{E}, \mathcal{R}, \pi)$ satisfies the condition $\mathcal{P}_1(\pi)$. With this criterion, it follows that a simultaneous rigorous measurement of two non-commuting quantities is not possible. But this impossibility concerns the statistical level of observation, it does not impose a positive restriction on the contents of the set $\mathcal{E} \in (\mathcal{E}, \mathcal{R}, \pi)$ of elementary events into which is rooted the probability measure $\pi \in (\mathcal{E}, \mathcal{R}, \pi)$;

it concerns π directly, not via $\mathcal{E}$. This becomes clearer if it is noted that (1) yields also the definition of a "good" apparatus for the measurement of Q : a system O is a "good" apparatus (observer) for Q if, in case that the object-system S is initially in an eigenstate of Q, the total system (S + O) performs the Schrödinger observation evolution (1) ($\mathcal{P}_1(\pi)$ is realized for the distribution of the outcomes). This reformulation of (1) brings into evidence the fact that (1) does not exclude the possibility that one given device be "good" at the same time for the observation of two non-commuting object-system quantities Q_1, Q_2, hence able also to yield a "good" individual joint registration of the observable values ${}^1q^0$ and ${}^2q^0$ of two apparatus variables characterizing the values ${}^1q$ and ${}^2q$ of Q_1 and Q_2. (Whether this individual joint registration can or cannot be rigorous, is a distinct problem to be examined by aid of a specific individual level criterion for the rigor of a measurement).

However, while (1) does not exclude the possibility of an individual joint observation of two non-commuting quantities, it does not take it constructively into consideration either, it simply ignores the problem of this possibility, giving no criterion for such an eventual good individual joint observation, and making no assertion concerning the conceivability of an acceptable criterion of this type. Thus (1) leaves in fact untouched the question whether it is possible or not to associate (O_Q, Ψ) coherently also with a definition of a good individual joint measurement of non-commuting quantities.

In these circumstances the exclusivity of the constructive use of (1) as a definition of a good observation, throughout the EWGD theory, is equivalent to an independent a priori elimination of the event "individual joint outcome $({}^1q^0, {}^2q^0)$ of two non-commuting quantities" from the set $\mathcal{E}$ of elementary events taken into account in the probability space $(\mathcal{E}, \mathcal{R}, \pi)$ which includes the researched probability measure π concerning the outcomes of measurements performable on microsystems. But this also is a probability postulate :

$$(\mathcal{P}_2(\pi)) \qquad \{\xi = ({}^1q^0, {}^2q^0)\} \notin \{\mathcal{E}\,\varepsilon\,(\mathcal{E}, \mathcal{R}, \pi)\}$$

The objection (b(1)) applies to $\mathcal{P}_2(\pi)$ also.

(d(1)) - WIGNER[12], YANASE and YARAKI[13,14] and STEIN and SHIMONY[15] have established rather severe limitations on the existence - inside the descriptive structure generated by (O_Q, Ψ) - of a Schrödinger evolution operator able to insure rigorously the total state transition symbolized by the arrow in (1). Thereby the utilizability of (1) in the very basis of the EWGD derivation, appears as objectionable from a new point of view, to be adjoined to $[b(1)]$ and $[c(1)]$.

B. <u>Concerning (2) and (3)</u>

The $|\Psi>$ vector in (O_Q, Ψ) possesses the definitory characteristic of being at all times and everywhere solution of a linear equation of evolution, hence the $|\Psi>$'s are submitted to an unrestricted superposition principle. Then, if (1) is accepted, (O_Q, Ψ) + (1) entails indeed (2) and (3). However, the limitations $[d(1)]$ on the arrow in (1) are transmitted to the arrows in (2) and (3).

C. <u>Concerning the interpretation of the final states in (2) and (3), and the qualitative formulation of the internal interpretation of the reduction problem.</u>

Here the following remarks, vital for this analysis, are to be made :

$[a(2'+3')]$ - Let us now make tentatively abstraction of $[b(1)]$, $[c(1)]$ and $[d(1)]$ and admit fully (1) as forming with (O_Q, Ψ) an acceptable basis for the EWGD derivation. Then (1) and the superposition principle to which are submitted by their definition the linearly evolving $\Psi \varepsilon (O_Q, \Psi)$, entail the form of the final total states in (2) and (3). Furthermore, the formal decomposability of these final states in elements (2') and (3'), is an incontestable mathematical fact. So far only correct mathematical manipulation has been done, no physical interpretation has yet intervened.

But the situation changes as soon as the EWGD physical significance attributed to the symbols q_i^O of (2') and $(q_i^{O,1} \ldots q_k^{O,r})$ of (3') is introduced. The assertion that the symbol q_i^O, in an element (2') <u>which in (2) is included in the final superposition state of (S+O)</u>, $\Sigma_i\, c_i\, \Phi^s_{q_i}\, \Psi^O\, (q_i^O \ldots)$, describes the observer as possessing the physically realized characteristic of definitely "perceiving" the particular relative system state $c_i\, \Phi^s_{q_i}$, and the similar assertion that the symbol $(q_i^{O,1} \ldots q_k^{O,r})$, in an element (3') <u>which in (3) is included in the final superposition state of (S+O)</u>,

$$\sum_{i,j\ldots k} c_i \ldots c_k\, \Phi^{S_1}_{q_i} \ldots \Phi^{S_r}_{q_k}\, \Psi^{S_{r+1}} \ldots \Psi^{S_N} \times \Psi^O\, (\ldots q_i^{O,1},\, q_j^{O,2}, . q_k^{O,r}, .$$

describes a physically realized "memory configuration" of O, do <u>not</u> follow <u>mathematically</u> from (1) and from the superposition principle for the $\Psi \varepsilon (O_Q, \Psi)$. The symbols $(q_i^O \ldots)$ and $(q_i^{O,1} \ldots q_k^{O,r})$ included in the final superpositions (2) and (3) possess the same type of merely conceptual existence, as the frequencies ν_i possess in the writing of the Fourier structure of an electromagnetic signal : no mathematical reasons exist which would require to associate to this graphic presence, a <u>physical</u> individualization also. The following considerations will add relief to this remark :

The EWGD interpretation of (2') and (3') is equivalent to the following assertion, which will be indicated by the notation (justified later) $\wp$(smr) :

($\wp$(smr)) : During a Q - observation Schrödinger evolution of (S+O), every observable effect of the object system S having an initial state representable by $\Sigma_i\ c_i\ \Phi^S_{q_i}$, on an apparatus O good for the measurement of Q (see [d(1)]), is identical to one of the effects which would be produced on O by microsystems S having initially an eigenstate $\Phi^S_{q_i}$ of Q.

This is equivalent to the alternative formulation :

($\wp$'(smr)) : During a Q - observation Schrödinger evolution of (S+O), the mathematical interference (the phase relations) in the representation $\Sigma_i\ c_i\ \Phi^S_{q_i}$ of the initial object system state, between the orthogonal elements $c_i\ \Phi^S_{q_i}$ associated to O_Q, has no observable counterpart.

Now, in order that the EWGD interpretation of (2') and (3') "follows mathematically" from $(O_Q\ ,\ \Psi)$ + (1), it would be necessary that the Schrödinger evolution symbolized by the arrows in (2) and (3) be such as to insure the formal basis for the physical assertion $\wp$(smr) ≡ $\wp$'(smr), for any initial object state and for any dynamical quantity Q. That is to say, it ought to be such as to insure the suppression of any overlapping of the spatial domains of existence of the orthogonal elements $c_i\ \Phi^S_{q_i}$ of the object system initial superposition state $\Sigma_i\ c_i\ \Phi^S_{q_i}$. But it is well known that no existing Schrödinger evolution operator insures a rigorous realization of this condition at a finite time, not even in the most favorized cases, of conservative quantities and discrete spectra (like in Stern-Gerlach measurements of a given spin component). Moreover, for certain initial object system states and certain quantities, even an approximate fulfilment of the formulated condition is impossible (consider total energy measurements in bound states, or a momentum vector measurement in a free Young-interference state)*. Hence the assertion [$\wp$(smr) ≡ $\wp$'(smr)] implied in the EWGD interpretation of (2') and (3') cannot be obtained deduc-

---

* Problem : is there a relation between this fact and the WIGNER[12] limitations on measurement ?

tively from $(O_Q , \Psi) + (1)$. It follows that it is an additional implicit independant postulate.

But this additional independent postulate $\mathcal{P}(smr)$ has the same nature as the quantum mechanical reduction postulate, $\mathcal{P}_{QM}(r)$, which EWGD intended to disolve in an "internal interpretation". Its content obviously justifies the denomination of "postulate of simultaneous multiple reduction", indicated by the notation adopted for it. Indeed, while the quantum mechanical reduction postulate $\mathcal{P}_{QM}(r)$ asserts the reduction by measurement, of the initial object system superposition state $\Sigma_i\ c_i\ \Phi^S_{q_i}$, to <u>one</u> of its elements $c_i\ \Phi^S_{q_i}$, $(\mathcal{P}(smr) \equiv \mathcal{P}'(smr))$ postulates the reduction by measurement, of $\Sigma_i\ c_i\ \Phi^S_{q_i}$, to all of its orthogonal elements $c_i\ \Phi^S_{q_i}$ simultaneously (to the unweighted (for the moment) mixture corresponding to $\Sigma_i\ c_i\ \Phi^S_{q_i}$).

This EWGD simultaneous multiple reduction of $\Sigma_i\ c_i\ \Phi^S_{q_i}$, is a weaker assumption than the quantum mechanical reduction, so that one can write the logical inclusion

(7) $$\mathcal{P}_{QM}(r) \subset \mathcal{P}(smr)$$

The whole subsequent EWGD branching Universes interpretation of the final states (3) (based-via $\mathcal{P}(smr)$-on the EWGD interpretation of (3')) has precisely the role to compensate the gap left between $\mathcal{P}(smr)$ and the more radical orthodox reduction postulate.

$[b(2'+3')]$ - The postulate $(\mathcal{P}(smr) \equiv \mathcal{P}'(smr))$ and (1) entail together the conclusion that each observable value q^O_i of a variable of an observer system good for the observation of a quantity Q,is numerically identical to an eigenvalue q_i of O_Q. This means, moreover, that - degeneracy neglected - each one of the observable values q^O_i is related to one eigenvector Φ_{q_i} of O_Q, by the equation $O_Q\ \Phi_{q_i} = q_i\ \Phi_{q_i}$ generated by (O_Q, Ψ), where the Φ_{q_i} are considered in the same limiting conditions as the object system state vector. But the propositions just mentioned constitute the quantum mechanical eigenvalues postulate $\mathcal{P}_{QM}(ev)$: thus $\mathcal{P}_{QM}(ev)$ is logically equivalent with the association of (1) with the EWGD postulate $\mathcal{P}(smr)$

(8) $$(\mathcal{P}(smr) + (1)) \sim \mathcal{P}_{QM}(ev)$$

$[c(2'+3')]$ - Now the eigenvalues postulate $\mathcal{P}_{QM}(ev)$ also is a probability postulate, a very strong one in fact ; it determines the positive contents assigned to the set of elementary events, in the probability space associated with the outcomes of measurements of Q on the object systems S

$$\mathcal{P}'_1(\pi)\quad \mathcal{E}^Q = \{q^O_i \equiv q_i\ (\Phi_{q_i})\} = \mathcal{E}^Q(O_Q)$$

(q_i^0 and q_i can be considered here as concerning a whole complete family of commuting quantities).

The objection [b(1)] applies to $\mathcal{P}'_1(\pi)$ also, hence via $\mathcal{P}(smr)$ to the EWGD interpretation of (2') and (3').

The probability postulate $\mathcal{P}'_1(\pi)$ has simultaneously the role of a rigourous definition of the semantical assumption contained in ($\mathcal{C}$), that O_Q "represents" Q :

(8') $$\mathcal{P}'_1(\pi) \sim \mathcal{D}(O_Q \longleftrightarrow Q)$$

[d(2'+3')] - The relations (7) and (8) entail

(9) $$[\mathcal{P}_{QM}(r) + (1)] \subset \mathcal{P}_{QM}(ev)$$

This stresses the fact that the quantum mechanical eigenvalues postulate is <u>derivable</u> from the quantum mechanical reduction postulate, once (1) is accepted : the separate assertion in QM of both these postulates is redundant.

To conclude this point, a synopsis followed by a general comment :

$\{[O_Q, \Psi] + (1)\} \not\to$ {EWGD interpretation of (2') and (3')}
{EWGD interpretation of (2') and (3')} $\sim \{[O_Q, \Psi] + (1) + \mathcal{P}(smr)\}$
$\mathcal{P}(smr) \supset \mathcal{P}_{QM}(r)$
$[\mathcal{P}(smr) + (1)] \sim \mathcal{P}_{QM}(ev) \sim \mathcal{P}'_1(\pi) \sim \mathcal{D}(O_Q \longleftrightarrow Q)$
$[\mathcal{P}_{QM}(r) + (1)] \subset \mathcal{P}_{QM}(ev)$

I find it very remarkable that according to $\mathcal{P}'_1(\pi)$ the possible elementary events $(q_i \in \mathcal{E}) \in [\mathcal{E}, \mathcal{R}, \pi]$ <u>are independent of the initial object-system state described by Ψ^S</u>, being determined exclusively by the measured quantity O_Q, via $O_Q \Phi_{q_i} = q_i \Phi_{q_i}$ (solved in the same limiting conditions as the evolution equation which yields Ψ^S). In QM the state vector Ψ^S (as it will appear) acts exclusively on the weighting of the elements attributed to $\mathcal{E}$, on the functional form and argument of the probability measure rooted into $\mathcal{E}$. The analysis of the EWGD interpretation of the final states in (2) and (3), concentrated in the above synopsis, shows that this situation is by no means a logical necessity entailed alone by the acceptance of $[O_Q, \Psi]$ (as a convenient descriptive basis) and of (1) : one has to add $\mathcal{P}(smr)$ to $[O_Q, \Psi] + (1)$ in order to obtain $\mathcal{P}_{QM}(ev)$, hence $\mathcal{P}'_1(\pi)$. Indeed if $\mathcal{P}(smr)$ is not introduced, there exists no reason whatever for which the values q^0 of Q resulting from the interactions between the object-system S in a state $\Psi^S = \Sigma_i c_i \Phi^S_{q_i}$ and the observer-system O in a state Ψ^O should be conceived as depending only on the <u>separated</u> Φ_{q_i} from Ψ^S, so that one be permitted to write $\mathcal{E}^Q = \{q^O \equiv q_i(\Phi_{q_i})\} = \mathcal{E}^Q(O_Q)$; in absence of $\mathcal{P}$(smr) these values q^O are to be conceived as depending also on the mathematical interferences of the $\Phi^S_{q_i}$ from Ψ^S, hence on the state Ψ^S in its integrality, so that in absence of $\mathcal{P}$(smr) one has to write in general

$$\mathcal{E}^Q = \{q^O \neq q_i(\Phi_{q_i})\} = \mathcal{E}^Q(O_Q, \Psi^S)$$

It is the reduction postulate $\mathcal{P}$(smr) which interposes a screen between the characteristics of the state of the object-system (described by Ψ^S) and the contents assigned to $\mathcal{E}^Q$. Thus $\mathcal{P}$(smr) appears to be the minimal assumption necessary in addition to $[O_Q, \Psi] + (1)$ in order to insure the assertability of $\mathcal{P}_{QM}$(ev), hence of $\mathcal{P}'_1(\pi)$ - vital in QM. As to $\mathcal{P}_{QM}(r) \subset \mathcal{P}$(smr), it is a stronger assumption which has in QM the same role as $\mathcal{P}$(smr) in EWGD. Thus it is for $\mathcal{P}'_1(\pi)$ that QM pays the heavy price of the well known "interpretation" problem of the direct physical significance assignable to $\mathcal{P}_{QM}(r)$. As to the EWGD theory, it is in order to insure reobtention of the quantum mechanical propositions $\mathcal{P}_{QM}$(ev) (hence $\mathcal{P}'_1(\pi)$) while starting out from $\mathcal{P}$(smr) instead of $\mathcal{P}_{QM}(r)$, that it pays the price of the branching Universes interpretation, which - contrary to the authors' claim - is not entailed by $[O_Q, \Psi] + (1)$ alone, but stems precisely from the independent postulation $\mathcal{P}$(smr), weaker than $\mathcal{P}_{QM}(r)$, and therefore still more strange in its direct physical implications.

D. <u>Concerning the quantitative formulation of the internal interpretation of the reduction problem (the quantum mechanical probability laws)</u>

It can be shown that the final considerations by which the quantum mechanical probability laws are obtained in EWGD, merely develop quantitatively the probabilistic and semantic implications contained in the association of the descriptive structure generated by $[O_Q, \Psi]$, with the postulates $\mathcal{P}_2(\pi)$, $\mathcal{P}'_1(\pi)$ already admitted implicitely before and with a new postulation including in fact the previous ones.

Indeed, EVERETT researches directly the probability law concerning the outcomes of measurements, in the form of a function $\mu(|c_i|)$ of the amplitude of the complex coefficients of the orthogonal components $\Phi^S_{q_i}$ in the object system initial superposition state $\Psi^S = \Sigma_i c_i \Phi^S_{q_i}$. But <u>why</u> a function of the amplitude of these coefficients ? And why a function of these coefficients only, and not for instance a function $\mu(c_i, \Gamma)$ of these coefficients and also of some other descriptive elements Γ generated by (O_Q, Ψ) ? And why a function of only one of these coefficients, at a time, and not a function $\mu(c_i, c_j, \ldots c_k, \Gamma)$ of several such coefficients and some Γ ? All these conceivable forms are equally permitted a priori, as long as really the <u>unique</u> restriction imposed is to shape out a formal expression of the researched probability law, by using in a mathematically consistent way, only the descriptive structure generated by (O_Q, Ψ). This unique restriction, by itself, if stripped of any additional assumption, obviously has not the power to select deductively one of these possibilities, while eliminating all the others. But, in fact, this is not the unique restriction imposed, in EWGD there are also the additional restrictions implicitely introduced in the previous stages by the postulates $\mathcal{P}_2(\pi)$ and $\mathcal{P}'_1(\pi)$, as well as other further restrictions, as it will appear. As soon as all these restrictions are taken into account, the choice is directed indeed towards $\mu(|c_i|)$. To show this, we shall now bring into evidence the constitution of the whole EWGD chain which relates in fact (O_Q, Ψ) to the quantum mechanical probability law.

Let us distinguish between the direct symbolization $(\mathcal{E}, \mathcal{R}(\Delta^{\mathcal{E}}), \pi(\Delta^{\mathcal{E}}))$ of the probability space of the physical events consisting of the outcomes of measurements performable on microsystems, and the various conceivable formal measure spaces $(\mathcal{S}, \mathcal{R}(\Delta^{\mathcal{S}}), \mu(\Delta^{\mathcal{S}}))$ isomorphic to $(\mathcal{E}, \mathcal{R}(\Delta^{\mathcal{E}}), \pi(\Delta^{\mathcal{E}}))$. The problem to be solved now, is to determine such a formal measure space, expressed in terms of descriptive elements generated by (O_Q, Ψ).

Let us first consider $(\mathcal{E}, \mathcal{R}(\Delta^{\mathcal{E}}), \pi(\Delta^{\mathcal{E}}))$ itself. A priori the set of elementary events $\mathcal{E}$ could contain any group of apparatus indications ${}^1q^0$, ${}^2q^0$ ... ${}^Nq^0$ concerning the values ${}^1q$, ${}^2q$, ... ${}^Nq$ of measured object system quantities Q_1, Q_2 ... Q_N. But $\mathcal{P}_2(\pi)$ - acting directly on $\mathcal{E}$ - has eliminated from $\mathcal{E}$ the consideration of all the conceivable joint individual registrations concerning two or more non-commuting quantities (a first restriction for the admitted $\mathcal{E}$). Noting by Q_n a complete family of quantities commuting with a given Q and by ${}^nq^0$ the values of the corresponding apparatus variables, the effect of $\mathcal{P}_2(\pi)$ on $\mathcal{E}$ can be defined by saying that $\mathcal{E}$ is splitted into a multiplicity $\{\mathcal{E}^{Q_n}\}$ of distinct sets $\mathcal{E}^{Q_n}$ of elementary events (each one corresponding to a family Q_n of commuting quantities) admitted to be experimentally disconnected in time at the statistical level of observation (and consequently in the expressibility of their probability measures).

$$\mathcal{P}_2(\pi) \longrightarrow \quad \mathcal{E} = \{\mathcal{E}^{Q_n}({}^nq^0)\} = \{\mathcal{E}({}^nq^0)\}$$

Let us now consider the effect produced on one $\mathcal{E}({}^nq^0)$, by the eigenvalues postulate $\mathcal{P}'_1(\pi)$ (we drop from now on the index n). $\mathcal{P}'_1(\pi)$ asserts that each observable value of an apparatus observable is uniquely related (by a numerical identity) to an eigenvalue q_i of the operator O_Q corresponding to Q. $(\forall_q 0 \; \varepsilon \; \mathcal{E}(q^0)) \equiv q_i$, (degeneracy neglected). Whereas q^0 indicates directly a physical event, q^i is a formal element generated by (O_Q, Ψ), via the equation $O_Q \; \Phi_{qi} = q_i \; \Phi_{q2}$. Hence $\mathcal{P}'_1(\pi)$ operates a passage from the direct description of the contents of $\mathcal{E}$ in terms of the elementary physical events q^0 concerned by the theory, to an isomorphic basic set $\mathcal{S}$ of formal elements q_i generated by (O_Q, Ψ).

$$(\mathcal{P}'_1(\pi) \sim \mathcal{D}(O_Q <-> Q)) \rightarrow (\mathcal{E}(q^0) <-> \mathcal{S}(q_i))$$

(The presence of $\mathcal{D}(O_Q <-> Q)$ stresses here again the semantical counterpart (8') of $\mathcal{P}'_1(\pi)$).

Now $\mathcal{S}(q_i)$ has to be related to a formal measure space generated by (O_Q, Ψ) and isomorphic to $(\mathcal{E}(q^0), \mathcal{R}(\Delta q^0), \pi(\Delta q^0))$. The first step is possible by the use - once more- of the equation $O_Q \; \Phi_{qi} = q_i \; \Phi_{qi}$. This equation puts the $q_i \; \varepsilon \; \mathcal{S}(q_i)$ in bijective relation with eigenvectors of O_Q (degeneracy neglected). This permits an isomorphic mapping from $\mathcal{S}(q_i)$ to a new set $\mathcal{S}(\Phi_{qi})$ of formal elements generated by (O_Q, Ψ) and isomorphic to $\mathcal{E}(q^0)$

$$(O_Q \; \Phi_{qi} = q_i \; \Phi_{qi}) \quad \rightarrow \quad (\mathcal{S}(q_i) \rightarrow \mathcal{S}(\Phi_{qi}))$$

Given an arbitrary separable Hilbert space $\mathcal{H}$, the $\Phi_{qi} \varepsilon \; \mathcal{S}(\Phi_{qi})$ form an orthonormal basis in $\mathcal{H}$, and each Φ_{qi} spans a one-dimensional closed subspace $\underline{\Phi}_{qi}$ of $\mathcal{H}$, corresponding to a projector of the family of projectors equivalent to O_Q. Thus a new bijection $\Phi_{qi} <-> \underline{\Phi}_{qi}$ is defined, permitting a new isomorphic mapping

$$\mathcal{S}(\Phi_{qi}) \quad \rightarrow \quad \mathcal{S}(\underline{\Phi}_{qi})$$

Let us consider the measure space $(\mathcal{S}(\underline{\Phi}_{qi}), \mathcal{R}(\Delta\underline{\Phi}_{qi}), \mu(\Delta\underline{\Phi}_{qi}))$ based on $\mathcal{S}(\underline{\Phi}_{qi})$. In this measure space, the measure $\mu(\Delta\underline{\Phi}_{qi})$ has the significance of a measure on the closed subspaces of $\mathcal{H}$ defined by O_Q. But, in the arbitrary $\mathcal{H}$ considered so far, there is no reason whatever that $\mu(\Delta\underline{\Phi}_{qi})$ be isormophic to $\pi(\Delta q^0) \; \varepsilon \; (\mathcal{E}(q^0), \mathcal{R}(\Delta q^0) \; \pi(\Delta q^0))$ even <u>though</u> $\mathcal{S}(\underline{\Phi}_{qi}) \; \varepsilon \; (\mathcal{S}(\underline{\Phi}_{qi}), \mathcal{R}(\Delta\underline{\Phi}_{qi}), \mu(\Delta\underline{\Phi}_{qi}))$ is isomorphic to $\mathcal{E}(q^0) \; \varepsilon \; (\mathcal{E}(q^0), \mathcal{R}(\Delta q^0), \pi(\Delta q^0))$. Only some new restrictive assumption concerning $\mathcal{H}$ could insure such an isomorphism of the functions μ and π, entailing then also the global isomorphism of the physical probability space $(\mathcal{E}(q^0), \mathcal{R}(\Delta q^0), \pi(\Delta q^0))$ with a formal measure space $(\mathcal{S}(\underline{\Phi}_{qi}), \mathcal{R}(\Delta\underline{\Phi}_{qi}), \mu(\Delta\underline{\Phi}_{qi}))$.

In this stage of the analysis, the attention is naturally

directed upon the second EWGD semantical assertion contained in the fundamental claim ($\mathcal{C}$), that the object system state vector Ψ^S "describes" the isolated object-system state. It becomes soon obvious by reflection that the specific role of this assertion is precisely to insure the isomorphism of the probability measure $\pi(\Delta q^0)$, with a formal measure $\mu(\Delta\underline{\Phi}_{qi})$. Rather straightforwardly, the rigorous restatement of the semantical definition of Ψ^S is the following one :

$\mathcal{D}(\Psi^S$ <-> state of S) : If the set $\mathcal{S}(\underline{\Phi}_{qi})$ <-> $\mathcal{E}(q^0)$ is integrated as an orthonormal basis into the particular Hilbert space $\mathcal{H}^S$ of the initial object-system state vector Ψ^S, then the formal measure $\mu(\Delta\underline{\Phi}_{qi})$ "put" by Ψ^S on the $\Delta\underline{\Phi}_{qi} \in \mathcal{R}(\Delta\underline{\Phi}_{qi})$ (hence in particular on the $\underline{\Phi}_{qi} \in \mathcal{S}(\underline{\Phi}_{qi})$) and the probability measure $\pi(\Delta q^0)$ which weigths the $\Delta q^0 \in \mathcal{R}(\Delta q^0)$ (hence in particular the $q^0 \in \mathcal{E}(q^0)$), are isomorphic functions.

The semantical definition $\mathcal{D}(\Psi^S$ <-> state of S) is a probability assertion. Now, as it happens also for $\mathcal{D}(O_Q$ <-> Q), the specific semantical input $\mathcal{D}(\Psi^S$ <-> state of S) obviously is by no means entailed deductively by the acceptance of (O_Q, Ψ) as a basic system of mathematical beings convenient for a description of microsystems (<u>any</u> initial semantical input into a formal system is <u>essentially</u> descriptive, non-deductive). This input is only coherently associable with (O_Q, Ψ) by an independent act of thought which is necessary in order to specify the <u>way</u> in which (O_Q, Ψ) has to be used for the "description" of microsystems. Thus the probabilistic assertion $\mathcal{D}(\Psi^S$ <-> state of S) is a new postulate $\mathcal{P}_3(\pi)$:

$$\mathcal{P}_3(\pi) \sim \mathcal{D}(\Psi^S \text{ <-> state S}) :$$

$$\pi(\Delta q^0) \in \left(\mathcal{E}(q^0), \mathcal{R}(\Delta q^0), \pi(\Delta q^0)\right)^{(\text{state of S})} \text{<->} \mu(\Delta\underline{\Phi}_{qi}) \in \left(\mathcal{S}(\underline{\Phi}_{qi}), \mathcal{R}(\Delta\underline{\Phi}_{qi}) \mu(\Delta\underline{\Phi}_{qi})\right)^{\mathcal{H}^S}$$

The probability postulate $\mathcal{P}_3(\pi)$ is a very strong postulate. Indeed : the isomorphy of two measures included in two distinct measure spaces, <u>entails</u> the isomorphy (modulo an ensemble of measure zero) of the two corresponding basic sets of disjoint elements, hence also the global isomorphy of the two spaces. But in our case, the isomorphy $\mathcal{E}(q^0)$ <-> $\mathcal{S}(\underline{\Phi}_{qi})$ of the basic sets has been established by use of $\mathcal{P}_2(\pi)$ and $\mathcal{P}'_1(\pi)$. Consequently

$$\{\mathcal{E}(q^0) \text{ <-> } \mathcal{S}(\underline{\Phi}_{qi})\} \subset \{\mathcal{P}_2(\pi) + \mathcal{P}'_1(\pi)\}$$

It follows then that

(11) $\mathcal{P}_3(\pi) \subset \{\mathcal{P}_2(\pi) + \mathcal{P}'_1(\pi)\}$

and

(12) $\mathcal{P}_3(\pi) \rightarrow \left\{ \left(\mathcal{E}(q^0), \mathcal{R}(\Delta q^0), \pi(\Delta q^0)\right)^{(\text{state of S})} <-> \left(\mathcal{S}(\underline{\Phi}_{q_i}), \mathcal{R}(\Delta\underline{\Phi}_{q_i}), \mu(\Delta\underline{\Phi}_{q_i})\right)^{\mathcal{H}^S} \right\}$

A short reflection permits familiarization with the important relation (11), which deserves a very particular attention, I think. As to (12), it identifies finally the researched expression of the physical probability space to be described, in terms of descriptive elements generated by (O_Q, Ψ). (The upper indexes "state of S" and "$\mathcal{H}^S$" used in the formulation of $\mathcal{P}_3(\pi)$ and in (12) stress the specification of the compared spaces, with respect to the state of S, introduced by means of $\mathcal{D}(\Psi^S$ <-> state of S). In QM this specification is realized exclusively through the measures $\pi(\Delta q^0)$ <-> $\mu(\Delta\underline{\Phi}_{q_i})$, while the basic sets $\mathcal{E}(q^0)$ <-> $\mathcal{S}(\underline{\Phi}_{q_i})$ are independent of the state of S, in consequence of $\mathcal{P}'_1(\pi)$.

Now, with $\mathcal{P}_3(\pi)$, the probability measure $\pi(\Delta q^0)$ possesses indeed the form of a function $\mu(|c_i|)$, which moreover satisfies the EWGD additivity and normalization conditions. This can be seen by taking into account Gleason's theorem which establishes that, in a separable Hilbert space $\mathcal{H}$ of dimension at least three, the measure $\mu(\Delta\underline{\Phi}_{q_i}) \in \left(\mathcal{S}(\underline{\Phi}_{q_i}), \mathcal{R}(\Delta\underline{\Phi}_{q_i}), \mu(\Delta\underline{\Phi}_{q_i})\right)^{\mathcal{H}}$ necessarily is a regular frame function (a frame function for which a self adjoint positive semi-definite operator T ($\mathcal{H}$) of the trace class does exist) of weight 1 (if $\mathcal{H}$ is required with a unit norm) :

$$\mu(\Delta\underline{\Phi}_{q_i}) = \sum_i \mu_i\,(\underline{\Phi}_{q_i}) = \sum_i (T(\mathcal{H})\ \Phi_{q_i}, \Phi_{q_i}) =$$

$$= \sum_i |c_i|^2 = \mu(|\gamma|) = \sum_i \mu\,(|c_i|)$$

i = 1, 2... n and $\Delta\underline{\Phi}_{q_i}$ denotes the closed subspace of $\mathcal{H}$ spanned by $\{\Phi_{q_1} \ldots \Phi_{q_n}\}$ (in the last equalities the EWGD notation from (4) and (5) are reproduced). Then, by $\mathcal{P}_3(\pi)$, one obtains the quantum mechanical structure of $\pi(\Delta q^0)$.

The EVERETT[1] - GRAHAM[5] derivation is <u>included</u> in Gleason's demonstration. While the Everett-Graham demonstration does not justify the dependance $\mu(|c_i|)$ of the argument $|c_i|$ (it takes it for granted and then derives from it the functional form $\Sigma_i\ |c_i|^2$ for the function μ of this accepted argument $(|c_i|)$: $\mu(|c_i|)^2 = \Sigma_i\ |c_i|^2)$, Gleason's demonstration shows at the same time the necessity of the argument $|c_i|$ for a measure $\mu(\Delta\Phi_{q_i}) \in \left(\mathcal{S}(\Phi_{q_i}), \mathcal{R}(\Delta\underline{\Phi}_{q_i}), \mu(\Delta\Phi_{q_i})\right)^{\mathcal{H}}$ and the necessity of the specific functional form $\Sigma_i\ |c_i|^2 \equiv \Sigma_i(T\Phi_{q_i}, \Phi_{q_i})$ for $\mu(|c_i|) \equiv \mu(\Delta\underline{\Phi}_{q_i})$. Thereby Gleason's demonstration yields at the same time a basis and a confirmation of the Everett-Graham demonstration.

Now, Gleason's demonstration can be inserted <u>concerning</u> $\pi(\Delta q^O)$ only from the precise moment <u>on</u> at which $\mathcal{P}_3(\pi)$ has been accepted. If $\mathcal{P}_3(\pi)$ <u>is</u> taken into account, $T(\mathcal{R})$ becomes $T(\mathcal{R}^S) = T(\Psi^S)$ and one can write : $\pi(\Delta q^O) = \mu(\Delta\underline{\Phi}_{qi}) = \Sigma_i\ (T(\Psi^S)\ \Phi_{qi}, \Phi_{qi}) = \Sigma_i\ |c_i(\Psi^S)|^2$. It can be noted here that for $\Psi^S \equiv \Phi_{qi}$ one obtains

$$\pi(\Delta q^O) = 0, \text{ for } (q^O \equiv q_i) \notin \Delta q^O \text{ and}$$

$$\pi(\Delta q^O) = \pi(q^O \equiv q_i) = 1, \text{ for } (q^O \equiv q_i) \in \Delta q^O,$$

which is $\mathcal{P}_1(\pi)$. Thus $\mathcal{P}_1(\pi)$ is indeed a partial postulation of the probability law entailed by $\mathcal{P}_3(\pi)$ as the objection $\left(b(1)\right)$ for (1) asserted

(13) $$\mathcal{P}_3(\pi) \subset \mathcal{P}_1(\pi)$$

But in <u>absence</u> of $\mathcal{P}_3(\pi)$, Gleason's demonstration obviously has no kind of implication whatever concerning $\pi(\Delta q^O)$, since intrinsically it has a purely mathematical character, dealing in abstracto with the measure determined by orthogonal projectors on the closed subspaces of a Hilbert space. If this is taken into account together with (11) and (13), it follows that :

The semantico-probabilistic assumption

$$\mathcal{D}(\Psi^S \text{ <-> state of S}) \sim \left\{ \mathcal{P}_3(\pi) \subset \begin{cases} \mathcal{P}_2(\pi) + \mathcal{P}'_1(\pi) \\ \mathcal{P}_1(\pi) \end{cases} \right\}$$

is the <u>necessary</u> (and sufficient) postulational element which <u>has</u> to be adjoined to $\left(O_Q, \Psi\right)$, in order to realize a complete coherent syntactical connection with the quantum mechanical probability law and, more generally, with the quantum mechanical probability space and with the whole "geometry of QM".

The EWGD development remains unaware of this.But implicitely it does introduce $\mathcal{P}_3(\pi)$ in its integrality in the final stage, after having already introduced it partially in the earlier stages by its implications $\mathcal{P}_1(\pi)$, $\mathcal{P}_2(\pi)$ and $\mathcal{P}'_1(\pi)$, making thereby an even redundant use of it. Paradoxically, the source of this redundancy lies precisely in the desire to avoid the assertion of $\mathcal{P}_3(\pi)$ and in the belief that this is realizable by the adjunction to $\left(O_Q, \Psi\right)$ of the "definition" (1) and of the EWGD "interpretation" of the final states in (2) and (3), of which the probabilistic implications $\mathcal{P}_1(\pi)$, $\mathcal{P}_2(\pi)$ and $\mathcal{P}'_1(\pi)$ are rather hidden indeed.

2° From EWGD back to QM

Once the above analysis is closed the EWGD theory ceases to be interesting from my point of view, so that I shall drop it and return to usual QM. For this it suffices to replace the multiple simultaneous reduction postulate $\mathcal{P}(smr)$, by the quantum mechanical reduction postulate $\mathcal{P}_{QM}(r) \subset \mathcal{P}(smr)$. It can be easily seen from what precedes that this has no effect whatever on the structure obtained for the probability space. The effect is localized in the physical interpretation associable with the formalism of QM. Whereas in EWGD, with $\mathcal{P}(smr)$, the postulate $\mathcal{P}'_1(\pi) \supset \mathcal{P}_3(\pi)$ obliges to associate the formalism with a multiply branching physical reality, in usual QM, with $\mathcal{P}_{QM}(r) \subset \mathcal{P}(smr)$, even though the same $\mathcal{P}'_1(\pi) \supset \mathcal{P}_3(\pi)$ is obtained, the formalism indicates a one-branch physical reality.

3° Conclusion and perspectives

Thus the final conclusion is reached that the quantum mechanical probability law cannot be derived from the quantum mechanical descriptive basis (O_Q, Ψ). QM is a formal-semantical structure which can be symbolized by

$$(14) \qquad \left[(O_Q, \Psi) + \left\{ \begin{matrix} \mathcal{P}_3(\pi) \subset \mathcal{P}_2(\pi) + \mathcal{P}'_1(\pi) \\ \mathcal{P}_3(\pi) \subset \mathcal{P}_1(\pi) \end{matrix} \right. \right] \longrightarrow$$

$$\left[(\mathcal{E}(q^0), \mathcal{R}(\Delta q^0), \pi(\Delta q^0))^{(\text{state of } \mathbf{S})} \Longleftrightarrow (\mathcal{S}(\underline{\Phi}_{q_i}), \mathcal{R}(\Delta\underline{\Phi}_{q_i}), \mu(\Delta\underline{\Phi}_{q_i}))^{\mathcal{H}^S} \right].$$

of which a simpler expression is

$$(14') \qquad \left[(O_Q, \Psi) + \mathcal{P}_3(\pi) \right] \rightarrow (\mathcal{E}(q^0), \mathcal{R}(\Delta q^0), \pi(\Delta q^0))^{QM}$$

where the isomorphic spaces from (14) are indistinctly symbolized in the second term of the implication.

As soon as something is modified in $\{\mathcal{P}_3(\pi) \subset (\mathcal{P}_2(\pi) + \mathcal{P}'_1(\pi)$, $\mathcal{P}_3(\pi) \subset \mathcal{P}_1(\pi)\}$ the logical connection between (O_Q, Ψ) and $(\mathcal{E}(q^0), \mathcal{R}(\Delta q^0), \pi(\Delta q^0))^{QM}$ breaks down. This establishes the syntactical liberty to conceive the possibility of a whole class of non-quantum mechanical probability laws, included in non-quantum mechanical probability spaces, which could be knitted with (O_Q, Ψ) by some new convenient independent semantico-probabilistic assumptions.

Obviously however, the pragmatic value of this formal liberty can be decided only by a systematic confrontation of the various physical implications of $\{\mathcal{P}_3(\pi) \subset (\mathcal{P}_2(\pi) + \mathcal{P}'_1(\pi)), \mathcal{P}_3(\pi) \subset \mathcal{P}_1(\pi)\}$, with an a priori totally unrestricted examination of all the conceivable physical processes which could permit to assert that an information has been obtained concerning the "value" of some object-system "quantity" (all the conceivable "measurement processes"). Such an investigation can be now organized on the basis of the knowledge produced by the analysis of the EWGD theory.

This analysis has brought forth a detailed identification

(a) - of the internal syntactical structure of the postulational element $\mathcal{P}_3(\pi)$ which insures the link between (O_Q, Ψ) and $(\mathcal{E}(q^0), \mathcal{R}(\Delta q^0), \pi(\Delta q^0))^{QM}$ (the relations (11) and (13) between $\mathcal{P}_3(\pi)$ and $\mathcal{P}_2(\pi)$, $\mathcal{P}'_1(\pi)$, $\mathcal{P}_1(\pi)$, with their derivation, and the relations (8) and (9) between $\mathcal{P}'_1(\pi)$ and $\mathcal{P}_{QM}(ev)$, $\mathcal{P}_{QM}(r)$, (1)) ;

(b) - of the external syntactical bonds of $\mathcal{P}_3(\pi)$, with (O_Q, Ψ) ($\mathcal{P}_3(\pi) \sim \mathcal{D}(\Psi^S \longleftrightarrow$ state of S) itself, $\mathcal{P}'_1(\pi) \sim \mathcal{D}(O_Q \longleftrightarrow Q)$, the relation (12), and the Gleason connection of $\mathcal{P}_3(\pi)$ with the quantum mechanical form $\Sigma_i\ (T(\Psi^S)\ \Phi_{q_i}, \Phi_{q_i}) = \Sigma_i\ |c_i(\Psi^S)|^2$ of $\pi(\Delta q^0)$) ;

(c) - finally, of the semantical charge of $\mathcal{P}_3(\pi)$ (the assumptions $\mathcal{P}_2(\pi)$, $\mathcal{P}'_1(\pi) \supset (\mathcal{P}_{QM}(r) + (1))$, $\mathcal{P}_1(\pi)$).

This syntactical-semantical explicit knowledge offers now a rather elaborate structure of reference for a study of any conceptual or experimental situation which concerns the values of dynamical quantities of microsystems. In particular, when referred to this structure, the problems of joint probabilities and of "hidden" variables might obtain a clearer restatement entailing a definite answer.

Besides the mentioned informational results, the analysis of the EWGD theory permitted also a first perception of a methodologically important fact : in a probabilistic theory it is fertile to consider the probability space in its whole, as one conceptual entity which is more fundamental than the probability measure contained in it. Indeed, the probability measure alone expresses explicity only certain particular characteristics of a complex configuration which cannot be understood completely without a direct and systematic examination - separately and in relation to one another - of all the three elements which make up the probability space : $\mathcal{P}_2(\pi)$ and $\mathcal{P}'_1(\pi) \sim \{\mathcal{P}_{QM}(ev) \supset (\mathcal{P}_{QM}(r) + (1))\}$ are both vital assumptions which concern <u>directly</u> $\mathcal{E}(q^0) \in (\mathcal{E}(q^0), \mathcal{R}(\Delta q^0), \pi(\Delta q^0))^{QM}$ and the important relation (11) as well as the fundamental demonstration of Gleason, cannot be obtained and understood by consideration of $\pi(\Delta q^0)$ alone. This methodological remark possesses a remarkable power which will become perceptible progressively in the course of a subsequent research (possessing a constructive character).

BIBLIOGRAPHY

1. - H. EVERETT, Rev. Mod. Phys., 29, 454 (1957).

2. - J. WHEELER, Rev. Mod. Phys., 29, 463 (1957).

3. - B. DE WITT, Phys. Today, 23 (n° 9), 30 (1970).

4. - B. DE WITT, in Foundations of Quantum Mechanics, Varenna 1970, Acad. Press Inc., New York (1971).

5. - R. GRAHAM, Ph. D. Thesis, University of North Carolina.

6. - A. GLEASON, J. of Rat. Mech. and Analysis, 6, 885 (1957).

7. - E. WIGNER, in Foundations of Quantum Mechanics, Varenna 1970, Acad. Press. Inc., New York (1971).

8. - H. MARGENAU, The nature of Physical reality, Mc Graw Hill Book Co, New York (1950).

9. - L. BALLENTINE, Rev. Mod. Phys., 42, 358 (1970).

10. - J. ANDRADE E SILVA, Portgal. Phys., 4, 257 (1966).

11. - M. ROBINSON, Can. J. Phys., 47, 963 (1969).

12. - E. WIGNER, Z. Physik, 133, 101 (1952).

13. - H. ARAKI and M. YANASE, Phys. Rev., 129, 940 (1961).

14. - M. YANASE, in Foundations of Quantum Mechanics, Varenna 1970, Acad. Press Inc., New York (1971).

15. - H. STEIN and A. SHIMONY, in Foundations of Quantum Mechanics, Varenna 1970, Acad. Press. Inc., New York (1971).

MEAN ERGODIC SEMIGROUPS AND INVARIANT IDEALS IN ORDERED BANACH SPACES

Rainer J. Nagel

Fachbereich Mathematik der Universität Tübingen

Tübingen, Germany

Mean ergodic semigroups of linear operators in Banach spaces have been studied in [5]. Here, we recall the main results (section 1) and prove the following theorem:

Let S be a mean ergodic semigroup of positive operators in an ordered Banach space E such that $Tu = u$ for all $T \in S$ and some topological unit $u \in E_+$. There exists a canonical bijection between the maximal S-invariant fp-ideals in E and the extreme rays of the positive S'-fixed cone in E'.

This is a generalization to the non-lattice case of a result of S c h a e f e r ([6], where $S = \{T^n : n \in \mathbb{N}\}$, $E = C(X)$) and T a k a h a s h i ([9], where S is amenable and $E = C(X)$), and it yields a characterization of the ergodic states of a C^*-algebra.

1. Mean ergodic semigroups of linear operators

Let E be a Banach space and let S be a semigroup in $\mathcal{L}_s(E)$ the space of all bounded linear operators on E endowed with the strong operator topology. The closed convex hull $\overline{\text{co}}\, S \subset \mathcal{L}_s(E)$ also is a semigroup.

(1.1) DEFINITION: *The semigroup* S *is called mean ergodic if* $\overline{co}\, S$ *contains a zero* P (i.e., *an element* P *such that* $TP = PT = P$ *for all* $T \in \overline{co}\, S$).

In the following we restate briefly the main properties of mean ergodic semigroups and refer the reader to [5] for proofs and a more explicit discussion.

Remarks: 1. The zero P of the mean ergodic semigroup S is a (uniquely determined) projection onto the fixed space F of S in E . The adjoint P' projects E' onto the dual fixed space F' and $(PE)'$ is norm-isomorphic to $P'E'$.

2. For each $x \in E$, $x' \in E'$ we have $Px \in \overline{co}\, Sx = \overline{co}\{Tx: T \in S\}$ and $P'x' \in \overline{co}\, S'x'$.(On E' , we consider always the $\sigma(E',E)$-topology) . Hence, for a mean ergodic semigroup S it follows that $\overline{co}\, Sx$ and $\overline{co}\, S'x'$ always contain a fixed point of S and S' , respectively.

3. The following result of D a y [1] is essential for the applications of the above concept: Let S be a bounded semigroup in $\mathcal{L}_s(E)$. If S is amenable (i.e. if there exists an invariant mean on the space $C_b(S)$ of all bounded continuous functions on S), then $\overline{co}\, S'x' \subset (E',\sigma(E',E))$ and $\overline{co}\, Sx \subset (E'',\sigma(E'',E'))$ contains a fixed point for each $x' \in E'$, resp. $x \in E$. Every abelian semigroup and every compact group is amenable.

(1.2) THEOREM (see [5], 1.2 and 1.7): *For a bounded semigroup* S *in* $\mathcal{L}_s(E)$, *the following conditions are equivalent:*

(a) S *is mean ergodic.*

(b) $\overline{co}\, Sx$ *and* $\overline{co}\, S'x'$ *contains a fixed point of* S *resp.* S' , *for each* $x \in E$ *and each* $x' \in E'$.

If S *is amenable, this is equivalent to:*

(c) *The fixed space* F *of* S *in* E *separates the fixed space* F' *of* S' *in* E' .

(d) $\overline{co}\, Sx$ *contains a* S-*fixed point for each* $x \in E$.

(1.3) COROLLARY (see also [5]): *Each of the following conditions on the space* E *or on the semigroup* S *implies* S *to be mean ergodic:*

(i) E *is a Hilbert space and* S *is contractive* (use 1.2.b).

(ii) S *is amenable and* Sx *is* $\sigma(E,E')$-*relatively compact for all* $x \in E$ (use Remark 3 and 1.2.d).

(iii) S *is a compact group in* $\mathcal{L}_s(E)$ (use 1.3.ii).

(iv) S *is a compact abelian semigroup in* $\mathcal{L}_s(E)$ (use 1.3.ii).

(v) E *is reflexive and* S *is amenable* (use 1.3.ii).

(vi) E *is a Banach lattice with order continuous norm,* S *is a semigroup of positive operators and there exists a topological unit* [1] $u \in E_+$ *and a strictly positive linear form* $\mu \in E'_+$ *such that* $Tu \leq u$ *and* $T'\mu \leq \mu$ *for all* $T \in S$ (see [5]).

(1.4) The ergodic theorem of Kovács - Szücs ([2], see also [8]) has recently been playing a considerable role in mathematical physics. Formulated as follows it fits exactly in our theory.

THEOREM ([2]): *Let* E *be a base norm Banach space such that* E', *as an ordered Banach space, is isomorphic to the self-adjoint part of a* C^*-*algebra* A. *Let* S *be a semigroup of positive contractions on* E *such that* S' *is a group of automorphisms of* A. *If* $Tu = u$ *for all* $T \in S$ *and some topological unit* $u \in E_+$, *then* S *is mean ergodic*.

Questions: 1. If E is an AL-space, the result (1.3.vi) shows that the above assumption on S' may be dropped. Can this be done also in the "non-commutative" case?

2. What are the base norm spaces for which a mean ergodic theorem of the above type holds?

2. Ideals in ordered Banach spaces

In this section we recall briefly some of the results of [4]. To this end we assume that E is an ordered Banach space with closed and generating positive cone E_+.

---

1) $u \in E_+$ is a topological unit (or quasi-interior point) if $\bigcup_{n\in\mathbb{N}} n[-u,u]$ is dense in E.

(2.1) DEFINITION: A closed subspace I of E is called an fp-ideal if one of the following equivalent conditions is satisfied:

(a) $I = \underline{H}(I) := \{x \in E : \forall\, n \in \mathbb{N} \quad \exists\, \mu, \nu \in E$ and $y \in I$ such that $\|\mu\|, \|\nu\| \leq n^{-1}$ and $-\mu - y \leq x \leq \nu + y\}$.

(b) $I^o \cap E'_+$ is a face of E'_+ and $\sigma(E',E)$-total in I^o.

The equivalence of (a) and (b) is proved in [4], from which we also quote the following two propositions (see [4], 3.2 and 3.3).

(2.2) PROPOSITION: If the dimension of E is greater than one, then there exists a fp-ideal I such that $o \neq I \neq E$.

(2.3) PROPOSITION: A subspace I of E is a maximal fp-ideal if and only if $I = f^{-1}(o)$ where f generates an extreme ray of E'_+.

(2.4) Finally, the following property will be needed in the proof of our main theorem in section 3 .

LEMMA: Let $T \in \mathcal{L}(E)$ be a positive operator. If I is an fp-ideal in E , then $T^{-1}I$ contains a uniquely determined greatest fp-ideal.

Proof: $T^{-1}I$ is a full ideal in E (see [4], 2.1) and $\underline{H}(o)$ is contained in $T^{-1}I$. By a maximality argument one proves the existence of maximal fp-ideals contained in $T^{-1}I$ (using [4], lemma 2.4). With the aid of [4], 2.6.iii , one shows uniqueness.

Examples: 1. If E is a Banach lattice, the fp-ideals are exactly the closed lattice ideals.

2. If E is the subspace of all self-adjoint elements of a C^*-algebra A , then there is a one-to-one correspondence between the fp-ideals in E and the closed left (or right) ideals in A .

3. Let E be an order unit space and let I be a closed subspace of E . I is an fp-ideal iff $I^o \cap K$, K the state space of E , is a face of K . In particular: I is a maximal fp-ideal iff $I = f^{-1}(o)$ for f an extreme point of K .

3. Invariant ideals

Our aim in this section is to generalize Theorem 2 of [6] and Theorem 4 of [9] to the case where E is not necessarily a space C(X) or even a Banach lattice. Hence, we assume $E = E_+ - E_+$ to be an ordered Banach space. Let S be a bounded semigroup of positive continuous linear operators on E . For the following definition compare [6].

(3.1) DEFINITION: A fp-ideal I in E is called an S-ideal if $TI \subset I$ for each $T \in S$. An S-ideal I is maximal if J = E for each S-ideal J properly containing I . Finally, S is called irreducible if (o) is a maximal S-ideal.

(3.2) The following theorem is motivated by the theory of "ergodic measures". It is due to Schaefer [6] for the case S = $= \{T^n : n \in \mathbb{N}\}$, $T \in \mathcal{L}(E)$ and E = C(X) , to Takahashi [9] for S an amenable semigroup and E = C(X) , and to Lotz (unpublished) for an arbitrary semigroup S and a Banach lattice E .

THEOREM: Let E be an ordered Banach space and let $S \subset \mathcal{L}_s(E)$ be a mean ergodic semigroup of positive operators with the corresponding projection $P > o$. Assume $Pu = u$ for one topological unit $u \in E_+$. Then

$$I \longrightarrow I^{o} \cap P'E'_+$$

is a bijection between the maximal S-ideals and the extreme rays of the dual positive fixed cone $P'E'_+$.

LEMMA: (PE)' is (norm and) order isomorphic to P'E' .

Proof: In view of 1.1, remark 1, the norm isomorphism is readily established. Concerning the order isomorphism we have to show that every $x' \in (PE)'$ which is positive on PE , has a positive continuous extension to all of E : Let U be the unit ball in E and take $x \in E$ such that $Px \leq y \in U$. Then $Px \leq Py$ and

$$\langle Px, x' \rangle \leq \langle Py, x' \rangle \leq \|P\| \, \|x'\| \quad .$$

Now apply Bauer's extension theorem ([7], V.5.4) .

<u>Proof of the theorem</u>: We show that I is a maximal S-ideal if and only if $PI = I \cap PE$ is a maximal fp-ideal in the ordered Banach space PE.

Let J be a maximal fp-ideal in PE and define k(J) to be the (uniquely determined) greatest fp-ideal contained in $P^{-1}J$ (apply 2.4). k(J) is a proper fp-ideal since $Pu = u \notin J$.

k(J) is an S-ideal: For $T \in S$ we get

$$T(k(J)) \subset \underline{H}(T(kJ))) \subset P^{-1}J \ .$$

We conclude $T(k(J)) \subset k(J)$ by [4], 2.6.iii and the maximality of k(J).

$k(J) \cap PE = J$: One inclusion is trivial. For the other take $x = Px \in k(J)$. Then $x \in P(k(J))$ and $x \in \underline{H}(J) = J$.

k(J) is maximal: Let L be an S-ideal such that $L \cap PE = J$. Then $x \in L$ implies $Px \in J = \underline{H}(J)$, hence $x \in k(J)$.

Now the theorem follows from the above lemma and from the proposition 2.3.

(3.3) COROLLARY: <u>Let</u> E <u>and</u> S <u>be as in the above theorem and assume that every non-trivial</u> fp-<u>ideal in</u> E <u>contains non-zero positive elements</u> (e.g., Simplex spaces, C^*-algebras). S <u>is irreducible iff the fixed space</u> PE <u>is</u> 1-<u>dimensional and</u> P <u>is strictly positive</u>.

<u>Proof</u>: One implication is trivial. For the other assume $\{o\}$ to be a maximal S-ideal in E. Then $\{o\}$ is also a maximal fp-ideal in the ordered Banach space PE. By 2.2, we conclude that the dimension of PE is one. The remainder is obvious.

(3.4) Finally, let E be an orderunit Banach space with unit u. Let $S \subset \mathcal{L}_s(E)$ be a semigroup of positive operators with $Tu = u$ for all $T \in S$. The extreme points of the invariant states

$$K_S := \{ f \in E'_+ : f(u) = 1 \text{ and } T'f = f \text{ for all } T \in S \}$$

are called the <u>ergodic states</u>. The application of the following corollary to the case of a C^*-algebra may be left to the reader (e.g. see [3]).

COROLLARY: *Let* E *be an order unit Banach space and* $S \subset \mathcal{L}_s(E)$ *a mean ergodic semigroup of positive operators such that* $Tu = u$ *for all* $T \in S$. *The maximal* S-*ideals in* E *correspond bijectively to the ergodic states.*

references

[1] Day, M.M.: Fixed point theorems for compact convex sets. Ill. Journ.Math. 5, 585-59o (1961).

[2] Kovács, I.; Szücs, J.: Ergodic type theorems in von Neumann algebras. Acta Sci.Math. 27, 233-246 (1966).

[3] Nagel, Bengt: Some results in non-commutative ergodic theory. Commun.math.Phys. 26, 247-258 (1972).

[4] Nagel, R.J.: Ideals in ordered locally convex spaces. Math.Scand. 29, 259-271 (1971).

[5] Nagel, R.J.: Mittelergodische Halbgruppen linearer Operatoren. to appear in: Annales Inst. Fourier.

[6] Schaefer, H.H.: Invariant ideals of positive operators in C(X), I. Ill.Journ.Math. 11, 7o3-715 (1967).

[7] Schaefer, H.H.: Topological Vector Spaces, 3rd print. Berlin-Heidelberg-New York: Springer 1971.

[8] Størmer, E.: Asymptotically abelian systems. Cargese lectures in physics, 4 (197o).

[9] Takahashi, W.: Invariant ideals for amenable semigroups of Markov operators. Kodai Math.Sem.Rep. 23, 121-126 (1971).

THE REPRESENTATION OF CLASSICAL SYSTEMS IN QUANTUM MECHANICS

Holger Neumann
Fachbereich Physik der Universität Marburg
Marburg, Germany

In the frame of an axiomatic foundation of quantum mechanics classical statistical theories and quantum theory can be treated on an equal footing. It is assumed that the physical system can be described by a base normed Banach space B and its dual B' satisfying axioms V1a,b, V2 and VId of [1] . A system is called classical if the lattice G of decision effects is Boolean. In this case B and B' are vector lattices. B is a complete separable L-space (see [2] , [3]). A bipositive linear isometry onto a space $L^1(S,\Sigma,\mu)$ of a finite measure space (S,Σ,μ) is called a representation of B. A representation of B provides a σ-isomorphism of G onto the quotient algebra $\Sigma/\mathfrak{J}_\mu$ where $\mathfrak{J}_\mu$ denotes the σ-ideal of μ-measure zero. On the other hand every σ-homomorphism φ of a Boolean σ-algebra Σ of subsets of a set S onto G determines a representation of B by $L^1(S,\Sigma,\mu)$, where μ is an appropriate measure vanishing on the cernel of φ .

The complete separable L-space B has a standard representation in an disconnected compact space $S_{B''}$, the state space of B', where $\Sigma_{B''}$ is the Boolean σ-algebra of Borel sets of $S_{B''}$. Since the topology on $S_{B''}$ is rather pathological this standard representation of B is certainly an "unphysical" one.

We shall show that the axiomatic approach of Ludwig sketched in [1] suggests a physically distinguished topological representation space. According to [1] the set L of effects contains a countable weakly dense subset $\underline{L}$. The norm closed linear hull of $\underline{L}$ has the properties:

$$D = \overline{\mathrm{lin}}^n \underline{L} \text{ is separable, } 1 \in D, \ \overline{L \cap D}^\sigma = L, \qquad (1)$$

where $\overline{L \cap D}^\sigma$ is the weak closure of $L \cap D$. If the subscript u denotes the unit sphere the equation $\overline{L \cap D}^\sigma = L$ is equivalent to $\overline{D}_u^\sigma = B'_u$. The set $L \cap D$ can be considered as the set of physically relevant effects.

To point out the physical meaning of D let $\sigma(K,\underline{L})$ denote the initial topology generated by the set $\underline{L}$ of real valued mappings on K. $\sigma(K,\underline{L})$ serves to describe the inaccuracy of the correspondence between reality and the set K of ensembles. It is not hard to prove

$\sigma(K,\underline{L}) = \sigma(K,D\cap L) = \sigma(K,D)$.

In order to see that $\underline{L}$ or equivalently D determines a distinguished representation of B consider the norm closed vector lattice D_ℓ generated by $\underline{L}$. D_ℓ also satisfies the statements (1) and is a separable Banach lattice with an order unit norm. Hence D_ℓ has a representation by the space of continuous functions on its state space S [2]. S is the set of extreme points of the base $K_{D_\ell'}$ of D_ℓ' equipped with the weak topology $\sigma(S,D_\ell)$. The separability of D_ℓ is equivalent to the metrizability of S. D_ℓ' can be identified with the space of all signed Borel measures on S.

Consider now the embedding $D_\ell \to B'$. The adjoint mapping is the canonical surjection $B'' \to D' = B''/D_\ell^\perp$. Since B can be identified with a subspace of B" there is defined a mapping $k:B \to D'$ and the following lemma can be proved [4]:

Lemma. The mapping $k:B \to D_\ell'$ is an isometric lattice isomorphism of B into D_ℓ' such that k(B) is a band in D_ℓ'.

Corollary. If D_ℓ' is identified with the space of all regular signed Borel measures on the state space S of D_ℓ there is a finite Borel measure μ such that $k(B) = L^1(S,\Sigma,\mu)$.

This corollary shows that the subset $\underline{L}$ of B' or equivalently D determine a representation of B in a metrizable compact space S. While $B' = L^\infty(S,\Sigma,\mu)$ the subsets $\underline{L}$ and D of B' generate the sublattice of continuous functions on S. We emphasize that the topology on S is independent of generating D_ℓ by $\underline{L}$.

Lemma. $\sigma(S,\underline{L}) = \sigma(S,D) = \sigma(S,D_\ell)$

Proof. It is sufficient to show that all functions of D_ℓ are $\sigma(S,\underline{L})$ continuous. However, the set of all $\sigma(S,\underline{L})$-continuous functions is a norm closed vector lattice containing $\underline{L}$.

The construction of D_ℓ is only necessary to find all points of S. As the example of the phase space will show it can be even of advantage not to consider all points of S.

In the sequel we shall discuss the representation of the classical system associated with the phase space of an n-particle system. Before entering into the discussion of this example it is necessary to give some general remarks on group representations by isomorphisms of B or B B and B' describing a not necessarily classical system. A particular transformation property of D will be exhibited.

Consider a locally compact group H. We assume for simplicity that the left and right uniform structures coincide. This uniform structure will be denoted by $\mathfrak{u}$. A physically meaningful metric on H should decrease when a pair of group elements is translated to infinity. According to this idea it will be assumed that there is another uniform structure $\mathfrak{u}_p < \mathfrak{u}$ on H which is suggested by physics and generates the same topology as $\mathfrak{u}$. As an example consider the 3-dimensional translation group T. A uniform structure $\mathfrak{u}_p$ is introduced on T according to the idea just explained if the one-point-compactification $\hat{T}$ of T is constructed and $\mathfrak{u}_p$ is defined to be the unique uniform structure induced by $\hat{T}$ on T. Then elements outside of circles centered at zero are collected into neighbourhoods of infinity.

Consider now a representation of H by mixture isomorphisms, i.e. order and norm preserving isomorphisms of B. These isomorphisms will be denot by the same symbols as the group elements. If one considers $L \cap D$ as the set of physically relevant effects the following assumption concerning the representation of H is reasonable:

The probability $\mu(gv,f)$ is $\mathfrak{u}_p$-uniformly continuous on H for all $v \in K$, $f \in L \cap D$. $L \cap D$ is invariant under the adjoint transformation g' for all $g \in H$.

(This assumption remains valid if K is replaced by $\underline{K}$ and $L \cap D$ is replaced by $\underline{L}$.)

This assumption together with the separability of B and the properties (1) of D imply [5] :

The mappings $H \longrightarrow B$ defined by $g \longmapsto gx$ are norm continuous for all $x \in B$. (2)

The mappings $H \longrightarrow D$ defined by $g \longmapsto g'y$ are norm continuous for all $y \in D$.

(2) corresponds to the usual continuity assumption for a group representation. If we define

$\Delta = \{ y \in B' \,/\, \mu(gx, y)$ is u_p-uniformly continuous for all $x \in B$ and $g \longmapsto g'y$ defines a norm continuous mapping $H \longrightarrow B' \}$

then Δ is an invariant norm closed linear subspace of B' containing D.

In concrete examples of spaces B and B' no subspace D is given. D is only known to exist on axiomatical grounds. If in an example there is a group representation in B such that Δ has the properties (1) one may put $D = \Delta$.

In the most important example of an irreducible quantum mechanical system, where B and B' are the well known operator spaces in a Hilbert space and H is the Galileian group without time translation, attempts to give a handy characterization of Δ failed so far.

We shall discuss now the example of the phase space Γ of an n-particle system: $B = L^1(\Gamma, \Sigma_\Gamma, \mu_L)$ and $B' = L^\infty(\Gamma, \Sigma_\Gamma, \mu_L)$ where μ_L denotes the Lebesgue measure on the Borel algebra Σ_Γ of Γ . We choose the physical group H to be the n-fold product of the Galileian group without time translation $H = \prod_{i=1}^{n} H_o^i$. An element $g_i \in H_o^i$ is of the form

$g_i = (R_i, a_i, v_i)$ where R_i denotes a rotation in Euklidean 3-space, a_i a space translation and v_i a velocity translation. A group element $g \in H$ acts on $s \in \Gamma$ such that $g_i \in H_o^i$ acts on the phase coordinates of the i-th particle as follows: $g_i(q_i, p_i) = (R_i q_i + a_i, R_i p_i + m_i v_i)$ where m_i is the mass of the i-th particle. A representation of H by mixture isomorphisms of B is defined by $gx(s) = x(g^{-1}s)$ where $s \in \Gamma$, $x \in B$.

The aim of our reasoning is to show that the representation of B in Γ corresponds to a representation constructed according to the methods developed in the beginning if $D = \Delta$ is chosen and Δ is determined by this group representation in B.

At first we have to define the uniform structure u_p on H. H_o^i is the topological product $H_o^i = SO(3) \times T \times T$. We define $\hat{H}_o^i = SO(3) \times \hat{T} \times \hat{T}$ where $\hat{T}$ is the one-point-compactification of T discussed above. Let $\hat{H} = \prod_{i=1}^{n} \hat{H}_o^i$ and the uniform structure u_p on H is defined to be the structure induced by the unique uniform structure on $\hat{H}$.

We shall define a corresponding compactification and uniform structure of Γ . Let $\hat{\Gamma}$ be the compactification of Γ achieved by the one-pointcompactifications of the space coordinates and the momentum coordinates of all particles. Thus $\hat{\Gamma}$ is homeomorphic to $\prod_{k=1}^{2n} \hat{T}_k$. The uniform structure induced by $\hat{\Gamma}$ on Γ will be denoted by $\mathcal{U}_{\hat{\Gamma}}$.

The space $C(\hat{\Gamma})$ of continuous functions on $\hat{\Gamma}$ can be identified with the space of $\mathcal{U}_{\hat{\Gamma}}$-uniformly continuous functions on Γ. Then the following theorem can be proved [4] :

Theorem. For $y \in B'$ we have $y \in \Delta$ if and only if there is a $\mathcal{U}_{\hat{\Gamma}}$-uniformly continuous function in the equivalence class of bounded measurable functions determined by y. In short $\Delta = C(\hat{\Gamma})$.

Since Δ contains the continuous functions with compact support $\overline{\Delta}_u^{\sigma} = B'_u$. Thus Δ has the properties (1) required for D and we may put $\Delta = D$. The state space S of $D_\ell = D = C(\hat{\Gamma})$ consists of the point measures on $\hat{\Gamma}$ and is homeomorphic to $\hat{\Gamma}$. Since D is invariant with respect to the group representation of H in B' a representation of H by homeomorphisms in S is defined.

According to the corollary of the first lemma there is a representation of B by a space $L^1(S,\Sigma,\mu)$, μ being a finite Borel measure on the Borel algebra Σ of S. The group representation of H in S determines an open subset $S' \subset S$ on which H acts transitively and the complement of which is of μ-measure zero. $L^1(S,\Sigma,\mu)$ has a natural restriction to the locally compact space S' and B is represented by $L^1(S',\Sigma',\mu')$. The above mentioned homeomorphism $\varphi : \hat{\Gamma} \longrightarrow S$ maps Γ onto S' and induces an isomorphism of $L^1(\Gamma,\Sigma_L,\mu_L)$ onto $L^1(S',\Sigma',\mu')$.

Thus the representation of the classical system in the phase space is obtained by constructing a representation by means of a subspace $D \subset B'$ along the lines developed in the beginning.

References

[1] H. Neumann, The Structure of Ordered Banach Spaces in Axiomatic Quantum Mechanics, this volume

[2] A. Goullet de Rugy, Representation of Banach Lattices, this volume.

[3] H.H. Schaefer, Topological Vector Spaces, New York, Heidelberg, Berlin 1971.

[4] H. Neumann, Classical Systems in Quantum Mechanical Spaces and Their Representations in Topological Spaces. Notes in Math.Phys. 10, Marburg (1972).

[5] G. Ludwig, Darstellung von Gruppen und Halbgruppen durch Transformationen von Gesamtheiten und Effekten, unpublished.

EXTENDED HILBERT SPACE FORMULATION OF DIRAC'S BRA AND KET FORMALISM AND ITS APPLICATIONS TO ABSTRACT STATIONARY SCATTERING THEORY

Eduard Prugovečki

Department of Mathematics
University of Toronto
Toronto, Canada

I. INTRODUCTION

The bra and ket formalism, in its original heuristic form introduced by Dirac [3], has become since its inception a standard tool of the physicist. In the last ten years there have been several attempts of providing a rigorous basis for this formalism (for a review see [1]). These attempts rely heavily on concepts first developed within the context of the theory of distributions. This theory necessitates the introduction of the concept of derivative of a distribution which in its turn dictates the choice of the space of test-functions and the topology on that space. Since Dirac's δ-symbol was first introduced in the context of quantum mechanics and later on was interpreted by L. Schwartz as a distribution, it would seem on first glance that topological vector spaces originating from the theory of distributions (such as rigged Hilbert spaces) would be the natural choice in the search for spaces of bra and ket vectors.

Let us ignore, however, this historical development and look instead at those problems in quantum mechanics where the bra and ket formalism seems most deeply entrenched and quite indispensible. Such problems appear in scattering theory and in quantum field theory. One of their outstanding mathematical features is that they are perturbation problems in the general sense of the word, namely in that all the important aspects of the problem are easy to understand and basic quantities easy to compute for an "unperturbed" free case and the task is to compute corresponding "perturbed" quantities. For such quantities as eigenfunction expansions or other similar non-Hilbert space objects the need of having derivatives to all orders rarely occurs in computations, if at all. Hence, the topologies introduced with differential equations in mind do not represent an asset. In fact, since they have not been designed with the specific problem under consideration in mind, they turn out to be a handicap, due to their relative intricacy and the unnatural restrictions they impose on basic quantities on the problem (e.g., limitations of the theory to everywhere infinitely many times differentiable potentials under which the Schwartz $\mathcal{S}$-space is stable).

In view of these remarks it should be clear that an approach in which the mathematics is chosen to fit the problem rather than trying to fit the problem to the mathematics by imposing artificial restrictions might have a better chance of success when attempting to solve realistic problems in quantum physics. In the search for suitable mathematics simplicity in the mathematical structure should be strived after rather than avoided. The formalism based on the concept of equipped Hilbert space developed by Berezenskiĭ [2] certainly meets the simplicity requirement. However, since the problems in scattering theory are of a different nature than those treated by Berezanskiĭ, a different approach to this formalism is desirable, as will be pointed out in Sec. II. This will make apparent the need for further expanding the scope of the equipped Hilbert space formalism. Thus we are led naturally to a structure which we call an extended Hilbert space and introduce in Sec. III. Some typical applications of the resulting formalism are then presented in the last section. Let us point out, however, that we do not attach any direct physical significance to the non-Hilbert space objects in this formalism since this might be in conflict with certain basic properties of quantum mechanical phenomena [4].

II. EQUIPPED HILBERT SPACE

We base the construction of equipped Hilbert space on the notion of equipping operator in a separable Hilbert space $\mathcal{H}$. An operator D in $\mathcal{H}$ is an equipping operator iff D and D* are densely defined and their inverses D^{-1} and D^{*-1}, respectively, are Hilbert-Schmidt operators of bound one.

In the sequel $\langle\cdot|\cdot\rangle$ denotes the inner product in the complex separable Hilbert space $\mathcal{H}$ chosen to be linear in the right variable and antilinear in the left variable, and $\|\cdot\|$ the corresponding norm.

For any given equipping operator D the sesquilinear form $\langle\cdot|\cdot\rangle_D$

(1) $$\langle f|g\rangle_D = \langle Df|Dg\rangle , \qquad f, g \in \mathcal{D}(D)$$

is positive definite on $\mathcal{D}(D) \times \mathcal{D}(D)$, and actually the domain of definition $\mathcal{D}(D)$ of D can be made [5] into a Hilbert space $\mathcal{K}_D$ (called the space with positive norm corresponding to D) if the above form is chosen as an inner product. On the other hand, the completion of $\mathcal{H}$ with respect to the positive-definite sesquilinear form $(\cdot|\cdot)_D$

(2) $$(f|g)_D = \langle D^{*-1}f|D^{*-1}g\rangle , \qquad f, g \in \mathcal{H} ,$$

is a Hilbert space $\mathcal{K}_D^{\dagger}$, called the space with negative norm corresponding to D. Furthermore, D* can be extended uniquely to a unitary transformation $\mathbb{D}$ of H on $K_D^\dagger$ [4].

The above terminology [2] reflects the fact that due to the conditions imposed on the bound of D^{-1} we have

$$\|f\|_D = \sqrt{\langle f|f\rangle_D} \leq \|f\| , \quad f \in K_D , \tag{3}$$

$$\|f\| \leq \|f\|_{\mathbb{D}^{-1}} = \sqrt{(f|f)_D} , \quad f \in H . \tag{4}$$

Since K_D is actually a subset of H and H is embedded in $K_D^\dagger$, we are dealing with an ordered triple $K_D \subset H \subset K_D^\dagger$ called an equipped Hilbert space. The Hilbert-Schmidt property of D implies that the embedding of K_D in H is quasi-nuclear. This feature is actually not sesential to the above construction. Its importance becomes apparent, however, as soon as generalized expansions are considered since it represents a necessary and sufficient condition for the existence of such an expansion corresponding to an arbitrary set of commuting operators [2].

Lemma 1. There is a unique sesquilinear form $(\cdot|\cdot\rangle_D$ on $K_D^\dagger \times K_D$ continuous in the respective norm topologies of $K_D^\dagger$ and K_D and such that $(f|g\rangle_D = \langle f|g\rangle$, $(\mathbb{D} f|g\rangle_D = \langle f|Dg\rangle$ for all $f \in K_D$ and all $g \in K_D$. The mapping $M : K_D^\dagger \to K_D'$ which takes $f \in K_D^\dagger$ into the functional $(f|\cdot\rangle_D$ is a bijective strongly continuous antilinear mapping between $K_D^\dagger$ and the dual K_D' of K_D.

The first part of this lemma easily follows from the observation that $\langle f|g\rangle = \langle D^{-1*} f|Dg\rangle$, $f \in H$, $g \in K_D$, while the second part is a straightforward consequence of Riesz' theorem [5].

It is important to note that although $K_D^\dagger$ can be identified with the dual of K_D as in the case of rigged Hilbert spaces, the construction of $K_D^\dagger$ does not rely on duality. This opens the possibility of a more general construction of extensions $A^\dagger$ of adjoints A^* of operators in H than the one based on duality.

We introduce now the concept of generalized expansions [2] for a set of commuting operators. Since in practice expansions corresponding to a complete set of observables are of almost exclusive importance, we limit our considerations to this case. In this context a set $\{A_1,\ldots,A_\nu\}$ of self-adjoint commuting operators in H is said to be complete if there it has a cyclic vector [6]. By a generalized expansion $\{\phi(\alpha)\}$ of a complete set $\{A_1,\ldots,A_\nu\}$ with respect to some measure $\rho(\Delta)$ having support on the Cartesian product $\sigma(A_1)\times\ldots\times(A_\nu)$ of the spectra of $A_1,\ldots,A_\nu$ we mean a family of vectors $\phi(\alpha) \in K_D^\dagger$ defined ρ-almost everywhere in $\mathbb{R}^\nu$ for which

$$(5)\qquad \langle f|E^{A_1,\ldots,A_\nu}(\Delta)g\rangle = \int_\Delta {}_D\langle f|\phi(\alpha))\,(\phi(\alpha)|g\rangle_D \, d\rho(\alpha)$$

for all $f, g \in K_D$. In the above relation $E^{A_1,\ldots,A_\nu}(\Delta)$ is the joint spectral measure of $A_1,\ldots, A_\nu$ and we define ${}_D\langle f|g) = \overline{(g|f\rangle_D}$. We note that at the present stage we do not require $\phi(\alpha)$ to be eigenvectors of some "generalized" adjoint $A_j^\dagger$ of A_j since this would restrict the scope of the formalism by artificial requests such as stability [1] of K_D under A_j, etc. We shall see in Sec. III that such a condition becomes a consequence of (7) if we regard $K_D^\dagger$ as being in some sense a part of a larger bra-space B obtained by "fusing together" spaces $K_D^\dagger$ corresponding to different equipping operators D.

Theorem 1. For every complete set $\{A_1,\ldots,A_\nu\}$ there is a generalized expansion $\phi(\alpha)$ with respect to the measure

$$(6)\qquad \rho_D(\Delta) = \mathrm{Tr}\{D^{-1*}E^{A_1,\ldots,A_\nu}(\Delta)D^{-1}\} \quad .$$

Proof: The existence of a cyclic vector for $\{A_1,\ldots,A_\nu\}$ implies the existence of a spectral representation space $L^2_\mu(\mathbb{R}^\nu)$ for these operators [6]. This means that there is a unitary transformation $\hat{U}$ of H onto $L^2_\mu(\mathbb{R}^\nu)$ for which $\left(\hat{U}E^{A_1,\ldots,A_\nu}(\Delta)f\right)(\alpha) = \chi_\Delta(\alpha)(\hat{U}f)(\alpha)$, $\forall f \in H$. The operator $\hat{U}D^{-1}\hat{U}^{-1}$ is an integral operator on $L^2_\mu(\mathbb{R}^\nu)$ with a Hilbert-Schmidt kernel $K_{D^{-1}}(\alpha,\beta)$ and consequently $\hat{\psi}_\alpha(\beta) = K_{D^{-1}}(\alpha,\beta)$ belongs to $L^2_\mu(\mathbb{R}^\nu)$ for μ-almost all $\alpha \in \mathbb{R}^\nu$. It is easy to verify [5,7] that the measures μ and ρ_D are equivalent and that the family $\left\{\phi(\alpha) = \frac{d\rho_D(\alpha)}{d\mu(\alpha)}\, \mathbb{D}\,\hat{U}^{-1}\,\hat{\psi}_\alpha\right\}$ satisfies (5) with $\rho \equiv \rho_D$.

By taking advantage of the existence of a cyclic vector for $\{A_1,\ldots,A_\nu\}$ and the fact that

$$(7)\qquad \langle f|g\rangle = \int_{\mathbb{R}^\nu} {}_D\langle f|\phi(\alpha))\,(\phi(\alpha)|g\rangle_D \, d\rho_D(\alpha)$$

we easily establish the following useful result [5].

Lemma 2. The mapping $U_D : f \to \hat{f}(\alpha) = (\phi(\alpha)|f\rangle_D$ extends to a unique unitary transformation of H onto $L^2_{\rho_D}(\mathbb{R}^\nu)$.

It should be emphasized at this point that the elements of $K_D^\dagger$ are equivalence classes of sequences $\{f_n\}$ in H which are D^*-convergent in the sense that $D^{*-1}f_n$ is a Cauchy sequence in the norm-topology of H ; the equivalence relation $\{f_n\} \sim \{g_n\}$ between two sequences holds iff $D^{*-1}(f_n - g_n) \to \mathbb{0}$. Hence, if H is an $L^2_\mu(\mathbb{R}^k)$ space of functions, this is not true of $K^\dagger_D$. However, an obvious identification of at least some elements of $K^\dagger_D$ with functions can be achieved if the space P_D of Borel-measurable functions $f(x)$ for which $\underset{\Delta \to \mathbb{R}^k}{\text{l.i.m.}} \int K(x,y) f(y) d\mu(y)$ exists is introduced, where $K(x,y)$ is the kernel of the operator D^{*-1} represented as an integral operator on $L^2_\mu(\mathbb{R}^k)$. This explains the existence of such objects as free and distorted plane waves, spherical waves, etc. as representations of generalized eigenvectors of different complete sets of operators with non-empty continuous spectra. In fact, it should be realized that P_D contains many additional functions besides those in $L^2_\mu(\mathbb{R}^k)$ For example, witness the case of $L^2(\mathbb{R}^3)$ with the choice $D = (-\nabla^2 + 1)\exp(\vec{x}^2)$ for which $K(\vec{x},\vec{y}) = \frac{1}{4\pi} |\vec{x}-\vec{y}|^{-1} \exp\{-\vec{x}^2 - |\vec{x}-\vec{y}|\}$. In this case P_D contains functions which are polynomially divergent at infinity.

III. EXTENDED HILBERT SPACE

From the point of view of applications it is important to be able to extend the adjoint A^* of an operator A in a unique manner to the "bra-space" (in our case to $K^\dagger_D$) with a minimum of restrictions on A . The most evident procedure is to define the generalized adjoint $A^\dagger$ of A on $K^\dagger_D$ by the relation

$$(8) \qquad (A^\dagger f | g\rangle_D = (f | Ag\rangle_D , \qquad \forall g \in D(A) \cap K_D .$$

If A maps K_D into itself and is bounded in the norm-topology of K_D then $(f|A(\cdot)\rangle_D$ is a continuous linear functional on K_D , and consequently we can take advantage of the existence of the mapping M in Lemma 1 to infer the existence of a unique $f^\dagger \in K^\dagger_D$ such that $(f^\dagger|\cdot\rangle_D = (f|A(\cdot)\rangle_D$. We can define $A^\dagger$ by setting $f^\dagger = A^\dagger f$.

The above procedure is essentially the same as the one used in the rigged Hilbert space approach [1]. From a practical point of view, the conditions of stability and boundedness in K_D imposed on A limit the scope of the construction of $A^\dagger$ and can be difficult to verify even for the rather restricted class of operators for which they are satisfied. It is therefore conducive to a more general approach to realize that $A^\dagger$ in the above construction is an extension to $K^\dagger_D$ of A^* in H , and that actually $f^\dagger = A^\dagger f$ is represented by the equivalence class of D^*-convergent sequences $\{A^* f_n\}$ if $f \in K^\dagger_D$ is represented by sequences $\{f_n\}$. This suggests that even if

$\{A^*f_n\}$ is not D*-convergent, it might be D'*-convergent for some other equipping operator $D' \neq D$. For example, if A is bounded and has a bounded inverse, then the choice $D' = \|(DA)^{-1}\|(DA)$ would be suitable, since it is easy to see that under these conditions D' is also an equipping operator.

The above argument indicates that it might be profitable to look upon $A^\dagger$ as a mapping of $K_D^\dagger$ into $K_{D'}^\dagger$. Difficulties are encountered, however, when we are dealing with families of parameter-dependent operators A_s and we have to investigate the limit of $A_s^\dagger$ as s approaches a certain value since in this case $K_{D'_s}^\dagger$ changes with s . Another difficulty occurs if A^{-1} does not exist or is not bounded. Then we could apply the same procedure to $A - \zeta\mathbb{1}$ instead, with ζ in the resolvent set of A , and define $A^\dagger$ by

(9) $$A^\dagger = (A-\zeta\mathbb{1})^\dagger + \bar{\zeta}\,\mathbb{1}^\dagger$$

except that we are faced with the problem of $(A-\zeta\mathbb{1}^\dagger)$ having range in $K_{D'}^\dagger$, where $D' = \|[D(A-\zeta\mathbb{1})]^{-1}\|D(A-\zeta\mathbb{1})$ while $\mathbb{1}^\dagger$ is the identity in $K_D^\dagger$, so that (9) is undefined. These and other similar reasons indicate the desirability of having different spaces $K_D^\dagger$ somehow fitted together in some kind of a superstructure.

The solution to this problem is to consider the family B of all D-convergent sequences $\{f_n\}$ corresponding to a given family E of equipping operators in which two sequences $\{f_n\}$ and $\{f'_n\}$ are considered equivalent iff $\|f_n - f'_n\| \to 0$ (it is trivial to verify that we are indeed dealing with an equivalence relationship). We shall call the resulting set of equivalence classes the <u>bra-space over E corresponding to H</u> . We embed H in B by identifying each element f in H with the element of B containing the sequence $\{f_n \equiv f\}$, thus obtaining a subset K of B which we call the <u>ket-space corresponding to H</u> . The ordered pair $K \subset B$ will be called <u>an extended Hilbert space</u>. Multiplication of $\{f_n\} \in B$ by $a \in \mathbb{C}$ is simply defined as $\{af_n\}$. Furthermore, $\{h_n\} \in B$ will be said to be the sum of $\{f_n\}$ and $\{g_n\}$ in B iff $\|h_n - (f_n+g_n)\| \to 0$ as $n \to \infty$.

The bra-space B is in general not linear since $\{f_n+g_n\}$ might not be D-convergent for any $D \in E$ when $\{f_n\}$ is D_1 convergent and $\{g_n\}$ is D_2-convergent for two distinct $D_1, D_2 \in E$. On the other hand B consists of linear spaces B_D , $D \in E$, which can be defined as being the domains of definition of the linear mappings $b_D : \{f_n\} \to \text{s-lim}\, D^{-1} f_n$ from B to K (since $D^{-1} f_n \in K$ and K is identified with H , "s-lim" stands for the limit in the norm - topology of H) .

The topology in $\mathcal{B}$ for which the family of all sets

(10) $\mathcal{V}(f;D_1,\ldots,D_n;\varepsilon) = \left\{g \mid \|b_{D_i}(f-g)\| < \varepsilon, \quad i=1,\ldots,n\right\}$

corresponding to all choices of $f \in \mathcal{B}$, $\varepsilon > 0$ and $D_1,\ldots,D_n \in \mathcal{E}$, $n=1,2,\ldots$, is a basis for the topology is called the <u>strong topology in $\mathcal{B}$</u> [note that in (10) for any given $f \in \mathcal{B}$ we implicitly restrict ourselves to such $D_i \in \mathcal{E}$ for which $f \in \mathcal{B}_{D_i}$].

In general this topology is not Hausdorff for obvious reasons. Had we introduced in $\mathcal{B}$ equivalence classes in which $\{f_n\} \sim \{g_n\}$ if $D^{-1}(f_n-g_n) \to \mathbf{0}$ for all $D \in \mathcal{E}$ we would have obtained a space of equivalence classes in which the above topology is Hausdorff. Actually this is the manner in which a bra-space $\mathcal{L}$ is defined in [5], and in addition to making the above topology Hausdorff it also has the advantage of yielding the ordered pair $\mathcal{H} \subset \mathcal{K}_D^\dagger$ for the special choice $\mathcal{E} = \{D^*\}$. On the other hand, there is the disadvantage of having to require that $\mathcal{E}$ is A-stable [5] in order to have $A^\dagger$ well-defined. Although this requirement is easy to satisfy in the applications to scattering theory treated in Sec. IV, it might prove to be a nuisance in the long run, and therefore we deem it desirable to investigate at present the alternative in which such a decomposition of $\mathcal{E}$ in equivalence classes is not carried out.

<u>Lemma 3</u>. For a given extended Hilbert space $\mathcal{K} \subset \mathcal{B}$ over $\mathcal{E}$ there is a unique complex functional $(\cdot|\cdot\rangle$ <u>in</u> $\mathcal{B} \times \mathcal{K}$ with the property $(f|g\rangle = \langle f|g\rangle$, $f, g \in \mathcal{K}$ and strongly continuous on $\mathcal{B}$ for each fixed $g \in \mathcal{K}$. Its restriction $(\cdot|\cdot\rangle_D$ to $\mathcal{B}_{D^*} \times \mathcal{K}_D$ is continuous in both arguments in the respective strong topologies.

<u>Proof</u>: A simple way of constructing $(\cdot|\cdot\rangle$ is by setting $(f|g\rangle = (m_{D^*}f|g\rangle_D$, $f \in \mathcal{B}_{D^*}$, $g \in \mathcal{K}_D$, where m_{D^*} is the mapping of $\mathcal{B}_{D^*}$ onto $\mathcal{K}_D^\dagger$ which to each equivalence class in $\mathcal{B}$ containing the D^*-convergent sequence $\{f_n\}$ assigns the equivalence class in $\mathcal{K}_D^\dagger$ containing the same sequence. Since this mapping is linear and continuous from $\mathcal{B}_{D^*}$ to $\mathcal{K}_D^\dagger$ supplied with the respective strong topologies, the continuity properties of $(\cdot|\cdot\rangle$ are established on $\mathcal{B}_{D^*} \times \mathcal{K}_D$. The possibility of any ambiguity arising when $f \in \mathcal{B}_{D_1^*} \cap \mathcal{B}_{D_2^*}$ and $g \in \mathcal{K}_{D_1} \cap \mathcal{K}_{D_2}$ for $D_1 \neq D_2$ is easily eliminated [5] by using Lemma 1 and the obvious fact that $\mathcal{K}$ is dense in $\mathcal{B}$ in the strong topology of $\mathcal{B}$.

We can now define the <u>bra-adjoint</u> $B^\dagger$ of B as an extension of B^* from $\mathcal{K}$ to $\mathcal{B}$ for which $(B^\dagger f|g\rangle = (f|Bg\rangle$, $g \in \mathcal{D}(B)$, and prove its existence by the procedure suggested earlier, i.e., by setting $B^{\dagger}f = \left\{B^{*}f_n\right\}_{n=1}^{\infty}$, $f = \left\{f_n\right\}_{n=1}^{\infty}$. Thus, if A , A^{-1} and A^{*-1} exist and are bounded then $A^{\dagger}$ is defined by for all $f \in B$ However, in general we have to consider $B = A - \zeta 1\!\!1$ instead. Since $(A - \zeta 1\!\!1)^{-1}$ and $(A - \zeta 1\!\!1)^{*-1}$ exist and are bounded if ζ is in the resolvent set of A , we can define $(A - \zeta 1\!\!1)^{\dagger} = B^{\dagger}$ as above at least for all $D^{*} \in E$ for which $\|(DB)^{-1}\|(DB)^{*} \in E$. If A is bounded this will be the case for all $D^{*} \in E$, but even for unbounded A we can expect to find some D for which this is true if $D(A)$ is dense in H and has an adjoint A^{*} . The definition of $A^{\dagger}$ then proceeds by the use of (9), which is now well-defined for all $f \in B$ for which f and $(A - \zeta 1\!\!1)^{\dagger}f$ belong to a common $B_{D^{*}}$.

In the above presented formalism most of the formal manipulations with Dirac's formalism become theorems. In particular we have [5]:

<u>Theorem 2</u>. If the closed operator B in K is a function $B = F(A_1,\ldots,A_{\nu})$ of the complete set $\{A_1,\ldots,A_{\nu}\}$ then for almost all $\alpha \in \mathbb{R}^{\nu}$ we have $\langle\phi_D(\alpha)|Bf\rangle = F(\alpha)\langle\phi_D(\alpha)|f\rangle$ for any $f \in K$.

IV. PERTURBATIONS OF BRA-EXPANSIONS IN STATIONARY SCATTERING THEORY

In this section we shall apply the preceding results to two-body scattering theory for short-range forces. The same procedures can be extended with very minor modifications to the long-range and the multichannel case by resorting to the two-Hilbert space formulation [8].

Let H and H_0 be the total and free Hamiltonians, respectively, acting in the Hilbert space H, and let $E^{H_0}(\Delta)$ be the spectral measure of H_0 . We shall assume that H_0 is a function $h(A_1,\ldots,A_{\nu})$ of a complete set $\{A_1,\ldots,A_{\nu}\}$ of observables, so that

(11) $$E^{H_0}(\Delta) = E^{A_1,\ldots,A_{\nu}}(h^{-1}(\Delta)) .$$

According to Theorem 2 we shall have for ρ_D-almost all $\alpha \in \mathbb{R}^{\nu}$ $H_0^{\dagger}\phi_D(\alpha) = h(\alpha)\phi_D(\alpha)$ for any bra-expansion $\phi_D(\alpha)$ of $\{A_1,\ldots,A_{\nu}\}$. We assume the existence of wave operators $\Omega_{\pm}$ such that

(12) $$\Omega_{\pm}^{*}f = \operatorname*{s-lim}_{\varepsilon\to+0} \Omega_{\pm\varepsilon}^{*} f$$

for f in the ranges $R(\Omega_{\pm})$ of $\Omega_{\pm}$. The operators $\Omega_{\pm\varepsilon}^{*}$ are defined in a time-independent manner by means of strong Riemann-Stieltjes integrals [5,6]:

(13) $$\Omega_{\pm\varepsilon}^{*} = \int_{-\infty}^{+\infty} E^{H_0}(d\lambda)\, \frac{\pm i\varepsilon}{H - \lambda \pm i\varepsilon} .$$

From time-dependent theory we have that $\Omega_{\pm}$ are partial isometries and therefore the sets $\left\{A_1^{(\pm)}, \ldots, A_{\nu}^{(\pm)}\right\}$ with

$A_i^{(\pm)} = \Omega_\pm A_i \Omega_\pm^*$ are complete sets of self-adjoint operators on $R(\Omega_\pm)$. Using the intertwining properties of $\Omega_\pm$ we obtain

(14) $$H\Big|_{R(\Omega_\pm)} = \Omega_\pm H_0 \Omega_\pm^* = h\left(A_1^{(\pm)}, \ldots, A_\nu^{(\pm)}\right) .$$

<u>Theorem 3.</u> The two families of all bra-vectors $\phi_D^{(\pm)}(\alpha) = \Omega_\pm^{*\dagger}\phi_D(\alpha)$ are bra-expansions of $\left\{A_1^{(\pm)}, \ldots, A_\nu^{(\pm)}\right\}$ with respect to the measure ρ_D. If $D \in E$ is such that at some $f \in K$

(15) $$\lim_{\lambda \to h(\alpha)} \left(\phi_D(\alpha) \,\middle|\, (H-\lambda \pm i\varepsilon)^{-1} f\right\rangle = \left(\phi_D^{\circ}(\alpha) \,\middle|\, (H-h(\alpha) \pm i\varepsilon)^{-1} f\right\rangle$$

for ρ_D-almost all $\alpha \in \mathbb{R}^\nu$ and (12) holds then

(16) $$\left(\phi_D^{(\pm)}(\alpha) \,\middle|\, f\right\rangle = \rho_D\text{-}\underset{\varepsilon \to +0}{\text{l.i.m.}} \left(\phi_D(\alpha) \,\middle|\, \frac{\pm i\varepsilon}{H-h(\alpha) \pm i\varepsilon} f\right\rangle$$

for the ket-vector $f \in K$ under consideration.

<u>Proof</u>: The first statement is a trivial consequence of the observation that $\left\langle f \,\middle|\, \Omega_\pm E^{A_1,\ldots,A_\nu}(\Delta)\Omega_\pm^* g\right\rangle = \int \left\langle \Omega_\pm^* f \,\middle|\, \phi_D(\alpha)\right) \left(\phi_D(\alpha) \,\middle|\, \Omega_\pm^* g\right\rangle d\rho_D(\alpha)$ and the fact that $\Omega_\pm E^{A_1,\ldots,A_\nu}(\Delta)\Omega_\pm^*$ are the spectral measures of $\left\{A_1^{(\pm)}, \ldots, A_\nu^{(\pm)}\right\}$.

To prove the second statement, we choose a sequence $\{\pi_n\}_{n=1}^\infty$ of subdivisions $-a_n = \lambda_0 < \lambda_1 < \ldots < \lambda_n = a_n$ of $[-a_n, +a_n]$ (where $a_n \to \infty$ as $n \to \infty$) with corresponding subdivision points $\lambda_j' \in (\lambda_{j-1}, \lambda_j)$ for which the corresponding Riemann-Stieltjes sums

(17) $$Q_{\pm\varepsilon_n}(\pi_n) = \sum_{k=1}^{n} E^{H_0}\left((\lambda_{k-1}, \lambda_k]\right) \frac{\pm i\varepsilon_n}{H - \lambda_k' \pm i\varepsilon_n} f$$

strongly approximate $\Omega_{\pm\varepsilon_n}^* f$ better than ε_n (i.e.,

$\| \Omega_{\pm\varepsilon_n}^* f - Q_{\pm\varepsilon}(\pi_n) f \| < \varepsilon_n$) for some sequence $\varepsilon_1, \varepsilon_2, \ldots \to +0$.

We show now that

(18) $$\left(\phi_D(\alpha) \,\middle|\, Q_{\pm\varepsilon_n}(\pi_n) f\right\rangle = \begin{cases} \left(\phi_D(\alpha) \,\middle|\, \dfrac{\pm i\varepsilon_n}{H - \lambda_k' \pm i\varepsilon_n} f\right\rangle & \text{if } \lambda_{k-1} < h(\alpha) \le \lambda_k \\ 0 & \text{if } |h(\alpha)| > a_n \end{cases}$$

for ρ_D-almost all $\alpha \in \mathbb{R}^\nu$. In fact, for any given Borel set $\Delta_k \subset \mathbb{R}^\nu$ and any orthonormal basis $\{u_n\}_{n=1}^\infty$ in H we have

$$\int_{\Delta} \langle u_n|\phi_D(\alpha)\big)\big(\phi_D(\alpha)|E^{A_1,\ldots,A_\nu}(\Delta_k)g_k\rangle\, d\rho_D(\alpha)$$

$$= \langle u_n|E^{A_1,\ldots,A_\nu}(\Delta\cap\Delta_k)g_k\rangle = \int_{\Delta\cap\Delta_k} \langle u_n|\phi_D(\alpha)\big)\big(\phi_D(\alpha)|g_k\rangle\, d\rho_D(\alpha)$$

for all Borel sets $\Delta \subset \mathbb{R}^\nu$. Since the set $\Delta^{(0)}$ on which $\langle u_n|\phi_D(\alpha)\big) = 0$ for all $n=1,2,\ldots$ has ρ_D measure zero, we conclude that

(19) $$\big(\phi_D(\alpha)|E^{A_1,\ldots,A_\nu}(\Delta_k)g_k\rangle = \chi_{\Delta_k}(\alpha)\big(\phi_D(\alpha)|g_k\rangle$$

for ρ_D-almost all $\alpha \in \mathbb{R}^\nu$. Thus (18) is obtained by observing that due to (11) we can apply (19) to each term in the sum (17) by taking $\Delta_k = h^{-1}\big[(\lambda_{k-1}, \lambda_k]\big]$ and $g_k = \pm i\varepsilon_n(H-\lambda_k' \pm i\varepsilon_n)^{-1}f$.

In view of the relation

$$\|[Q_{\pm\varepsilon_m}(\pi_m)-Q_{\pm\varepsilon_n}(\pi_n)]f\|^2$$

$$= \int_{\mathbb{R}^\nu} |\big(\phi_D(\alpha)|Q_{\pm\varepsilon_m}(\pi_m)f\rangle - \big(\phi_D(\alpha)|Q_{\pm\varepsilon_n}(\pi_n)f\rangle|^2\, d\rho_D(\alpha)$$

we conclude that $\hat{f}_\pm(\alpha) = \rho_D\text{-}\underset{n\to\infty}{\text{l.i.m.}}\ \big(\phi_D(\alpha)|Q_{\pm\varepsilon_n}(\pi_n)\,f\rangle$ exist iff the strong limits of $\Omega^*_{\pm\varepsilon_n} f$ exist as $\varepsilon_n \to +0$. If (15) is satisfied then by (18) we have

$$\rho_D\text{-}\underset{n\to\infty}{\text{l.i.m.}}\,|\big(\phi_D(\alpha)|Q_{\pm\varepsilon_n}(\pi_n)f\rangle - \big(\phi_D(\alpha)|\,\frac{\pm i\varepsilon_n}{H-h(\alpha)\pm i\varepsilon_n}f\rangle| = 0$$

and consequently (16) follows.

If (12) is true for any $f \in R(\Omega_\pm)$ then (16) is equivalent to the statement

(20) $$\phi_D^{(\pm)}(\alpha) \overset{w}{=} \rho_D\text{-}\underset{\varepsilon\to+0}{\text{l.i.m.}}\left(\frac{\pm i\varepsilon}{H-h(\alpha)\pm i\varepsilon}E_\pm\right)^\dagger \phi(\alpha)$$

where $E_\pm$ are the projectors onto $R(\Omega_\pm)$, while "$\overset{w}{=}$" indicates that the convergence is in the weak topology of B; in this topology the family

$$W(f;h_1,\ldots,h_n) = \{g|\ \big(f-g|h_i\rangle| < 1,\quad i=1,\ldots,n\}$$

corresponding to all $f \in K$ and $h_1, \ldots, h_n \in K$, $n=1,2,\ldots$, constitutes a basis for the topology. An analogous procedure in which the roles of H and H_0 are reversed yields

$$\phi(\alpha) \overset{w}{=} \lim_{\varepsilon\to+0} \left[\frac{\pm i\varepsilon}{H_0 - h(\alpha) \pm i\varepsilon}\right]^\dagger \phi^{(\pm)}(\alpha) \quad . \tag{21}$$

The above two sets of equations are equivalent to the Lippmann-Schwinger equations.

An expression relating the T-matrix to the T-operator can be obtained in a similar manner. We start by using (17), (18) and the intertwining property of Ω_- to derive [7]

$$\langle f|Tg\rangle = \lim_{\varepsilon\to+0} \frac{\varepsilon}{\pi} \int_{\mathbb{R}^\nu} \langle f|\phi_D(\alpha)\rangle \left(\phi_D(\alpha)\left|V\Omega_- \frac{1}{(H_0-h(\alpha))^2+\varepsilon^2} g\right.\right\rangle d\rho_D(\alpha) \quad .$$

where $V = H - H_0$. If $(V\Omega_-)^\dagger \phi_D(\alpha) \in K$ we can expand again to obtain

$$\langle f|Tg\rangle = \lim_{\varepsilon\to+0} \frac{\varepsilon}{\pi} \int d\rho_D(\alpha)\langle f|\phi_D(\alpha)\rangle \int d\rho_D(\beta) \frac{\langle (V\Omega_-)^\dagger\phi_D(\alpha)|\phi_D(\beta)\rangle}{(h(\alpha)-h(\beta))^2 + \varepsilon^2} (\phi_D(\beta)|g$$

Suppose now that the energy variable η is one of the components of $\alpha \in \mathbb{R}^\nu$ so that $\alpha = (\eta,\alpha')$ where $\alpha' \in \mathbb{R}^{\nu-1}$, and that the continuous spectrum of H_0 is absolutely continuous. Then the above limit $\varepsilon \to +0$ can be taken provided that the integrand is continuous in a neighborhood of the considered value of $\eta \in \mathbb{R}^1$. This yields the well-known formula for the T-matrix on the energy shell,

$$\langle \eta,\alpha'|T|\eta,\beta'\rangle = \langle (V\Omega_-)^\dagger\phi_D(\eta,\alpha')|\phi_D(\eta,\beta')\rangle \frac{\partial\rho_D(\eta,\alpha')}{\partial\eta} \frac{\partial\rho_D(\eta,\beta')}{\partial\eta}$$

where the above partial derivatives are Radon-Nikodym derivatives resulting from the transition to Lebesgue integration over the continuous spectrum of H_0 .

In conclusion, we would like to point out that one of the main advantages of the above considered abstract approach to stationary scattering theory lies in its flexibility and versality in applications. For example, this approach applies equally well to scattering theory for wave-packets and for statistical operators. In the later case the role of the Hilbert space $\mathcal{H}$ is played by Liouville space and instead of H and H_0 we are dealing with the transformers $\underline{H}$ and $\underline{H}_0$, respectively [8], where $\underline{H}\rho = [H,\rho]^{**}$ and $\underline{H}_0\rho = [H_0, \rho]^{**}$. Of course, it might seem more natural to consider $\underline{H}$ and $\underline{H}_0$ as transformers on the trace-class rather than Liouville space, but such an approach is precluded by the fact that $\underline{H}$ and $\underline{H}_0$ are not spectral operators on the trace-class except in the trivial case when H and H_0 have only a finite pure point spectrum [9].

References

[1] Antoine, J.P.: Dirac Formalism and Symmetry Problems in Quantum Mechanics I, General Dirac Formalism, J. Math. Phys. 10, 53-69 (1969).

[2] Berezanskiĭ, Ju. M.: Expansions in eigenfunctions of self-adjoint operators, American Mathematical Society, Rhode Island, 1968.

[3] Dirac, P.A.M.: The principles of quantum mechanics, Oxford University Press, Oxford, 1930.

[4] Prugovečki, E.: Fuzzy sets in the theory of measurement of incompatible observables, Found. Phys. 3, No. 4 (1973) (in press).

[5] __________ : The bra and ket formalism in extended Hilbert space, J. Math. Phys. 14, No. 9-10 (1973) (in press).

[6] __________ : Quantum mechanics in Hilbert space, Academic Press, New York and London, 1971.

[7] __________ : Eigenfunction expansions for stationary scattering theory in spaces with negative norm, (University of Toronto preprint).

[8] __________ : Multichannel stationary scattering theory in two-Hilbert space formulation, J. Math. Phys. 14, 957-964 (1973).

[9] __________ and Tip, A.: Semi-groups of rank-preserving transformers on minimal norm ideals in $B(H)$ (preprint).

Projections on Orthomodular Lattices

G.T. Rüttimann

Universität Bern, Institut für
Theoretische Physik, Bern,
(Switzerland)

Abstract

In terms of c-closure operators we give a necessary and sufficient condition for an orthocomplemented poset to be an orthomodular lattice. C-closure operators are closely related to projections and appear as a generalization of symmetric closure operators.

We show how a projection can be represented as a product of a SASAKI-projection and a symmetric closure operator. Finally, starting with a subset of an orthomodular lattice, we construct explicitly the symmetric closure operator that maps the lattice onto the commutant of that subset.

1. BACKGROUND MATERIAL

Let T be a poset with largest element 1. A subset $A \subseteq T$ is called weakly meet-complete whenever $\{ x \mid z \leq x \,;\, x \in A \}$ has a smallest element for every $z \in T$. An orthocomplementation on T is a mapping $x \in T \rightarrow x' \in T$ such that i) $x'' = x$, ii) $x \leq y \Rightarrow y' \leq x'$ and iii) $x \vee x'$ exists and is equal to 1. A poset T admitting an orthocomplementation is called orthocomplemented ($o := 1'$ is the smallest element; $x \wedge x'$ exists and is equal to o). $x, y \in T$ are said to be orthogonal (denoted: $x \perp y$) if $x \leq y'$.

If an orthocomplemented poset is indeed a lattice, then it is called <u>orthocomplemented lattice</u>. An orthocomplemented lattice L is an <u>orthomodular lattice</u> whenever $x \leq y$ ($x, y \in L$) $\Longrightarrow$ there exists a $z \in L$ such that $x \perp z$ and $x \vee z = y$. Two elements $x, y \in L$ are said to be <u>compatible</u> (denoted $x \leftrightarrow y$) when there exist three pairwise orthogonal elements $u, v, w \in L$ such that $x = u \vee v$ and $y = u \vee w$ (eg.: $x \leq y \Rightarrow x \leftrightarrow y$). $x \leftrightarrow y$ if and only if $L\{x, y, x', y'\}$ (generated sublattice) is BOOLEAN.

A <u>BEAR*-semigroup</u> or <u>FOULIS semigroup</u> B is an involution semigroup with zero and a mapping $a \in B \rightarrow a' \in P \subseteq B$, $P := \{ p \in B \mid p = pp^* = p^* \}$ such that $\{ b \mid ab = o \} = \{ c \mid a'c = c \}$. An element of P is called <u>projection</u>. $p_1 \leq p_2 :\Longleftrightarrow p_1 p_2 = p_1$ ($p_1, p_2 \in P$) makes P into a partially ordered set. If $p \in P$ then $p \leq p''$ is valid. If $p = p''$ then the projection is said to be <u>closed</u>. The poset P' of all closed projections is a lattice with smallest and largest element. The mapping $p \in P' \rightarrow p' \in P'$ is an orthcomplementation that makes P' into an orthomodular lattice.

Let L be an orthomodular lattice. A <u>hemimorphism</u> Ω of L is a mapping $L \rightarrow L$ with the following properties :

i) $\Omega o = o$, $\Omega(x \vee y) = \Omega x \vee \Omega y$;

ii) there exists a mapping $\Omega^* : L \rightarrow L$ such that

$\Omega^* o = o$, $\Omega^*(x \vee y) = \Omega^* x \vee \Omega^* y$

and $\Omega^*(\Omega x')' \leq x$, $\Omega(\Omega^* x')' \leq x$ ($x \in L$).

Clearly, Ω^* is a hemimorphism too, called the <u>adjoint</u> hemimorphism of Ω. A hemimorphism has exactly one adjoint. The set of hemimorphisms B(L) is an involution semigroup with zero due to "function composition"as multiplication and the mapping $\Omega \rightarrow \Omega^*$ as involution. $\phi_a z := (a' \vee z) \wedge a$ ($a, z \in L$) is a projection of B (L) (<u>SASAKI-projection</u>). The mapping $\Omega \in B(L) \rightarrow \Omega' := \phi_{(\Omega^* 1)'}$ makes B (L) into a Baer*-semigroup. The SASAKI-projections are exactly the closed projections of B (L) (notation : P (L) projections, P'(L) closed projections). The mapping $a \in L \rightarrow \phi_a \in P'(L)$ is an ortho-order isomorphism between the orthomodular lattices L and P'(L).

The following Lemma is often useful :

<u>Lemma 1</u> : A mapping Γ is a projection if and only if

i) $z_1 \leq z_2 \Longrightarrow \Gamma z_1 \leq \Gamma z_2$,

ii) $\Gamma \circ \Gamma = \Gamma$,

iii) $\Gamma(\Gamma z)' \leq z'$ $(z, z_1, z_2 \in L)$.

For basic results in the theory of BAER*-semigroup see [2] , [3] and for their application [5] , [6].

2. C-CLOSURE OPERATORS AND PROJECTIONS

A closure operator Ψ on a poset T (with 1) is a mapping $T \rightarrow T$ for which

i) $z \leq \Psi z$,

ii) $z_1 \leq z_2 \Longrightarrow \Psi z_1 \leq \Psi z_2$,

iii) $\Psi \circ \Psi = \Psi$ $(z, z_1, z_2 \in T)$.

The range of a closure operator Ψ on T is weakly meet-complete. To every weakly meet-complete subset $\overline{T}$ there exists exactly one closure operator with range $\overline{T}$. In an orthocomplemented poset T a closure operator satisfies

$$\Psi(\Psi(\Psi z)')' \leq \Psi z \quad (z \in T).$$

Those closure operators for which the equality

$$\Psi(\Psi(\Psi z)')' = \Psi z \quad (z \in T)$$

is valid are of special interest. As we will see later, they are closely related to projections and determining for the lattice and the orthomodular structure of T. We call these operators c-closure operators [7]. The set C (T) of c-closure operators is obviously not empty since the mapping $z \rightarrow 1$ $(z \in T)$ is a c-closure operator.

The following lemma will be used in the sequel :

Lemma 2 : Let T be an orthocomplemented poset and $\Psi \in C(T)$. Then

i) $\Psi 1 = 1$,

ii) $\Psi z \vee \Psi(\Psi z)' = 1$,

iii) $\Psi z \wedge \Psi(\Psi z)' = \Psi o$.

Proof : i) $1 \leq \Psi 1 \leq 1$.

ii) Of course $\Psi z \leq 1$ and $\Psi(\Psi z)' \leq 1$. If there is an $x \in T$ such that $\Psi z \leq x$ and $\Psi(\Psi z)' \leq x$ then also $(\Psi z)' \leq x$ since $(\Psi z)' \leq \Psi(\Psi z)'$. But $\Psi z \vee (\Psi z)' = 1$, hence $1 \leq x$. This proves that $\Psi z \vee \Psi(\Psi z)' = 1$.

iii) By monotony $\Psi o \leq \Psi z$ and $\Psi o \leq \Psi(\Psi z)'$. Let $x \in T$ be such that $x \leq \Psi z$ and $x \leq \Psi(\Psi z)'$. By monotony and idempotence of the closure operator we get $\Psi x \leq \Psi z$ and $\Psi x \leq \Psi(\Psi z)'$. Orthocomplementation gives $(\Psi z)' \leq (\Psi x)'$ and $(\Psi(\Psi z)')' \leq (\Psi x)'$. Again by monotony and using the fact that Ψ is a c-closure operator, we get $\Psi(\Psi z)' \leq \Psi(\Psi x)'$ and $\Psi z = \Psi(\Psi(\Psi z)')' \leq \Psi(\Psi x)'$. Using part ii) of this lemma, we conclude $\Psi(\Psi x)' = 1$. Then $(\Psi(\Psi x)')' = o$ and finally $x \leq \Psi x = \Psi(\Psi(\Psi x)')' = \Psi o$. QED

Now the theorem.

Theorem 3 : Let T be an orthocomplemented poset. T is an orthomodular lattice if and only if every segment $[a, 1] := \{x \in T \mid a \leq x \leq 1\}$ is the range of a c-closure operator.

Proof : a) Assume T to be an orthomodular lattice. The mapping $z \rightarrow a \vee z$ is a closure operator that maps T onto $[a, 1]$. But $z \rightarrow a \vee z$ is also a c-closure operator:

By orthomodularity, there is an $x \in T$ such that $x \perp a$ and $a \vee z = a \vee x$. Since $a \leftrightarrow x$, we get $a \vee (a \vee (a \vee z)')' = a \vee (a \vee (a \vee x)')' = a \vee (a' \wedge (a \vee x)) = a \vee x = a \vee z$.

b) Let $a, b \in T$, then there is a c-closure operator Ψ that maps T onto $[b, 1]$. Clearly $a \leq \Psi a$ and $b = \Psi o \leq \Psi a$. Let $x \in T$ such that $a \leq x$ and $b \leq x$. Now, from $a \leq x$ we get $\Psi a \leq \Psi x = x$, Ψ being a closure operator. Thus $a \vee b$ exists in T and is equal to Ψa. As T is an orthocomplemented poset, it follows that T is also a lattice. Orthomodularity of the lattice T is proved as follows : Let $a \leq b$ and Ψ be the c-closure operator that maps onto $[a, 1]$. Again we have $a = \Psi o$ and $\Psi b = b$. By lemma 2 we get now $a = \Psi o = \Psi b \wedge \Psi (\Psi b)' = b \wedge \Psi b' = b \wedge (a \vee b')$. QED

Remark: C (T) can be made into a poset with largest and smallest element by means of the ordering relation

$$\Psi_1 \leq \Psi_2 \quad :\Leftrightarrow \quad \Psi_2 z \leq \Psi_1 z \quad (\forall z \in T).$$

Let L be an orthomodular lattice and P (L) its poset of projections. Let $\Psi \in C(L)$, then $\Omega(\Psi) z := (\Psi(\Psi z)')'$ is a projection on L and the mapping $\Psi \rightarrow \Omega(\Psi)$ is an order-isomorphism between the poset of c-closure operators and the poset of projections. It is easily seen that $\Omega(\Psi)$ is a SASAKI-projection if and only if Ψ maps onto a segment $[a, 1]$.

3. SYMMETRIC CLOSURE OPERATORS

In this section we are going to characterize a class of projections that will be important for the decomposition of arbitrary projections.

Theorem 4 : The following statements are equivalent.

i) $\Psi \in C(L)$ with $\Psi o = o$.

ii) $\Psi \in P(L)$ with $\Psi 1 = 1$.

iii) Ψ is a closure operator with the property
$\Psi z = z \Rightarrow \Psi z' = z'$.

Proof : i) $\Rightarrow$ ii) By lemma 2 we have $o = \Psi o = \Psi z \wedge \Psi (\Psi z)'$. Since

$(\Psi z)' \leq \Psi(\Psi z)'$ we get by orthomodularity $(\Psi z)' = \Psi(\Psi z)'$. But $(\Psi z)' \leq z'$, hence $\Psi(\Psi z)' \leq z'$. Thus by lemma 1, Ψ is a projection and of course $\Psi 1 = 1$. ii) $\Longrightarrow$ iii) Ψ is a projection, thus $1 = \Psi 1 =$ $\Psi(\Psi z \vee (\Psi z)') = \Psi z \vee \Psi(\Psi z)'$ and $\Psi(\Psi z)' \leq (\Psi z)'$. Again by orthomodularity we get $\Psi(\Psi z)' = (\Psi z)'$. But $\Psi(\Psi z)' \leq z'$, thus $z \leq \Psi z$ which proves that Ψ is a closure operator. Let now $\Psi z = z$, then $\Psi z' = \Psi(\Psi z)' = (\Psi z)' = z'$.

iii) $\Longrightarrow$ i) By idempotence $\Psi(\Psi z) = \Psi z$, thus $\Psi(\Psi z)' = (\Psi z)'$. Then $(\Psi(\Psi z)')' = \Psi z$ and finally $\Psi(\Psi(\Psi z)')' = \Psi z$, hence $\Psi \in C(L)$. Since for a closure operator $\Psi 1 = 1$ is valid, we get $\Psi o = o$. QED

A closure operator Ψ is called <u>symmetric</u> [4], whenever $\Psi z = z \Rightarrow \Psi z' = z'$ (S (L) denotes the set of symmetric closure operators on L). Using theorem 4, we get $S(L) = P(L) \cap C(L)$. One easily sees that $S(L) \cap P'(L) = \{ I \}$ where $I z := z$.

<u>Remark</u> : The symmetric closure operators are exactly the fixed elements of the mapping $\Psi \in C(L) \longrightarrow \Omega(\Psi) \in P(L)$. Another interesting feature of symmetric closure operators is shown by the following theorem.

<u>Theorem 5</u> : The range of a symmetric closure operator is a weakly meet-complete orthomodular sublattice.

<u>Proof</u> : $\Psi \in S(L)$ being a closure operator, ΨL is clearly weakly meet-complete. Since also $\Psi \in P(L)$, we get for $z_1, z_2 \in \Psi L$ $z_1 \vee z_2 = \Psi z_1 \vee \Psi z_2 =$ $\Psi(z_1 \vee z_2) \in \Psi L$. Furthermore ΨL is closed under orthocomplementation. QED

We have now exhibited two classes of projections : SASAKI-projections and symmetric closure operators. SASAKI-projections map the orthomodular lattice L onto segments, whereas symmetric closure operators map L onto certain orthomodular sublattices. It is natural to ask whether it is possible to represent an arbitrary projection in terms of SASAKI-projections and symmetric closure operators.

<u>Theorem 6</u> : (<u>Decomposition Theorem</u> [8]). Every projection Γ on an orthomodular lattice L is a product of a SASAKI-projection ϕ and a symmetric closure operator Ψ : $\Gamma = \phi \circ \Psi$.

<u>Remark</u> : This decomposition of an arbitrary projection Γ is not unique. However, the SASAKI-projection is uniquely determined ($\phi = \phi_{\Gamma 1}$) and among the symmetric closure operators which decompose Γ in this way, there exists a smallest one ($\Psi_s z = \bigwedge \{ x \mid z \leq x, x \in \Gamma L \cup (\Gamma L)' \}$).

4. SYMMETRIC CLOSURE OPERATORS MAPPING ONTO VON NEUMANN SYSTEMS

In the sequel we assume that L is a complete orthomodular lattice. Let A be a non-empty subset of L. $A^0 := \{ x \mid x \leftrightarrow a \text{ for all } a \in A \}$ is the commutant of A. A^0 is a complete orthomodular sublattice of L. Clearly $A_1 \subseteq A_2 \Rightarrow A_2^0 \subseteq A_1^0$, $A^{000} = A^0$ and the mapping $A \to A^{00}$ is a closure operation. A VON NEUMANN-system N is a subset of L for which $N = N^{00}$. The notion of a VON NEUMANN system introduced here is a straight forward generalization of the notion of VON NEUMANN algebra when only considering its (orthomodular) lattice of projections [1].

A subset $V \subseteq L$ is a VON NEUMANN system if and only if V is the commutant of a subset. Hence one gets all VON NEUMANN systems by taking the commutants of subsets.

It is the goal of this section to give a characterization of those symmetric closure operators that map onto VON NEUMANN systems. Starting with a subset $A \subseteq L$ we construct explicitly the symmetric closure operator with range A^0. The mapping

$$z \in L \longrightarrow \Psi\{A\} z := \bigvee_{a \in A} [(a \vee z) \wedge (a' \vee z)] \in L$$

has the properties:

i) $\Psi\{A\}\, o = o$,

ii) $z \leq \Psi\{A\} z$,

iii) $z_1 \leq z_2 \Rightarrow \Psi\{A\} z_1 \leq \Psi\{A\} z_2$,

iv) $\Psi\{A\} \bigvee_i z_i = \bigvee_i \Psi\{A\} z_i$.

A simple lemma.

<u>Lemma 7</u> : $\Psi\{A\} z = z$ if and only if $z \in A^0$.

<u>Proof</u> : If $\Psi\{A\}\, z = z$, then $z = \bigvee_{a \in A} [(a \vee z) \wedge (a' \vee z)] \geq$ $\geq (\bar{a} \vee z) \wedge (\bar{a}' \vee z)$ for all $\bar{a} \in A$. Thus $z = (\bar{a} \vee z) \wedge (\bar{a}' \vee z)$ or $z \leftrightarrow \bar{a}$ for all $\bar{a} \in A$. Conversely, let $z \leftrightarrow a \in A$, then $z = (a \vee z) \wedge (a' \vee z)$ for all $a \in A$, hence $\Psi\{A\}\, z = z$. QED

As a consequence, the range of $\Psi\{A\}$ is in general larger than the commutant of A. But in the case that A is contained in a BOOLEAN sublattice we have $A^0 = \Psi\{A\} L$.

Consider now the mapping $\Pi\{A\} : L \to L$ defined as follows

$$\Pi\{A\}\ z := \bigvee_{n=1}^{\infty} \Psi^n\{A\} z .$$

As is easily seen $\Pi\{A\}$ inherits properties i) - iv) .

Theorem 8 : $\Pi\{A\}$ is a closure operator on L .

Proof : It remains to show that $\Pi\{A\}$ is idempotent. $\Pi\{A\}(\Pi\{A\}z) =$
$= \bigvee_{n=1}^{\infty} \Psi^n\{A\}(\bigvee_{m=1}^{\infty} \Psi^m\{A\}z) = \bigvee_{n=1}^{\infty}\bigvee_{m=1}^{\infty} \Psi^{n+m}\{A\}z =$
$= \Pi\{A\}z$ since $\Pi\{A\}$ preserves arbitrary joins and $\{\Psi^n\{A\}z\}$
(n = 1, 2, 3 .....) is a monotone sequence. QED

Theorem 9 : $\Pi\{A\}z = z$ if and only if $z \in A^0$.

Proof : If $z \in A^0$, then $\Psi\{A\}z = z$ and $\Psi^n\{A\}z = z$
by lemma 7 , thus $\bigvee_{n=1}^{\infty} \Psi^n\{A\}z = z$.
Conversely, if $\Pi\{A\}z = z$, then $\Psi\{A\}z \leq z$;
hence $\Psi\{A\}z = z$. Again by lemma 7 it follows that $z \in A$. QED

Theorem 10 : $\Pi\{A\}$ is a symmetric closure operator and has range A^0 .

Proof : $\Pi\{A\}$ is a closure operator on L (theorem 8).
By theorem 9 $A^0 \subseteq \Pi\{A\}L$, but if $y = \Pi\{A\}z$
then $\Pi\{A\}y = \Pi^2\{A\}z = \Pi\{A\}z = y$, hence
$y \in A^0$. Finally, the range A^0 is closed under orthocomplementation. Thus $\Pi\{A\}$ is a symmetric closure operator. QED

References

[1] J. Dixmier : "Les algèbres d'opérateurs dans l'espace Hilbertien". Paris, Gauthier-Villars (1957).

[2] D.J. Foulis : "Baer*-semigroups". Proceedings of the American Mathematical Society 11, 648 (1960).

[3] D.J. Foulis : "A Note on Orthomodular Lattices". Portugaliae Mathematica 21, 65 (1962).

[4] M.F. Janowitz : "Residuated Closure Operators". Portugaliae Mathematica 26, 221 (1967).

[5] J.C.T. Pool : "Baer*-semigroup and the Logic of Quantum Mechanics". Communications of Mathematical Physics 9, 118 (1968).

[6] J.C.T. Pool : "Semimodularity and the Logic of Quantum Mechanics". Communications of Mathematical Physics 9, 212 (1968).

[7] G.T. Rüttimann : "Closure Operators and Projections on Involution Posets". To appear in The Journal of the Australian Mathematical Society.

[8] G.T. Rüttimann : "Decomposition of Projections on Orthomodular Lattices". To appear in Canadian Mathematical Bulletin.

See also

M.F. Janowitz : "Equivalence Relations induced by Baer*-semigroups". Journal of Natural Sciences and Mathematics 11, 83 (1971). p. 98, theorem 4.7 .

THE ŠILOV BOUNDARY OF A CONVEX CONE

H.H. Schaefer
Mathematisches Institut der Universität Tübingen
Tübingen, Germany

If C denotes a cone of continuous real functions on a compact space K, a well-known theorem of H. Bauer [2] (see Alfsen [1] concerning the history of the subject) asserts that C possesses a Šilov boundary P in K; that is, there exists a unique minimal closed subset P of K on which each $f \in C$ attains its maximum. In the theory of ordered vector spaces, it is frequently helpful to consider the given space as a vector space of continuous (in general, affine) functions on certain compact subsets of its dual; then the question of an identification of the Šilov boundary of the positive cone arises naturally (cf. [1]). It is the purpose of the present note to give a short, intuitively clear presentation of the subject; the essential lemma is due to M. Powell (University of Maryland). Notation and terminology follows [3] .

LEMMA. Let H be an ordered vector space over $\mathbb{R}$ with positive cone H_+, and suppose H_+ to be $\sigma(H,G)$-closed where G is a H-separating subspace of the algebraic dual H^* . Let K be a $\sigma(H,G)$-compact convex subset of H and suppose x_0 is an extreme point of K which is maximal in K with respect to the ordering induced by H.

For every $\sigma(H,G)$-neighborhood U of x_0 , there exists a linear form $f \in G$ which is ≥ 0 on H_+ and such that $\sup_{x \in K \setminus U} f(x) < f(x_0)$.

Proof. By definition of the weak topology $\sigma(H,G)$, there exist closed semi-spaces $H_i := \{x \in H : f_i(x) \leq \alpha_i\}$, where $f_i \in G$ $(i = 1,\ldots,n)$ and $\alpha_i \in \mathbb{R}$, such that

$$x_0 \in H \setminus \left(\bigcup_1^n H_i\right) \subset U$$

Define $K_i := K \cap H_i$ and denote by K_0 the convex hull of $\bigcup_1^n K_i$; then K_0 is compact [3,II.10.2] . By Milman's theorem [3,II.10.5] we must have $x_0 \notin K_0$, since otherwise it would follow that $x_0 \in \bigcup_1^n K_i$ which is impossible. Since x_0 is maximal in K by hypothesis, it follows that

$(x_0 + H_+) \cap K = \{x_0\}$ and hence, $(x_0 + H_+) \cap K_0 = \emptyset$.

By a well-known separation theorem [3,II.9.2] , there exists a $\sigma(H,G)$-closed hyperplane $\{x \in H : f(x) = \alpha\}$ strictly separating K_o from $x_o + H_+$; without loss of generality we can assume that

$$\sup_{x \in K_o} f(x) < \alpha < f(x_o).$$

Then $f \in G$ is positive on H_+, since it is bounded below on H_+, and the assertion follows because of $K \setminus U \subset K_o$.

The following result (due to Choquet) is obtained by considering the trivial ordering with positive cone $\{0\}$.

COROLLARY. Let K be a convex compact subset of a locally convex space E. For each extreme point x_o of K and each weak neighborhood U of x_o, there exists an $f \in E'$ such that

$$\sup_{x \in K \setminus U} f(x) < f(x_o)$$

The main result is as follows.

THEOREM. Let E be an ordered locally convex space such that the linear hull of the positive cone E_+ is dense in E, and let K denote a convex, $\sigma(E',E)$-compact subset of the dual E'. Then the Šilov boundary of E_+, E_+ being considered as a cone of affine continuous functions on K, exists and is identical with the closure of the set of those extreme points which are maximal in K for the canonical order of E'.

Proof. Recall that a <u>boundary</u> of E_+ in K is a closed subset $Q \subset K$ such that each $x \in E_+$ assumes its maximum on Q; if P_o denotes the set of extreme points of K maximal in K with respect to the canonical order of E' (defined by : $x' \geq 0$ iff $\langle x,x' \rangle \geq 0 \;\; \forall x \in E_+$), the preceding lemma shows that $Q \supset P_o$ for every boundary Q of E_+. Thus it remains to show that the ($\sigma(E',E)$-) closure $\overline{P}_o$ is a boundary of E_+.

Let $x_o \in E_+$ be given; we show there exists a point $x'_o \in P_o$ such that $\langle x_o, x'_o \rangle = \sup \{ \langle x_o, x' \rangle : x' \in K \}$; we follow [1,§6]. A face F ($\neq \emptyset$) of K is called <u>hereditary</u> (with respect to $\geq$) if $x' \in F$, $y' \in K$ and $y' \geq x'$ implies $y' \in F$; denote by Φ the family of all closed hereditary faces of K. It is clear that Φ is ordered inductively under downward inclusion; therefore, each $F \in \Phi$ contains a minimal $F_o \in \Phi$ (Zorn's lemma). Let $\alpha := \sup\{ \langle x_o, x' \rangle : x' \in K\}$ and consider the set $F := \{ y' \in K : \langle x_o, y' \rangle = \alpha \}$; F is a closed face of K and evidently hereditary ($\geqslant$), so $F \in \Phi$. Now let F_o be a minim element of Φ contained in F; since F_o is convex compact, it contains an extreme point x_o' which, since F_o is a face of K, is extreme in K. Now if x_o' were not maximal in K there would exist $y' \in K$ satisfying $y' > x_o'$; since F_o is hereditary, we would have $y' \in F_o$. But since E_+ separates E', there must exist $y \in E_+$ such that $\langle y, y' \rangle > \langle y, x_o' \rangle$; if we let

$$F_y := \{ y' \in F_o : \langle y, y' \rangle = \sup_{z' \in F_o} \langle y, z' \rangle \} ,$$

then a purely computational verification shows that $F_y \in \Phi$. But $x_o' \notin F_y$ and so F_y must be properly contained in F_o, which contradicts the minimality of F_o. It follows that $x_o' \in P_o$ and, in particular, that $F_o = \{ x_o' \}$.

Therefore, $\overline{P}_o$ is a boundary of E_+ and the proof is complete.

Recall that a <u>Banach lattice</u> is a vector lattice and Banach space E su that $|x| \leq |y|$ implies $\|x\| \leq \|y\|$ $(x, y \in E)$. Applying the theorem to $K := \mathcal{U}^\circ \cap E_+'$ (the positive part of the dual unit ball) and keepin in mind that $\|x\| = \| |x| \|$ for all $x \in E$, we obtain this corollary.

COROLLARY. Let E denote a Banach lattice with unit ball U, and denc by P the $\sigma(E',E)$-closure of the set of those extreme points of $\mathcal{U}^\circ \cap E_+'$ which are maximal in $\mathcal{U}^\circ \cap E_+'$ for the canonical order of E'. Then for each $x \in E$, we have

$$\|x\| = \sup_{x' \in P} \langle |x| , x' \rangle .$$

<u>References</u>

[1] Alfsen, E., Convex Compact Sets and Boundary Integrals. Springer, Berlin-Heidelberg-New York 1971.

[2] Bauer, H., Schilowrand und Dirichletsches Problem. Ann.Inst. Fourier <u>11</u>, 89-136 (1961)

[3] Schaefer, H.H., Topological Vector Spaces. 3rd printing, Springer, Berlin-Heidelberg-New York 1971.

A RADON-NIKODYM-THEOREM FOR OPERATORS WITH AN APPLICATION TO SPECTRAL THEORY

M. Wolff

Abteilung Mathematik der Universität Dortmund
Dortmund, Germany

O. Introduction: E.B. Davies, H.P. Lotz, and A. Goullet de Rugy (see [4, 15]) gave independently of one another a representation of a Banach lattice E with quasiinterior points in its positive cone E_+ as a space of continuous numerical functions on a compact space Z. In applications, however, other representations are sometimes quite more natural than this abstract one; e.g. let X be a compact space, C(X) be the Banach lattice of all realvalued continuous functions on X, and μ be a positive linear form on C(X). Then the well-known theorem of Radon-Nikodym establishes a representation of the band B_μ, generated by μ in $C(X)' =: M(X)$, as the space $L^1(X,\mu)$, where now the basic space X has quite often only a little to do with that one given by E.B. Davies and H.P. Lotz.

It is the aim of our paper to give a general integration theoretic approach for representations of Banach lattices, as lattices of functions (mod. null-functions) on (clearly not uniquely determined) compact spaces, regaining besides others the theorem of Davies and Lotz as a special case. This will be done in sect. 2. In sect. 1 we outline a special case of the theory of abstract integration, taken from [1], while in sect. 3 we derive a general Radon-Nikodym-type theorem for bands of regular linear mappings from one Banach lattice to another (for a different R-N-Theorem see [5]). Finally we apply this R-N-theorem to prove a generalization of a theorem of H.H. Schaefer on the spectrum of real matrices ([14]) to quite arbitrary Banach lattices (even to simplex spaces).

I wish to thank Dr. R.J. Nagel and Professors Dr. H.H. Schaefer and Dr. G. Wittstock for valuable discussions, leading to easier proofs in sect. 3 and to an elimination of a minor mistake.

1. Abstract integration spaces: Following [7] we understand by a Banach lattice E over $\mathbb{C}$ the complexification $E = E_1 \oplus i E_1$ of a real Banach lattice E_1, satisfying $|x+iy| = \sup\{x \cos\varphi + y \sin\varphi : 0 \leq \varphi \leq 2\pi\}$, and $\|x+iy\| = \| |x+iy| \|$, the latter taken in E_1 (for an axiomatic treatment of complex vector lattices see [9], where we have proved, that this is the only reasonable construction of complex Banach lattices).

In the following (adapted from [1]) we set $\mathbb{K}$ for $\mathbb{R}$ or $\mathbb{C}$. Let X, Y, Z always denote compact spaces, $\overline{\mathbb{R}} = \mathbb{R} \cup \{\pm\infty\}$, $\overline{\mathbb{R}}_+ = \{\alpha \in \overline{\mathbb{R}} : \alpha \geq 0\}$, $C_{\mathbb{K}}(X) = \{f \in \mathbb{K}^X : f \text{ is continuous}\}$, $M(X) = C(X)'$. If we do not fix the special field, we often omit the index.

1. 1) Definition: A mapping $N: \overline{\mathbb{R}}_+^X \to \overline{\mathbb{R}}_+$ is called an upper norm, if it satisfies the following conditions:

(i) $f \leq g \Rightarrow N(f) \leq N(g)$

(ii) $N(\lambda f) = \lambda N(f) \qquad (\lambda \in \overline{\mathbb{R}}_+)$

(iii) $N(f+g) \leq N(f) + N(g)$

(iv) If (f_n) is an increasing sequence, converging to f pointwise, then $N(f) = \sup N(f_n)$

(v) $N(1_X) < \infty$ [1]

Examples: a) If μ is a positive Radon measure on X, i. e. an element of $M(X)_+$, then the upper integral N_μ (encombrement), constructed by Bourbaki [3], is an upper norm.

b) If (N_α) is a family of upper norms with $\sup\{N_\alpha(1_X) : \alpha \in A\} < \infty$, then $N(f) = \sup\{N_\alpha(f) : \alpha \in A\}$ defines a new upper norm.

As usual one defines:

1. 2) Definition: $f \in \overline{\mathbb{R}}^X$ ($\in \mathbb{C}^X$, resp.) is called N-negligible (negligible, for short if $N(|f|) = 0$. $A \subset X$ is called negligible, if 1_A is negligible. A property E holds a. e. (N), if $\{x : E(x) \text{ does not hold}\}$ is negligible.

We get quickly the following lemma:

1. 3) Lemma: a) The set of all negligible sets is a σ-ideal in the Boolean algebra of all subsets in X.

b) The set $\mathcal{N}_{\mathbb{K}}(N) = \{f \in \mathbb{K}^X : N(|f|) = 0\}$ is a σ-complete lattice ideal in $\mathbb{K}^X$. [2]

c) $f \in \overline{\mathbb{R}}^X$ ($\in \mathbb{C}^X$, resp.) and $N(|f|) < \infty$ implies $|f| < \infty$ a. e.

---

[1] 1_A denotes the characteristic function of the set A.

[2] For expressions not explained here we refer to [13, 16].

Proof: Integration theoretic arguments, as usual.

We set now $\mathcal{F}_{\mathbb{K}}(X,N) =: \mathcal{F}_{\mathbb{K}} := \{f \in \mathbb{K}^X : N(|f|) < \infty\}$ and obtain the following theorem:

1.4) Theorem: $\mathcal{F}_{\mathbb{K}}$ is an ideal of $\mathbb{K}^X$ and the function $f \to N(|f|)$ (denoted by N again) is a lattice seminorm, for which $\mathcal{F}_{\mathbb{K}}$ is complete. $F_{\mathbb{K}}(X,N) = \mathcal{F}_{\mathbb{K}}/\mathcal{N}$ is therefore (equipped with the induced norm) a Banach lattice over $\mathbb{K}$.

Proof: Everything is rather easy to prove except perhaps completeness which is implied by the following lemma.

1.5) Lemma: Let (f_n) be a Cauchy sequence in $\mathcal{F}_{\mathbb{K}}$. Then there exists a subsequence (f_{n_k}) converging a.e. on X. If $f \in \mathbb{K}^X$ is an a.e. limit of (f_{n_k}), then $f \in \mathcal{F}_{\mathbb{K}}$ and $\lim_{n\to\infty} N(|f-f_n|) = 0$.

Proof: Choose (f_{n_k}) such that $N(|f_{n_{k+1}} - f_{n_k}|) < 2^{-(k+1)}$ holds, and argue as usual.

We come now to our basic definition, observing $C_{\mathbb{K}}(X) \subset \mathcal{F}_{\mathbb{K}}$ because of $N(1_X) < \infty$.

1.6) Definition: $\mathcal{L}_{\mathbb{K}}(X,N)$ is the closure of $C_{\mathbb{K}}(X)$ in $\mathcal{F}_{\mathbb{K}}$ and $L_{\mathbb{K}}(X,N) = \mathcal{L}_{\mathbb{K}}(X,N)/\mathcal{N}$ is called abstract integration space (on X w.r.t. N).

1.7) Remarks: a) Clearly $L_{\mathbb{K}}(X,N)$ is a Banach lattice over $\mathbb{K}$. Loosely spoken, every element f of $L_{\mathbb{K}}(X,N)$ can be approximated by an N-Cauchy-sequence of $C_{\mathbb{K}}(X)$ converging a.e. to f. If $f \geq 0$ the sequence can be taken from $C_{\mathbb{K}}(X)_+$.

b) If $N = N_\mu$ (see the example after 1.1), we get $L(X,N_\mu) = L^1(X,\mu)$ etc. If $N_1 \leq \rho N_2$ for a positive constant ρ, then we get $\mathcal{F}_{\mathbb{K}}(N_2) \subset \mathcal{F}_{\mathbb{K}}(N_1)$, $\mathcal{L}_{\mathbb{K}}(N_2) \subset \mathcal{L}_{\mathbb{K}}(N_1)$ and $L_{\mathbb{K}}(N_2)$ can be mapped into $L_{\mathbb{K}}(N_1)$ in a natural manner.

c) If $f \in \mathcal{L}(X,N)$ and $|f| \leq n \cdot 1_X$, for each $g \in \mathcal{L}(X,N)$ the product fg is again in this space, and is negligible, if g is; so the multiplication by f induces a continuous linear operator U_f in $L(X,N)$, called multiplication operator and needed in §§ 3 a. 4.

2. Representation theory: This theory is based on the following simple observation.

2.1) Lemma: Let q be a lattice seminorm on $C_{\mathbb{K}}(X)$. Then there exists an upper norm N_q, such that a completion of $C_{\mathbb{K}}(X)$ w.r.t. q is equal to $\mathcal{L}_{\mathbb{K}}(X,N_q)$, and the completion of the normed lattice $C_{\mathbb{K}}(X)/q^{-1}(o)$, equipped with the induced norm, is equal to $L(X,N_q)$.

Proof: Set $S = \{f \in C(X): q(f) \leq 1\}$, $S'_+ = \{\mu \in M(X)_+ : \sup\{\langle\mu, |f|\rangle : q(f) \leq 1\} \leq 1\}$, and $N_q(h) = \sup\{N_\mu(h): \mu \in S'_+\}$ (see the ex. after 1.1). Since $q(1_X) < \infty$ we have N_q to be an upper norm, and $q(f) = N_q(|f|)$ for all $f \in C(X)$. The remainder is now obvious.

This lemma implies the representation theorem:

2.2) Theorem: Let $(E, \|\ \|)$ be a Banach lattice over $\mathbb{K}$, and let $j: C_{\mathbb{K}}(X) \to E$ be a lattice isomorphism onto a dense sublattice of E. Then there exists an upper norm N on X, such that j is extendable to an isometric lattice isomorphism from $L(X, N)$ onto E.

Proof: For $f \in C_{\mathbb{K}}(X)$ set $q(f) = \|j(f)\|$ and apply 2.1.

2.3) Corollary: (see [4]) Let E be a Banach lattice over $\mathbb{K}$, and u be a quasi-interior point of E_+, furthermore $E_u = \bigcup_{n\in\mathbb{N}} \{x \in E: |x| \leq nu\}$, and p_u the gauge function of $\{z: |z| \leq u\}$. Then there exists a compact space Z, an upper norm N on Z and an isometric lattice isomorphism T from $L_{\mathbb{K}}(Z, N)$ onto E, which maps 1_Z onto u and C(Z) onto E_u.

Proof: Since $J = \{x \in E: |x| \leq u\}$ is closed in E and since p_u induces on E_u a topology finer than that of E, (E_u, p_u) is an AM-space with unit u, and thus lattice isomorphic to some space C(Z) (Kakutani's theorem). Now apply 2.2.

3. Representation of operator lattices: Let E, F be vector lattices over $\mathbb{C}$, and F be order complete. A linear mapping $S: E \to F$ is called regular or orderbounded, if for all $z \in E_+$ the set $\{|Sy| : |y| \leq z\}$ is order bounded in F. By $|S|(z) = \sup\{|Sy| : |y| \leq z\}$ and linear extension we get a positive linear operator $|S|$. The set $\mathcal{L}^r(E, F)$ turns out to be a linear subspace of the space of all linear mappings, spanned by the cone $\mathcal{L}^r(E, F)_+$ of all positive mappings, this cone inducing a lattice ordering, for which the modulus of an element S is determined by the formula given above (see [8] sect. 3, and [9] for details). Moreover, if E and F are Banach lattices, then $\|T\|_r = \|\,|T|\,\|$ (operator norm of $|T|$) defines a lattice norm on $\mathcal{L}^r(E, F)$ w.r.t. which the space is complete.

Let us quickly recapitulate some notions on order convergence: A net $(x_\alpha)_{\alpha\in A}$ of a vector lattice E is said to converge in order to an element x (notation: $x = \text{o-lim}\, x_\alpha$), if $\overline{\lim}\,(|x - x_\alpha|) = \inf_\alpha (\sup_{\beta\geq\alpha} |x - x_\beta|) = 0$. If F is another vector lattice, a positive sublinear mapping $p: E \to F$ is called order-continuous, if for every decreasing net (x_α) in E with $\inf x_\alpha = 0$ $\inf(p(x_\alpha)) = 0$ holds to be true.

Because of the sublinearity we get at once, that such a map satisfies o–lim $p(y_\beta) = p(\text{o–lim } y_\beta)$ for arbitrary order convergent nets.

A positive bijective mapping T is a lattice isomorphism, if T^{-1} is positive, too.

Let now X, Y be compact, $E = C(X)$, $F = C(Y)$.

By a theorem of Hulanicki and Phelps ([6], Prop. 3.2) we get that each linear form μ on $E \otimes F$, which is positive on $K_p = \{\Sigma f_i \otimes g_i,\ f_i \geq 0,\ g_i \geq 0\}$, is uniquely extendable to an (equally denoted) positive Radon measure on $X \times Y$. This implies that each $\rho \in (E \otimes F)^*$ of the form $\rho = \mu_1 - \mu_2 + i(\mu_3 - \mu_4)$, the μ_i being positive on K_p, can be uniquely extended to an equally denoted continuous linear form on $C(X \times Y)$. So we get at once the following statement (which should already be well-known).

3.1) Theorem: The mapping $T \to \mu_T (\mu_T(f \otimes g) = \langle g, Tf \rangle)$ is an isometric, order–continuous lattice isomorphism from $\mathcal{L}^r(C(X), M(Y))$ onto $M(X \times Y)$. The inverse mapping is given by $\mu \to T_\mu\ (T_\mu(f) = \mu(f \otimes .))$.

Proof: Since every $T \in \mathcal{L}^r(C(X), M(Y))$ can be written as $T = T_1 - T_2 + i(T_3 - T_4)$ with $T_i \geq 0$, all statements follow by straightforward computation, except perhaps that concerning the norm, being established by the following inequality

$$\| T_\mu \|_r = \| \, | T_\mu | \, \| = \| T_{|\mu|} \| = \sup \{ \| \, |\mu| \, (f \otimes .) \| : |f| \leq 1_X \}$$
$$= |\mu| (1_X \otimes 1_Y) = \| \mu \| .$$

Unless otherwise stated we consider from now on only abstract integration spaces $L(X,N)$ with $C(X) \cap \mathcal{M}(N) = \{0\}$. In that case $C(X)$ is viewed as a dense sublattice of $L(X,N)$.

3.2) Corollary: Let $E = L(X,N)$, Y compact, and F be a lattice ideal of $M(Y)$. Then the mapping $\mathcal{L}^r(E,F) \ni T \to \overline{T} = T_{|C(X)} \to j \circ \overline{T} \to \mu_T$, where $j: F \to M(Y)$ denotes the inclusion map, is an order continuous lattice isomorphism onto a lattice ideal of $M(X \times Y)$.

Proof: The mapping $T \to j \circ \overline{T}$ is easily seen to be an order continuous lattice isomorphism onto an ideal in $\mathcal{L}^r(C(X), M(Y))$. Now apply 3.1.

Those abstract integration spaces $L(Y, N') = F$ which can be embedded as ideals of $M(Y)$, are easily characterized:

3.3) Lemma: (cf. [10], 3.4): For an abstract integration space $F = L(Y,N)$ the following statements are equivalent:

(i) There exists a lattice isomorphism j from F onto an ideal in $M(Y)$.

(ii) F is order complete, and there exists a strictly positive order continuous linear form on F.

Proof: (i) ⇒ (ii): Since each ideal in M(Y) is order complete, F has to be order complete. The linear form $u \to \langle 1_Y, j(u) \rangle$ is the desired one.

(ii) ⇒ (i): Let the linear form in question be ρ (viewed as a Radon measure on Y). Since ρ is continuous we have $N_\rho \leq \gamma \cdot N$ $(\gamma > 0)$ and therefore F = L(Y, N) as a dense sublattice of $L^1(Y, \rho)$ (see 1. 7b). Since this latter space is already known to be embeddable onto an ideal in M(Y), we only have to show that F is an ideal in $L^1(Y, \rho)$. Let $0 \leq v \leq u \in F$, $v \in L^1(Y, \rho)$. There exists an N_ρ-Cauchy-sequence (u_n) from F converging to v. Then the same is true for (w_n) with $w_n = \inf(u, u_n^+) \in F$. But (w_n) is order-bounded in F, so we get $v_1 = \underline{\lim}\, w_n \leq \overline{\lim}\, w_n = v_2$ in F. Since ρ is order continuous and (w_n) a ρ-Cauchy-sequence, we get $\rho(v_1) = \underline{\lim}\, \rho(w_n) = N_\rho(v) = \overline{\lim}\, \rho(w_n) = \rho(v_2)$, hence $\rho(v_2 - v_1) = 0$, which implies $v_1 = v_2$ because ρ is strictly positive on F. Necessarily we have $v_1 \leq v \leq v_2$ in $L^1(Y, \rho)$ i. e. $v = v_1 = v_2 \in F$, **q.e.d.**

In the situation of 3. 2, if T is a positive linear map, and S in the band B_T, generated by T in $\mathscr{L}^r(E, F)$, we get μ_S to be in the band generated by μ_T in M(X×Y). This latter being isomorphic to $L^1(X\times Y, \mu_T)$ we can assign to each $S \in B_T$ a function K_S, determined a. e. (μ_T) in $L^1(X\times Y, \mu_T)$. If we now have F as in 3. 3, we have for each $S \in \mathscr{L}^r(E, F)$ a measure μ_S on X×Y, defined by $\mu_{S_1}(f \otimes g) = \langle g, Sf.\rho \rangle = \int g(y)(Sf)(y)d\rho(y)$, Sf to be considered as an element in $L^1(Y, \rho)$. Combining all these mappings we get

<u>3. 4) Theorem (Radon-Nikodym-type theorem) : Let E = L(X, N), F = (Y, N') be abstract integration spaces, the latter being order complete and possessing a strictly positive, order continuous linear form ρ. Let $T \in \mathscr{L}^r(E, F)$, and B_T be the band, generated by T. Then if $\mu_T \in M(X\times Y)$ is defined by $\mu_T(f \otimes g) = \int g(y)(Tf)(y)d\rho(y)$ for all $(f, g) \in C(X)\times C(Y)$, for each $S \in B_T$ there exists a function K_S (determined uniquely a. e. (μ_T)) in $L^1(X\times Y, \mu_T)$, such that for all $f \in C(X)$, $g \in C(Y)$ the relation $\langle g, Sf.\rho \rangle = \int f(x)g(y)K_S(x, y)d\mu_T(x, y)$ holds. The mapping $\psi_T: S \to K_S$ is an order continuous lattice isomorphism of B_T onto an ideal R_T of $L^1(X\times Y, \mu_T)$, containing $L^\infty(X\ Y, \mu_T)$. Its inverse mapping is denoted by m_T.</u>

Proof: Obvious by 3. 2, 3. 3, and the preceding remarks. For T to be of rank one our result stands in special close connection with those given in [11]. But instead of going into details we give only three other applications, the first of it needed in § 4.

3.5) Corollary 1: For $f_o \in C(X)$, $g_o \in C(Y)$ and $K \in R_T$ we have $m_T(f_o \otimes 1_Y \cdot K \cdot 1_X \otimes g_o) = U_{g_o} \circ m_T(K) \circ U_{f_o}$, where U_{f_o} and U_{g_o} denote the multiplication operators (see 1.7 c).

Proof: Set $m_T(K) = S$. Clearly $U_{g_o} S U_{f_o} =: V \in B_T$. For arbitrary $f \in C(X)$, $g \in C(Y)$ we get $\iint f(x)g(y) K_V(x,y)d\mu_T(x,y) = \int g(y)(Vf)(y)d\rho(y) = \int g(y)g_o(y)(Sff_o)(y)d\rho(y) =$ $= \iint g(y)g_o(y)f(x)f_o(x)K(x,y)d\mu_T(x,y)$ which implies $K_V = f_o \otimes 1_Y \cdot K \cdot 1_X \otimes g_o$.

Finally let us discuss the problem, under which conditions B_T turns out to be an abstract integration space on $X\times Y$. This need not be the case in general as simple examples ($F = \mathbb{C}$) show. However, a necessary condition is that $C(X\times Y)$ is dense in R_T for the norm $\|K\| = \|m_T(K)\|_r$.

3.6) Corollary 2: Let the norm on B_T be order continuous. Then B_T is isomorphic to an abstract integration space $L(X\times Y, N_T)$ for a suitable upper norm N_T.

Proof: $C(X\times Y)$ is dense in R_T for the L^1-norm, thus order dense[3]. Now $m_T(C(X\ Y))$ is order dense in B_T, since m_T is order continuous. The norm on B_T being order continuous, $m_T(C(X\times Y))$ is normdense in B_T. Now apply 2.2.

3.7) Corollary 3: Let E and F be separable Banach lattices, F being order complete and having a strictly positive order continuous linear form; furthermore let $T \in \mathcal{L}^r(E,F)$ be positive, and the norm on B_T be order-continuous. Then there exists an $S \in B_T$, such that B_T is generated by S and T, i.e. B_T is the smallest closed vector sublattice containing S and T.

Sketch of a proof: On account of the separability one can represent E and F as abstract integration lattices over metrizable spaces X and Y; but then $L(X\times Y, N_T)$ has to be separable again, since $X\times Y$ is metrizable and $C(X\times Y)$ is separable. So B_T is separable and one can apply [17] Thms 1.9, 1.11 or 1.12, resp.

4. Application to spectral theory:

At first we prove some lemmata to establish the main theorems in an easy manner.

4.1) Lemma: Let E, F be Banach lattices, F being order complete, and u a quasi-interior point of E_+. If S, T are of $\mathcal{L}^r(E,F)$, $0 \leq |S| \leq T$ and $Su = Tu$, then $S = T$.

Proof: We have $S = S_1^+ - S_1^- + i S_2$, where S_1 and S_2 are real operators. $Su = Tu$

---

3) Every element of R_T is the limit a.e. of a sequence in $C(X\times Y)$, but this means the σ-limit of that sequence ([2], XV).

yields $S_2u = O$, therefore $Tu = S_1u \leq S_1^+u \leq |S|u \leq Tu$; this implies $(T-|S|)u = O$, and since $O \leq (T-|S|)$ and u is a quasiinterior point, we get $|S| = T$ and similar $S_1^+ = T$, hence the assertion.

4.2) Corollary: With the notation of 3.6 we have:
If $K \in R_T$, $|K| \leq 1_{X\times Y}$ and $\int K(x,y)d\mu_T(x,y) = \int d\mu_T(x,y)$, then $K = 1_{X\times Y}$ a.e. (μ_T).
Proof: Apply 4.1 and 3.4 to T and $m_T(K)$.

Let now E be a Banach lattice, F a vector sublattice (not necessarily closed), and f an element of F, $|f| = u$, p_u the gauge function of $\{z \in F: |z| \leq u\}$ on F_u, q_u that one of $\{z \in E: |Z| \leq u\}$ on E_u. The f-hull G_f is defined to be the smallest vector sublattice of E_u, generated by f and being closed for q_u.

4.3) Lemma: If (F_u, p_u) is complete, then G_f is already contained in F.

Proof: Represent the AM-space with unit u (E_u, q_u) as a space C(X). Then F_u turns out to be a closed sublattice, containing f and u; the assertion follows at once.

For our main theorem let now at first E be an order complete complex Banach lattice with strictly monotone, order continuous norm.

4.4) Theorem[4)]: Let S be a real operator on E into itself, and T a positive contraction, majorizing $|S|$, furthermore let λ be an eigenvalue of S of modulus 1. Then λ^{2k} is an eigenvalue of T and λ^{2k+1} such one of S for each integer **k**. Moreover, if f is an eigenvector of S for the eigenvalue λ, then to each λ^{2k} one eigenvector of T, and to each λ^{2k+1} one eigenvector of S are to be found in the f-hull G_f.

Proof: Let f be an eigenvector of S for λ; then by our assumptions and $|f| = |Sf| \leq |S|(|f|) \leq T|f|$ we get $\|f\| \leq \|T|f|\| \leq \|f\|$, hence $|f| = T|f|$, so we may restrict ourselves to the space $\overline{E_{|f|}}$, being invariant under S and T. With the aid of 2.3 we identify this latter space as an abstract integration space

---

4) H.H. Schaefer proved a similar theorem for n×n-matrices ([14]) as well as for AL-spaces and AM-spaces with unit (oral communication), cf. 4.5, 4.7 and 4.8.

$L(X,N)$ with $|f| = 1_X$ and $E_{|f|} = C(X)$. By a result of G. Lozanovski (see [1o], 4.8) $L(X,N)$ possesses a strictly positive order continuous linear functional (since the norm is order continuous and the space is order complete). So we are able to apply 3.4, 3.5, 4.1 and 4.2. Let K_S be $\psi_T(S)$. From $|f| = 1_X$ and $Sf = \lambda f$ we derive $(\bar{\lambda} U_{\bar{f}} S U_f) 1_X = 1_X = T 1_X$ [5], which gives us

(i) $\bar{\lambda}\, 1_X \otimes \bar{f} \cdot K_S \cdot f \otimes 1_X = 1_{X \times X}$ by 4.1, 4.2 and 3.5, or $K_S = \lambda(1_X \otimes f)(\bar{f} \otimes 1_X)$.

Since K_S has to be real this implies

(ii) $K_S^2 = 1_{X \times X}$ and (by induction on n and on −n) for alle $k \in \mathbb{Z}$

(iii) $\bar{\lambda}^{2k+1}(1_X \otimes \bar{f}^{2k+1}) K_S (f^{2k+1} \otimes 1_X) = 1_{X \times X}$ and

(iv) $\bar{\lambda}^{2k}(1_X \otimes \bar{f}^{2k})(f^{2k} \otimes 1_X) = 1_{X \times X}$. Applying 3.5 we get

$1_X = (\bar{\lambda}^{2n+1} U_{\bar{f}^{2n+1}} S U_{f^{2n+1}})(1_X)$ and

$1_X = (\bar{\lambda}^{2n} U_{\bar{f}^{2n}} T U_{f^{2n}})(1_X)$

which yields $Sf^{2n+1} = \lambda^{2n+1} f^{2n+1}$ and $Tf^{2n} = \lambda^{2n} f^{2n}$.

Furthermore G turns out to be the sup-norm-closure in $C(X)$ of the algebra generated by f and $\bar{f}$, hence the assertion follows from 4.3.

<u>4.5) Corollary:</u> <u>Let E be an arbitrary complex Banach lattice, S a real operator on T, T a positive operator majorizing S and −S, and μ a strictly positive linear form on E with $T'\mu \leq \mu$. Then for an eigenvalue λ of S of modulus one the same statement as in 4.4 holds to be true.</u>

<u>Proof:</u> Consider the completion E^{μ} of E with respect to the norm $p(f) = \langle |f|, \mu \rangle$. E^{μ} is an AL-space and S and T are continuously extendable to $\bar{S}$, $\bar{T}$ on E^{μ}, because of $T'\mu \leq \mu$. Now E^{μ}, $\bar{T}$, $\bar{S}$ satisfy the requirements of 4.4. Let $f \in E$ be an eigenvector of S for $\lambda(|\lambda| = 1)$. Then the lattice $G_f \subset (E^{\mu})_{|f|}$ has already to be contained in $E_{|f|}$ by 4.3, because E is a Banach lattice, and so $(E_{|f|}, p_{|f|})$ is already complete in E. The remainder is now an obvious consequence of 4.4.

The next results require deep methods in spectral theory; the easiest case was handled by H.P. Lotz [7]. The trick is to change the peripherical spectrum of an operator into the point spectrum of a suitable other operator on a suitable other space. Since we have to guarantee the new space to satisfy the requirements of 4.4, the most convenient method to use is nonstandard analysis. (The case, in

---

[5] The bars denoting the complex conjugate.

which E is an AL-space can be treated by a refinement of H.P. Lotz's method) This method being far beyond of the scope of this paper, we give only the results. Proofs of more general statements are to be found in [18].

Let us recall some further definition from [2].

4.6) Definition ([2], XV, § 14): A Banach lattice E is said to have uniformly monotone norm, if for all $\varepsilon > O$ there exists a $\delta > O$ such that for all $f, g \in E_+$ with $\|f\| \leq 1$ the relation $\|f + g\| \leq \|f\| + \delta$ always implies $\|g\| \leq \varepsilon$.

Examples are all uniformly convex Banach lattices as well as all AL-spaces.

4.7) Theorem: Let E be a Banach lattice with uniformly monotone norm, let S be a real operator on E, T a positive contraction, majorizing S and -S. If λ is in the spectrum of S and $|\lambda| = 1$, then the set $\{\lambda^{2k+1} : k \in \mathbb{Z}\}$ is in the spectrum of S and the set $\{\lambda^{2k} : k \in \mathbb{Z}\}$ is in the spectrum of T.

4.8) Corollary: The theorem holds to be true, if one requires E to be nothing more than a simplex space.

Proof: Consider E', S', T'. E' is an AL-space and the other conditions of 4.7 are also satisfied.

References

1. Bichteler, K.: Integration Theory (with special attentions to vector measures) Lecture Notes in Mathematics 315, Berlin-Heidelberg-New York: Springer Verlag 1973.
2. Birkhoff, G.: Lattice Theory, 3rd ed., AMS Coll. Publ. XXV (1967)
3. Bourbaki, N.: Éléments de Mathematique, Livre VI: Integration, chap. II-IV, Paris: Herman et Cie, 1956.
4. Goullet de Rugy, A.: Representation of Banach lattices, this volume.
5. Hackenbroch, W.: Zum Radon-Nikodymschen Satz in Vektorverbänden, Manuskript 1972.
6. Hulanicki, A, Phelps, R.R.: Some applications of tensor products of partially-ordered linear spaces, J. Funct. Anal. 2, 177-2o1 (1968).
7. Lotz, H.P.: Über das Spektrum positiver Operatoren, Math. Z. 1o8, 15-32 (1968).
8. Luxemburg, W.A.J., Zaanen, A.C.: The linear modulus of an order-bounded linear transformation, Indag. Math. 33, 422-437 (1971).

9. Mittelmeyer, G., Wolff, M.: Über ein einheitliches Axiomensystem für Vektorverbände über $\mathbb{R}$ und $\mathbb{C}$ und den Absolutbetrag ordnungsbeschränkter Operatoren, to be published.

1o. Nagel, R.J.: Ordnungsstetigkeit in Banachverbänden, Manuscripta math. 9, 9-27 (1973).

11. Nagel, R.J., Schlotterbeck, U.: Integraldarstellungen regulärer Operatoren auf Banachverbänden, Math. Z. 127, 293-3oo (1972).

12. Peressini, A.L., Sherbert, D.R.: Ordered topological tensor products, Proc. London Math. Soc. 19, 177-19o (1969).

13. Schaefer, H.H.: Topological vector spaces, 3rd printing, Berlin-Heidelberg-New York: Springer 1971.

14. Schaefer, H.H.: On the characteristic roots of real matrices, Proc. AMS 28, 91-92 (1971).

15. Schaefer, H.H.: On representation of Banach lattices by continuous numerical functions, Math. Z. 125, 215-232 (1972).

16. Schaefer, H.H.: Orderings of vector spaces, this volume.

17. Wolff, M.: Darstellung von Banach-Verbänden und Sätze vom Korovkin-Typ, Math. Annal. 2oo, 47-67 (1973).

18. Wolff, M.: Über das Randspektrum gewisser komplexer ordnungsbeschränkter Operatoren, to be published.

Lecture Notes in Physics

Bisher erschienen / Already published

Vol. 1: J. C. Erdmann, Wärmeleitung in Kristallen, theoretische Grundlagen und fortgeschrittene experimentelle Methoden. 1969. DM 20,–

Vol. 2: K. Hepp, Théorie de la renormalisation. 1969. DM 18,–

Vol. 3: A. Martin, Scattering Theory: Unitarity, Analyticity and Crossing. 1969. DM 16,–

Vol. 4: G. Ludwig, Deutung des Begriffs physikalische Theorie und axiomatische Grundlegung der Hilbertraumstruktur der Quantenmechanik durch Hauptsätze des Messens. 1970. DM 28,–

Vol. 5: M. Schaaf, The Reduction of the Product of Two Irreducible Unitary Representations of the Proper Orthochronous Quantummechanical Poincaré Group. 1970. DM 16,–

Vol. 6: Group Representations in Mathematics and Physics. Edited by V. Bargmann. 1970. DM 24,–

Vol. 7: R. Balescu, J. L. Lebowitz, I. Prigogine, P. Résibois, Z. W. Salsburg, Lectures in Statistical Physics. 1971. DM 18,–

Vol. 8: Proceedings of the Second International Conference on Numerical Methods in Fluid Dynamics. Edited by M. Holt. 1971. DM 28,–

Vol. 9: D. W. Robinson, The Thermodynamic Pressure in Quantum Statistical Mechanics. 1971. DM 16,–

Vol. 10: J. M. Stewart, Non-Equilibrium Relativistic Kinetic Theory. 1971. DM 16,–

Vol. 11: O. Steinmann, Perturbation Expansions in Axiomatic Field Theory. 1971. DM 16,–

Vol. 12: Statistical Models and Turbulence. Edited by M. Rosenblatt and C. Van Atta. 1972. DM 28,–

Vol. 13: M. Ryan, Hamiltonian Cosmology. 1972. DM 18,–

Vol. 14: Methods of Local and Global Differential Geometry in General Relativity. Edited by D. Farnsworth, J. Fink, J. Porter and A. Thompson. 1972. DM 18,–

Vol. 15: M. Fierz, Vorlesungen zur Entwicklungsgeschichte der Mechanik. 1972. DM 16,–

Vol. 16: H.-O. Georgii, Phasenübergang 1. Art bei Gittergasmodellen. 1972. DM 18,–

Vol. 17: Strong Interaction Physics. Edited by W. Rühl and A. Vancura. 1973. DM 28,–

Vol. 18: Proceedings of the Third International Conference on Numerical Methods in Fluid Mechanics, Vol. I. Edited by H. Cabannes and R. Temam. 1973. DM 18,–

Vol. 19: Proceedings of the Third International Conference on Numerical Methods in Fluid Mechanics, Vol. II. Edited by H. Cabannes and R. Temam. 1973. DM 26,–

Vol. 20: Statistical Mechanics and Mathematical Problems. Edited by A. Lenard. 1973. DM 22,–

Vol. 21: Optimization and Stability Problems in Continuum Mechanics. Edited by P. K. C. Wang. 1973. DM 16,–

Vol. 22: Proceedings of the Europhysics Study Conference on Intermediate Processes in Nuclear Reactions. Edited by N. Cindro, P. Kulišić and Th. Mayer-Kuckuk. 1973. DM 26,–

Vol. 23: Nuclear Structure Physics. Proceedings of the Minerva Symposium on Physics. Edited by U. Smilansky, I. Talmi, and H. A. Weidenmüller. 1973. DM 26,–

Vol. 24: R. F. Snipes, Statistical Mechanical Theory of the Electrolytic Transport of Non-electrolytes. 1973. DM 20,–

Vol. 25: Constructive Quantum Field Theory. The 1973 "Ettore Majorana" International School of Mathematical Physics. Edited by G. Velo and A. Wightman. 1973. DM 26,–

Vol. 26: A. Hubert, Theorie der Domänenwände in geordneten Medien. 1974. DM 28,–

Vol. 27: R. Kh. Zeytounian, Notes sur les Ecoulements Rotationnels de Fluides Parfaits. 1974. DM 28,–

Vol. 28: Lectures in Statistical Physics. Edited by W. C. Schieve and J. S. Turner. 1974. DM 24,–

Vol. 29: Foundations of Quantum Mechanics and Ordered Linear Spaces. Advanced Study Institute Held in Marburg 1973. Edited by A. Hartkämper and H. Neumann. 1974. DM 26,–